# Advances in Industrial Control

More information about this series at http://www.springer.com/series/1412

John T. Wen · Sandipan Mishra
Editors

# Intelligent Building Control Systems

## A Survey of Modern Building Control and Sensing Strategies

Springer

*Editors*
John T. Wen
Department of Electrical, Computer,
and Systems Engineering
Rensselaer Polytechnic Institute
Troy, NY
USA

Sandipan Mishra
Department of Mechanical, Aerospace,
and Nuclear Engineering
Rensselaer Polytechnic Institute
Troy, NY
USA

ISSN 1430-9491 ISSN 2193-1577 (electronic)
Advances in Industrial Control
ISBN 978-3-319-68461-1 ISBN 978-3-319-68462-8 (eBook)
https://doi.org/10.1007/978-3-319-68462-8

Library of Congress Control Number: 2017954468



Printed on acid-free paper

This Springer imprint is published by Springer Nature
The registered company is Springer International Publishing AG
The registered company address is: Gewerbestrasse 11, 6330 Cham, Switzerland

*To our students and families*

# Series Editors' Foreword

The series *Advances in Industrial Control* aims to report and encourage technology transfer in control engineering. The rapid development of control technology has an impact on all areas of the control discipline. New theory, new controllers, actuators, sensors, new industrial processes, computer methods, new applications, new philosophies, new challenges, etc. Much of this development work resides in industrial reports, feasibility study papers, and the reports of advanced collaborative projects. The series offers an opportunity for researchers to present an extended exposition of such new work in all aspects of industrial control for wider and rapid dissemination.

A module on "Building Services" was often an item in the curriculum of the Electrical Engineering undergraduate degree. It would deal with topics such as the power supply to commercial and residential buildings and the regulations in force for the electrical installations for these building types.

However, times change and new technologies are shaping the future of building construction and services. Furthermore, control systems theory, design, simulation, and implementation are all playing a role in developing this future.

Take building construction, for example. In regions of the world where earthquakes are a frequent occurrence, building contraction engineers have worked with control engineering specialists to investigate how to make buildings earthquake proof. This involves considering the "building" as a dynamic system and devising passive and active control mechanisms and schemes to absorb the shocks delivered by an earthquake. This merger of building science and control engineering is ongoing in countries like Japan and Chile.

As another example, take the electric power supply to a building. The drive toward the increased use of renewable sources of energy to mitigate the effects of climate change in some small way has had an impact too. Rooftop solar panels are widespread on commercial, public, and domestic residential buildings. Individual rooftop wind turbine installations are less well used but still a viable option for localized power generation. The electrical engineering aspect of these developments is how to integrate these intermittent energy supplies with the conventional grid supply of electric power to premises. Most recently there has been consideration of using battery storage to try and make premises complete independent of external power supplies, depending rather on self-generated supplies from the localized renewable energy sources. Control systems engineering has a significant role to play here in devising such a completely independent power supply system.

Development in the control of the interior environment of a building has come about in a piecemeal way with different environmental aspects being dealt with separately. Heating, ventilation, and air conditioning (HVAC) control is a well-known and major component of interior environment control; indeed it forms the key middle part of this edited volume *Intelligent Building Control Systems* drawn together by Editors John T. Wen and Sandipan Mishra. Their edited volume opens with the fundamental idea of a supervisory control structure for the control of a building's interior environment. The middle part of the volume concentrates on HVAC systems looking at topics such as interior multi-zone modeling, and the hardware technology of HVAC systems. Model predictive control makes a predictable appearance here. It is in Part III that the "intelligent" or "smart" integrated control system concept comes into its own. Smart lighting control and energy management systems are given in one chapter each. Thus the volume closes with a forward look at new emerging integrated intelligent building control systems.

The Editors John T. Wen and Sandipan Mishra are both at the Rensselaer Polytechnic Institute, Troy, NY, USA and have research and applications interests in a wide range of control fields including smart building control. The list of contributors is suitably international and shows there is global interest in developing intelligent building systems. It is a subject that is capable of exciting new developments for energy management systems particularly as localized renewable energy systems become standard for commercial, public, and residential buildings; and control engineers, researchers, and companies will have a significant role to play in developing this future. This is a very welcome contribution to the *Advances in Industrial Control* monograph series and to the wider control community and its literature.

Other volumes in the *Advances in Industrial Control* series that deal with applications of control in buildings and the interior environment include the following:

- *Control of Traffic Systems in Buildings* by Sandor A. Markon, Hajime Kita, Hiroshi Kise and Thomas Bartz-Beielstein (ISBN 978-1-84628-448-9, 2006); and

- *Modeling and Control of Greenhouse Crop Growth* by Francisco Rodríguez, Manuel Berenguel, José Luis Guzmán and Armando Ramírez-Arias (ISBN 978-1-84996-280-3, 2015).

M.J. Grimble
M.A. Johnson
Industrial Control Centre
Glasgow, Scotland, UK

# Preface

This book evolved over several years based on discussions and exchanges at various conferences. The motivation is compelling—commercial and residential building energy consumption constitutes a staggering percentage of the overall energy consumption in the United States and other countries. With the proliferation of sensors and embedded controllers, building control systems are becoming more "intelligent", offering a ripe opportunity for the control community to contribute to creating solutions for energy efficiency, occupant comfort, and building management. Building control systems also pose challenges as complex interconnected subsystems, with interactions between human occupants and automation subsystems, and time-varying operating conditions depending on usage and weather conditions. Attempts to solve these issues in building control have resulted in a thriving research community. In automation and control conferences, special sessions on intelligent building, smart building, building control, etc., are regular fixtures in the conference programs. We feel the time is right for a collection of papers from leading academic and industrial researchers, in U.S., Europe, and Asia, to capture the current state of building control research and development, ranging from the overall building management, software architecture, to HVAC, lighting and human comfort, and other building-related issues including integration with the smart grid and the so-called Internet-of-Things. This book presents multiple facets of intelligent building control. Chapters 2–3 present the overall building control system architecture. Chapters 3–7 focus specifically on the HVAC system from vapor compression cycle to temperature control. In Chap. 8, the role of human mediation in buildings is illustrated. Chapters 9–11 go beyond building HVAC to examine other aspects of intelligent building control, including lighting, power distribution, and network connectedness. We hope that this book will benefit those interested in gaining a broader perspective and contemporary approaches on intelligent building control systems.

Troy, NY, USA
March 2017

John T. Wen
Sandipan Mishra

# Acknowledgements

The editors would like to thank all the authors that contributed to the chapters of this book. This is truly a collaborative effort. Our students, Matt Minakais and Charles Okaeme, helped with reviewing and proofreading the book.

Our research on building control at Rensselaer Polytechnic Institute has been supported by the National Science Foundation, NSF-CNS 1230687 SEP Collaborative: A Unified Framework for Sustainability in Buildings through Human Mediation, 9/1/12-8/31/16, the Center for Light Enabled Systems and Applications (LESA) under the Engineering Research Centers Program (ERC) of the National Science Foundation, EEC-0812056, Center for Automation Technologies and Systems (CATS) under a block grant from the New York State Empire State Development Division of Science, Technology and Innovation (NYSTAR) under contract C090145, and the Hewlett Packard Labs Innovative Research Projects Award.

We gratefully acknowledge the professional support from the Springer Verlag and its staff, particularly the Editor for Engineering, Oliver Jackson.

# Contents

# Contributors


**Sina Afshari** University of Michigan, Ann Arbor, MI, USA

**Andrew Alleyne** University of Illinois at Urbana-Champaign, Champaign, IL, USA

**David M. Auslander** University of California, California, USA

**Francesco Borrelli** University of California, Berkeley, CA, USA

**Scott A. Bortoff** Mitsubishi Electric Research Laboratories, Cambridge, MA, USA

**Daniel J. Burns** Mitsubishi Electric Research Laboratories, Cambridge, MA, USA

**Stefano Di Cairano** Mitsubishi Electric Research Laboratories, Cambridge, MA, USA

**Frank Chuang** University of California, Berkeley, CA, USA

**Allan Daly** NEXTracker, Inc., Fremont, CA, USA

**Claus Danielson** Mitsubishi Electric Research Laboratories, Cambridge, MA, USA

**Steve Gabel** Honeywell Global Labs (retired), Golden Valley, MN, USA

**Santosh K. Gupta** ECSE Department, Rensselaer Polytechnic Institute, Troy, New York, USA

**M.H. Toufiq Imam** Rensselaer Polytechnic Institute, Troy, NY, USA

**Qing-Shan Jia** Center for Intelligent and Networked Systems, Department of Automation, Tsinghua National Laboratory for Information Science and Technology, Tsinghua University, Beijing, China

**Karl H. Johansson** ACCESS Linnaeus Centre, School of Electrical Engineering, KTH Royal Institute of Technology, Stockholm, Sweden

**Koushik Kar** ECSE Department, Rensselaer Polytechnic Institute, Troy, New York, USA

**Bryan Keating** University of Illinois at Urbana-Champaign, Champaign, IL, USA

**Sarah M. Koehler** University of California, Berkeley, CA, USA

**Justin Koeln** University of Illinois at Urbana-Champaign, Champaign, IL, USA

**Christopher R. Laughman** Mitsubishi Electric Research Laboratories, Cambridge, MA, USA

**Yudong Ma** NEXTracker, Inc., Fremont, CA, USA

**Sandipan Mishra** Rensselaer Polytechnic Institute, Troy, NY, USA

**Marco Molinari** School of Electrical Engineering, KTH Royal Institute of Technology, Stockholm, Sweden

**Alessandra Parisio** School of Electrical and Electronic Engineering, The University of Manchester, Manchester, UK

**Girija Parthasarathy** Honeywell Global Labs, Golden Valley, MN, USA

**Christopher Price** Texas A&M University, College Station, TX, USA

**Marco Pritoni** Lawrence Berkeley National Lab, California, USA

**Bryan P. Rasmussen** Texas A&M University, College Station, TX, USA

**Tariq Samad** Technological Leadership Institute, University of Minnesota, Minneapolis, MN, USA

**Petr Stluka** Honeywell Global Labs, Prague, Czech Republic

**Damiano Varagnolo** Department of Computer Science, Electrical and Space Engineering, Luleå University of Technology, Luleå, Sweden

**John T. Wen** Rensselaer Polytechnic Institute, Troy, NY, USA

**Yuanming Zhang** Center for Intelligent and Networked Systems, Department of Automation, Tsinghua National Laboratory for Information Science and Technology, Tsinghua University, Beijing, China

**Qianchuan Zhao** Center for Intelligent and Networked Systems, Department of Automation, Tsinghua National Laboratory for Information Science and Technology, Tsinghua University, Beijing, China

# Abbreviations

| | |
|---|---|
| 3R2C | Three Resistance and Two Capacitance |
| AAR | ARCnet to ARCnet Router |
| ADMM | Alternating Direction Method of Multipliers |
| AHC | Akademiska Hus Controller |
| AHU | Air Handling Unit |
| ALC | Automated Logic Corporation |
| AMIGO | Approximate M-constrained Integral Gain Optimization |
| ANSI | American National Standards Institute |
| API | Application Programming Interface |
| ARC | Advanced Rooftop Control |
| ARX | Autoregressive Exogenous |
| ASHRAE | American Society of Heating Refrigerating and Air-conditioning Engineers |
| AutoDR | Automated Demand Response |
| BACnet | Building Automation Control Network |
| BAS | Building Automation System |
| BASs | Building Automation Systems |
| BDDI | Building Data-Dependent Identification |
| BMS | Building Management Systems |
| BOS | Building Operating System |
| BRCM | Building Resistance–Capacitance Modeling |
| BTMS | Building Thermal Management Systems |
| CC | Cooling Capacity |
| CCT | Correlated Color Temperature |
| CF | Compressor Frequency |
| CFD | Computational Fluid Dynamics |
| CHP | Combined Heat and Power |
| CPP | Critical Peak Pricing |
| CRAC | Computer Room Air Conditioning |
| CRI | Color Rendering Index |

| | |
|---|---|
| DAE | Differential-Algebraic Equation |
| DDC | Direct Digital Control |
| DER | Distributed Energy Resource |
| DIC | Decentralized Integral Controller |
| DMPC | Distributed Model Predictive Control |
| DMPC | Decentralized Model Predictive Control |
| DMPC | Deterministic Model Predictive Control |
| DoD | Department of Defense |
| DOE | U.S. Department of Energy |
| DR | Demand Response |
| EBI | Enterprise Building Integrator |
| EC | Equipment Controller |
| EEV | Electronic Expansion Valve |
| EIA | U.S. Energy Information Administration |
| EKF | Extended Kalman Filter |
| EMCS | Energy Management and Control Systems |
| EMS | Energy Management System |
| ESS | Energy Storage System |
| EV | Electric Vehicle |
| FCV | Finite Control Volume |
| FEM | Finite Element Method |
| FERC | U.S. Federal Energy Regulatory Commission |
| FPM | Fault Propagation Metric |
| FTOC | Finite Time Optimal Control |
| HEMS | Home Energy Management System |
| HVAC | Heating, Ventilation, and Air conditioning |
| IAQ | Indoor Air Quality |
| IBA | Illumination Balancing Algorithm |
| IDF | Indoor Fan |
| IEC | International Electrotechnical Commission |
| IFTTT | If This Then That |
| IG | Internal Gains |
| IMC | Internal Model Control |
| IoT | Internet-of-Things |
| ISO | International Organization for Standardization |
| IT | Information Technology |
| LC-DMPC | Limited-Communication-Decentralized MPC |
| LDR | Light-Dependent Resistor |
| LED | Light-Emitting Diode |
| LESA | Lighting-Enabled Systems and Applications |
| LLMSE | Linear Least Mean Square Error |
| LMI | Linear Matrix Inequality |
| LPV | Linear Parameter Varying |
| LQ | Linear Quadratic |
| LTI | Linear Time Invariant |

| | |
|---|---|
| LTM | Light Transport Matrix |
| MAP | Maximum A-Posteriori |
| MB | Moving Boundary |
| MEM | Micro-Electro-Mechanical |
| ME-VCS | Multi-Evaporator Vapor Compression System |
| MILP | Mixed Integer Linear Program |
| MIMO | Multi-Input/Multi-Output |
| MMSE | Minimum Mean Square Error |
| MPC | Model Predictive Control |
| MPC | Model-Based Predictive Control |
| MPI | Message Passing Interface |
| MRF | Markov Random Field |
| OAT | Outdoor Air Temperature |
| ODE | Ordinary Differential Equation |
| ODF | Outdoor Fan |
| OLF | Optimality Loss Factor |
| OOP | Object-Oriented Programming |
| OpenADR | Open Automated Demand Response |
| PC | Plant Controller |
| PDE | Partial Differential Equation |
| PI | Proportional Integral |
| PID | Proportional–Integral–Derivative |
| PLCs | Programmable Logic Controllers |
| PMV | Predicted Mean Vote |
| PPD | Predicted Percent Dissatisfied |
| pQP | parametric Quadratic Program |
| PRV | Pressure Regulating Valve |
| PV | Photovoltaic |
| PWA | Piecewise Affine |
| PWM | Pulse Width Modulation |
| RC | Resistance–Capacitance |
| RF | Radio Frequency |
| RGB | Red–Green–Blue channel |
| ROI | Return-On-Investment |
| RTP | Real-Time Pricing |
| RTU | Rooftop Forced-Air Packaged Unit |
| SBPI | Simulation-Based Policy Improvement |
| SCADA | Supervisory Control And Data Acquisition |
| SCR | Smart Conference Room |
| SICS | Swedish Institute for Computer Science |
| SISO | Single-Input-Single-Output |
| sMAP | Simple Measuring and Actuation Profile |
| SMB | Switched Moving Boundary |
| SMPC | Stochastic Model-Based Predictive Control |
| SQP | Sequential Quadratic Programming |

| | |
|---|---|
| SVM | Support Vector Machine |
| TEDA | Tianjin Economic-Technological Development Area |
| TEV | Thermostatic Expansion Valve |
| TEV/TXV | Thermostatic Expansion Valve |
| TOU | Time-Of-Use |
| UKF | Unscented Kalman Filter |
| UPS | Uninterruptible Power Supply |
| VAV | Variable Air Volume |
| VCC | Vapor Compression Cycle |
| VCS | Vapor Compression System |
| VFD | Variable Frequency Drive |
| VRF | Variable Refrigerant Flow |
| ZC | Zone Controller |

# Chapter 1
# Introduction and Overview

**John T. Wen and Sandipan Mishra**

## 1.1 Overview

The built environment is an integral part of modern human society. Buildings offer shelter from the elements, security from predators, privacy from intrusion, and space for work, entertainment, and storage. Buildings require numerous resources to operate—natural gas for heating and cooling, electricity for lighting and equipment, water for drinking and cleansing, and fresh air for ventilation. This resource consumption also results in significant environmental footprint, with waste effluence and direct $CO_2$ emissions from heating and cooking. As buildings become larger, more complex and multifunctional, resource demand and environmental impact have dramatically escalated. In the most recent report from the U.S. Energy Information Administration (EIA), in 2015, about 40% of total U.S. energy consumption, and almost 75% of the electricity, was from residential and commercial buildings [1]. Despite intensive conservation effort, the projected consumption for the next two decades remain relatively flat. As shown in Fig. 1.1 [1], major energy consumers in buildings are heating, cooling, refrigeration, and lighting. The category "other" represents electronic devices and networking equipment. As these devices are not covered under energy efficiency regulations, their energy consumption is projected to have significant growth.

With typically large numbers of occupants and diverse usage of space in commercial buildings, automated building management involving coordination of multiple subsystems is becoming essential. These subsystems include heating, ventilation, and air conditioning (HVAC), lighting, power distribution, transportation, telecommunication, plumbing, access and evacuation control, security and monitoring, and

J.T. Wen (✉) · S. Mishra
Rensselaer Polytechnic Institute, Troy, NY 12180, USA
e-mail: wenj@rpi.edu

S. Mishra
e-mail: mishrs2@rpi.edu



J.T. Wen and S. Mishra (eds.), *Intelligent Building Control Systems*,
Advances in Industrial Control, https://doi.org/10.1007/978-3-319-68462-8_1

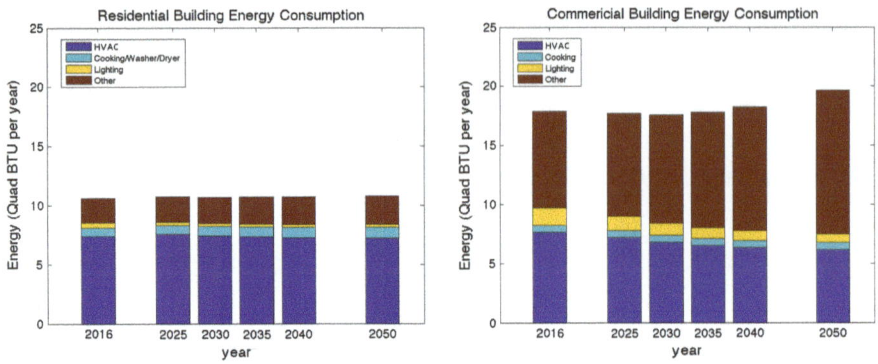


(a) Residential sector delivered energy consumption (quad BTU/year)

(b) Commercial sector delivered energy consumption (quad BTU/year)

**Fig. 1.1** Key energy consumers in residential and commercial buildings (data from [1])

fire safety. As sensors, embedded controllers, microprocessors, and network communication become more powerful and lower cost, it is now feasible, and even essential, to model the building dynamics and coordinate various building control subsystems to achieve energy efficiency, occupant comfort, and safety.

The need and opportunity for energy savings and streamlined building operations has spurred growing interest in building components, devices, and systems. There are now numerous journals dedicated to building systems, including significant emphasis on building management, such as top publications listed from Google Scholar: *Energy and Buildings*, *Building and Environment*, *Building Research and Information*, *Buildings*, and others. There are also annual conferences dedicated to building science and technology, such as *BuildSys, American Society of Heating, Refrigerating and Air-Conditioning Engineers (ASHRAE) Conferences, Greenbuild International Conference*, among others. Professional societies such as ASHRAE, ASME, and IEEE are leading the standard setting and technology dissemination.

The orchestration of multiple networked embedded processors, sensors, and actuators in a highly coupled system to achieve desired operation and performance is the essence of *intelligent building control*, the focus of this book. The need for such technology creates an ideal opportunity for control engineers and researchers. Indeed, the controls community has been gaining increasing interest in building control, bringing a variety of tools to building management. Major conferences such the American Control Conference (ACC), Conference on Decision and Control (CDC), International Federation on Automatic Control (IFAC) now regularly feature multiple sessions dedicated to building control systems.

Because of the large share of energy consumed by the HVAC subsystem, thermal management is typically the most well-studied aspect of building control. The basic formulation of the problem is on regulating air flow into each zone to maintain a user-specified temperature profile in the presence of ambient variability, including outside temperature, solar radiation, occupant, and equipment heat generation.

The system typically involves multiple "zones" equipped with thermostats and controller regulating heated or cooled air input. The heat transfer within and between the zones, including the outdoor as a zone, may involve conduction, convection, and radiation. Various levels of model complexity are possible, from the high fidelity, but time-consuming, finite element method (FEM) [7] to lumped thermal resistance–capacitance (RC) models [9]. For control analysis and synthesis, the RC model is typically used. The RC model is linear, but possibly time-varying because of window/door opening, and the environmental heat load would certainly change with time of day due to occupancy, outdoor condition variations. If the system input is considered to be heated into each zone (positive for heating, negative for cooling) and the system output is the (averaged) zone temperature, the control system is a coupled multi-input/multi-output (MIMO) linear system. Because of the RC approximation, the system is also strictly passive [14], meaning that the overall system is energy dissipating. If the input is the mass flow rate, it would introduce a nonlinear input-to-state map. There are also multiple constraints related to the air handling unit (bound on the maximum mass flow), supply air temperature, and outside air temperature. Many control strategies take advantage of the inherent passivity and introduce decentralized controllers, such as zone-based bang-bang or proportional-integral (PI) control, to more centralized optimal control, balancing temperature control with energy consumption. A particularly popular approach is the model predictive control (MPC) [6] which applies finite horizon optimization based on the current measured condition. Energy consumption may be additionally considered in an outer loop, from automatic temperature setpoint scheduling (e.g., as in night setback) to adjustment based on energy pricing as in demand response, which has been used extensively in smart grid [4]. There has also been explicit incorporation of user feedback/incentive to balance between user comfort and energy efficiency. This could range from presenting energy usage information to active user participation in providing comfort feedback [3].

As the thermal zones in buildings today could be far-flung, network implementation and communication delay are important considerations in control system implementation. The control architecture tends to contain tight distributed local loops (based on sensors and actuators in each zone) with an outer loop managing the overall system-wide coordination. The software architecture in such distributed control and communication network needs to be carefully designed to allow efficient information exchange and avoid delays and interruptions. There are low-level network communication protocols available such as BACnet, LonWorks, etc. But the critical middleware layer still tends to be proprietary (see the discussion in Chaps. 2 and 3). Some exceptions are from the automation, robotics, or smart grid community, e.g., Data Distribution Service (DDS) [2] and their commercial vendors, RTI [11] and PrismTech [8] and related work in Industrial Internet [5], Robot Operating System (ROS) [10], Volttron (led by Department of Energy for smart grid) [12], and our home-grown (at Rensselaer Polytechnic Institute) object-oriented system Robot Raconteur [13], etc. Some of these systems have been used in building automation and control architectures in various levels of sophistication. There is also a growing trend toward open-source software to encourage broad community involvement and

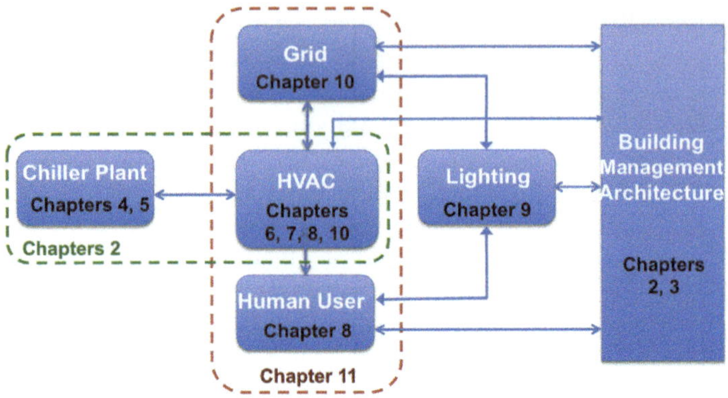


**Fig. 1.2** Structure and coverage of this book

adoption. With the advent of the Internet-of-Things (IoT), there are many important emerging issues in a distributed control architecture, issues such as plug-and-play and self-commissioning (e.g., smart lighting, cameras, and other sensors), and of course the increasingly ubiquitous issues of privacy and cybersecurity.

This book will present some important recent advances in intelligent building control, with both industrial and academic perspectives. Most of the authors are from the U.S., but perspectives from Europe and China are also included. The major emphasis is on thermal management, with consideration of the whole building operation, and some aspects beyond thermal system, including lighting, smart grid, and IoT. The structure of the book is shown in Fig. 1.2. Chapters 2 and 3 consider the overall building management architecture. Chapter 2 focuses on the HVAC and chiller plant, and Chap. 3 considers the software architecture. Chapters 4 and 5 address the modeling and control of the vapor compression cycle typically used in the chiller plant, which supplies the cooling water to the HVAC system, and the refrigeration system. Chapter 4 provides a detailed review of the modeling and surveys various control techniques. Chapter 5 considers the application MPC to the control of vapor compression cycles. Chapters 6–8 all address temperature control with building HVAC systems. Chapter 6 uses a graphic model to analyze multi-zone buildings and surveys various control techniques. Chapter 7 applies MPC to multi-zone temperature control. Chapter 8 explicitly considers human comfort in the optimization problem and uses a distributed optimization approach for its solution. Chapters 9–11 go beyond the building thermal control. Chapter 9 addresses distributed lighting control of the increasingly popular LED lighting. Chapter 10 addresses MPC-based coordinated HVAC and smart grid control. Chapter 11 examines the use of pervasive sensors to facilitate building management, from energy optimization to building evacuation.

## 1.2 The Building Control System

The first part of the book considers the building management architecture and the software platform.

Chapter 2 "Architectures and Algorithms for Building Automation – An Industry View" brings an industrial perspective on the state of the art in building automation. It focuses on the multifunctional aspect of commercial buildings and provides an overview of *Building Automation System (BAS)* as a commercial product. The chapter covers three examples of BAS, pointing out the opportunity and challenges in real-world implementations:

1. Adaptive control for HVAC based on model predictive control principles, with models automatically updated as more data are collected. The algorithm is executed in the cloud, enabling applications where on-site computing resources may be limited. The chapter also presents results from a pilot implementation.
2. Central plant optimization for campuses and large commercial buildings. To achieve significant energy savings, multiple models and forecasts are integrated into an optimization scheme. The solution has been implemented at a large military base, although issues with data quality did not permit reliable validation.
3. Automated demand response, leveraging connectivity with the smart grid. In particular, the chapter discusses the development of the OpenADR standard, which facilitates interoperability between applications worldwide.

Chapter 3 focuses on the software platform to integrate networked devices, such as smart thermostats, in new and existing buildings. As the market rapidly evolves, both commercial and open-source software platforms emerge. In the academic literature, this software is frequently called *Building Operating System (BOS)*. This chapter reviews previous recent work on BOS and describes one common architecture and its features. Since most of the research has not focused on controls, the authors propose an extension of existing BOS that aims at facilitating the task of writing, testing, and deploying control sequences. The chapter illustrates the design and development of such software, and presents preliminary test results. Finally, this chapter also discusses some of insights gained from implementation in a building, and the challenges and future work necessary to advance this area of research.

## 1.3 HVAC Control

The second part of the book focuses on the HVAC control system. Chapters 4 and 5 consider the vapor compression cycle (VCC) in the chiller plant that supplies the cooling water to the HVAC system, while Chaps. 6-8 consider HVAC control in multi-zone buildings.

Chapter 4 "HVAC System Modeling and Control: Vapor Compression System Modeling and Control" presents an overview of the modeling and control of VCC. A basic vapor compression cycle consists of two heat exchangers, an expansion valve, and a compressor. An ideal thermodynamic cycle utilizes four processes to transfer heat between the external environment and the conditioned space: heat absorption and evaporation in the evaporator, compression, heat rejection and condensation, and expansion. During steady-state operation, refrigerant flows through the system to efficiently transfer energy and maintain occupied spaces within desired temperature and humidity levels. This chapter presents dynamic modeling and control techniques to address transient operations when the system undergoes varying heat loads.

Chapter 5 illustrates an industrial multi-zone VCC control system. Controller requirements represent desired performance of production-scale equipment, including baseline requirements common in control literature such as constraint enforcement, reference tracking, disturbance rejection, and also extended requirements necessary for commercial application, including selectively deactivating zones, implementable on embedded processors with limited memory/computation, and compatibility with demand response events. This chapter presents a controller architecture based on model predictive control to meet the trade-offs inherent in this large set of requirements. Experiments are also presented demonstrating constraint enforcement and automatic deactivation of zones.

Chapter 6 "Multi-zone Temperature Modeling and Control" gives an overview of multi-zone building temperature modeling and control. It particularly focuses on model identification. Depending on the application, building modeling approaches range from data-based to physics-based. For models that require data for identification, the building dynamics must be sufficiently excited by system input signals. For models that employ physical parameter data, the model may not capture all of the important underlying mechanisms. Ideally, the entire system model would be easily reconfigurable in order to account for changes to the building geometry and the effects of switching off and replacing components for routine maintenance. This chapter is motivated by these requirements to generate a system modeling framework that is rooted in physical understanding of the system, but is general enough to account for unique features of buildings and flexible enough to be easily reconfigurable. Using a single-zone example, the chapter discusses three categories of modeling approaches: black-box, gray-box, and white-box.

Chapter 7 "Distributed Model Predictive Control for Forced-Air Systems" focuses on control design of forced-air systems. The chapter covers thermal modeling for controls, predictive control design, and implementation of such controllers in real-world buildings. The currently available computational platforms and communication systems are reported. Our main objective is to discuss how advanced control relates to the existing building practices; in particular, a distributed control logic "Trim and Respond" is described in detail. The "Trim and Respond" logic is shown to match a one-step explicit distributed model predictive controller. The chapter concludes with an algorithm for advanced distributed model predictive control that is implementable on existing distributed building control systems.

Chapter 8 "Human-in-the-loop Thermal Management for Smart Buildings" includes user feedback in HVAC control, which is absent in most building thermal control systems. Consider office floors with cubicles, conference rooms, student dorms, homes, and other multi-occupant spaces where temperature set points on thermostats are chosen irrespective of the number of occupants and their individual preferences. This existing approach is not only nonuser-centric but also suboptimal from both energy consumption and occupant satisfaction/productivity perspectives. Individual occupant's feedback and preference can be obtained through wearable sensors or through applications on personal devices such as smartphones. This chapter proposes algorithms that take into account each occupant's preferences along with the thermal correlations between different zones in a building, to arrive at optimal thermal settings for all zones of the building in a coordinated manner.

## 1.4 Beyond HVAC: Lighting, Power, and Network

Chapters 9-11 examine some of the issues beyond building HVAC control.

Chapter 9 "Smart Lighting Control Systems" considers illumination control for light-emitting diode (LED)-based lighting systems. Illumination is a key factor in creating a comfortable building interior and affects health, productivity, mood, comfort, security, and safety of the occupants. Moreover, about 404 billion kilowatt hours (kWh) of electricity were used for artificial lighting by the residential sector and the commercial sector in the United States in 2015, which is about 15% of the total electricity consumed by both of these sectors and about 10% of total U.S. electricity consumption. The new generation of lighting systems takes advantage of recent developments in solid-state lighting and the advances in the state-of-the-art spectral and occupancy sensing technology to deliver functionality beyond just illumination. These lighting fixtures can mix light from different sources to create efficient, high-quality, and healthy illumination. Moreover, such systems also aim to use the data acquired by various sensors in the illuminated space to estimate the number and location of the occupants, user preferences, available daylight, etc. The lighting control algorithm then uses this information to intelligently adjust the input signals to individual LEDs and achieve the desired control objective in an energy-efficient manner. This chapter presents a comparative study of several state-of-the-art feedback control algorithms for advanced lighting systems by experimental evaluation on a standard test bed. Four algorithms are selected and implemented in a standard test bed and experimentally evaluated using various metrics such as power consumption, color quality, etc.

Chapter 10 "Energy Management Systems for Intelligent Buildings in Smart Grids" describes energy management frameworks for buildings in a smart grid scenario. An energy management system (EMS) is responsible for optimally scheduling end-user smart appliances, heating systems, ventilation units, and local generation devices. This chapter discusses the performance and the practical implementation of novel stochastic MPC schemes for HVAC systems, and illustrates how these schemes take into account several sources of uncertainties, e.g., occupancy and weather conditions. It further shows how to integrate local generation capabilities and storage systems into a holistic building energy management framework.

Chapter 11 "Controlling the Internet of Things – from Energy Saving to Fast Evacuation in Smart Buildings" considers the application of IoT in smart buildings, in which occupant-oriented wireless sensor network collects information on both the supply and the demand. This chapter discusses how control formulation could simultaneously improve energy efficiency, comfort, and safety. The specific topics addressed include heterogeneous information fusion-based building occupant-level estimation, energy management of smart microgrid of buildings, and building-mounted wind turbines for electric vehicle charging and fast evacuation.

## References


1. Annual energy outlook (2017) with projection to 2050, U.S. Energy Information Administration, 2017. www.eia.go/aeo
2. DDS: The proven data connectivity standard for IoT, Object Management Group 2016. http://portals.omg.org/dds
3. Erickson VL, Cerpa AE (2012) Thermovote: Participatory sensing for efficient building HVAC conditioning. In: Proceedings of the Fourth ACM Workshop on Embedded Sensing Systems for Energy-Efficiency in Buildings, pp 9–16
4. Gatsis N, Giannakis GB (2011) Residential demand response with interruptible tasks: Duality and algorithms. In: 50th IEEE Conference on Decision and Control and European Control Conference (CDC-ECC). pp 1–6
5. Industrial Internet Consortium, Industrial Internet Consortium, (2016). retrieved Nov 2016. http://www.iiconsortium.org/
6. Ma Y, Borrelli F, Hencey B, Coffey B, Bengea S, Haves P (2012) Model predictive control for the operation of building cooling systems. IEEE Transact Control Syst Technol 20(3):796–803
7. McBee BK (2011) Computational approaches to improving room heating and cooling for energy efficiency in buildings. Ph.D. dissertation, Virginia Tech
8. Prismtech Vortex platform, PrismTech, (2016). retrieved Nov 2016. http://www.prismtech.com/vortex
9. Riederer P, Marchio D, Visier J, Husaunndee A, Lahrech R (2002) Room thermal modelling adapted to the test of HVAC control systems. Build Environ 37(8):777–790
10. ROS, Open Source Robotic Foundation (2014) Accessed 24 Nov 2014. https://ros.org
11. RTI Connext DDS platform, Real Time Innovation, Inc (2016) retrieved Nov 2016. https://www.rti.com/products/dds/
12. Volttron Github, Department of Energy, (2016). retrieved Nov 2016. https://github.com/VOLTTRON/volttron
13. Wason Technology, Robot Raconteur, (2014) Accessed 28 Jan 2017. https://robotraconteur.com
14. Wen JT, Mishra S, Mukherjee S, Tantisujjatham N, Minakais M Building temperature control with adaptive feedforward. In: 52nd Conference on Decision and Control (CDC), pp 4827–4832 2013

# Part I
# Building Level Design and Control Architectures

# Chapter 2
# Architectures and Algorithms for Building Automation—An Industry View

**Petr Stluka, Girija Parthasarathy, Steve Gabel and Tariq Samad**

## 2.1 Introduction

The importance of buildings in developed societies can hardly be overstated. Most of us live, work, and shop in buildings. When we travel, much of our time is spent in airports and hotels, restaurants, and museums. Our children are educated in schools and colleges. When we are ill, or for preventative purposes, we go to hospitals and other buildings housing healthcare services. Government offices, data centers, sports complexes, and law courts are other prominent examples.

If there is one parameter that highlights the prominence of buildings in society, it is energy. Altogether, buildings are responsible for over 40% of the energy, and almost 75% of the electricity consumed in the U.S. [1]. The consumption is marginally higher in residences versus commercial buildings. The energy footprint is correlated with the carbon footprint, with 39% of the nation's CO2 emissions coming directly (e.g., natural gas combustion for heating and cooking) and indirectly (emissions from fossil-fueled power plants that are generating electricity that is used in buildings) from buildings [2]. Both in terms of energy consumption and carbon dioxide emissions, buildings are the largest sector—more than industrial plants and transportation.

Much research in buildings is focused on energy efficiency and reducing energy use. Intelligent automation and control technologies in particular have garnered much attention. Specific topics of research include heating, ventilation, and cool-

P. Stluka (✉)
Honeywell Global Labs, Prague, Czech Republic
e-mail: petr.stluka@honeywell.com

G. Parthasarathy
Honeywell Global Labs, Golden Valley, MN, USA
e-mail: girija.parthasarathy@honeywell.com

S. Gabel
Honeywell Global Labs (retired), Golden Valley, MN, USA
e-mail: sgabel1974@gmail.com

T. Samad
Technological Leadership Institute, University of Minnesota,
Minneapolis, MN, USA
e-mail: tsamad@umn.edu



J.T. Wen and S. Mishra (eds.), *Intelligent Building Control Systems*,
Advances in Industrial Control, https://doi.org/10.1007/978-3-319-68462-8_2

ing (HVAC) control; automated demand response for smart grids; optimization of equipment such as compressors and chillers; advanced lighting and comfort control; new vapor compression systems; and thermal energy storage through building envelopes and radiant slabs. All of these topics, among others, are represented in this volume, and some of our own work, which has also focused on several building energy management solutions, is discussed in more detail later.

Before we review this work, there is an important point that bears emphasis. Despite the high energy, electricity, and greenhouse gas footprint of buildings, substantial obstacles exist in incorporating promising, and even validated, research advances into commercial practice. These include the following:

- In many cases, buildings are occupied by tenants who do not pay utility bills themselves; these bills are paid by the building owner or manager. Thus, tenants have little economic incentive to adopt efficiency measures.
- Return-on-investment (ROI) periods on innovative energy-saving technologies, especially where new equipment needs to be installed, are often as long as a few years. The ROI is, of course, exacerbated by the current low price of energy.
- Energy expenditure is often not a significant part of the operating budget. Consider a home, overall, a 5% reduction in electricity use in most residences would have a huge impact on the nation's consumption and emissions, but a $5 reduction in a monthly utility bill may not be sufficient incentive for many homeowners to make the time or financial investment involved.
- Advanced technologies often require appropriately trained technicians and engineers to use them, and most buildings do not have such staff available; hiring such staff would be a significant additional expense.
- Retrofitting new technologies into existing buildings, which were designed and built without the prospect of such future enhancements, is especially expensive. Building lifetimes are typically on the order of a few or several decades, and building equipment typically has a lifetime of a few decades as well; it will be a long while before the building stock can accommodate research advances easily.
- While energy management systems are widely used for large commercial buildings, they are rare in small commercial buildings. Brambley et al. [3] reported that the percentage of buildings equipped with energy management and control systems (EMCS) increases from about 5% for buildings of 1,000—10,000 sq. ft. to about 70% for buildings larger than 500,000 sq. ft.
- Government investment in applied research and technology transfer in building technologies is low. A recent report [4] notes that *the federal government spends more than 30 times as much on research for electricity generation as it does on research on the buildings that consume three-quarters of this electricity*.

It is certainly not the case that energy efficiency plays no role when building developers or refurbishers are specifying a building automation system. But energy is not the sole concern and indeed it is usually not the principal concern. According to one market report, the substantial growth in Building Automation Systems (BASs) compounding at an annual rate of 10.65% to reach over $100 billion by 2022 is being led by security and access control [5].

The multifunctional nature of building automation systems [6] will be evident in the next section of this chapter, where we provide an overview of BASs using a major commercial product as an example and we discuss some related technology trends.

With this background, we then discuss three topics in building automation that leverage some of these technology trends in BASs:

- An adaptive control strategy for HVAC based on model predictive control principles, with models automatically updated as more data is collected. The algorithm is executed in the cloud, enabling applications where onsite computing resources may be limited. Results of pilot implementations are presented.
- Central plant optimization for campuses and large commercial buildings. Multiple models and forecasts are integrated into an optimization scheme. The solution was implemented at a large military base although issues with data quality did not permit reliable validation.
- Automated demand response, leveraging connectivity with the smart grid. In particular, we discuss the development of the OpenADR standard, which is facilitating applications worldwide.

These and other applications have been successfully implemented, but the path from research to practice can be tortuous. Therefore, in the final section, we present a path for technology transfer to commercial product and we describe some of the challenges involved.

## 2.2 Building Automation Systems

BAS control and monitor mechanical and electrical equipment, such as HVAC, lighting, power systems, fire systems, and security systems. Over the years, BASs have advanced through several major evolution stages [7]. They initially relied on pneumatic controls with compressed air (starting in the 1950s) but later these systems were replaced by microprocessors and Direct Digital Controls (DDC) in the 1980s, subsequently leading to the introduction of standardized building protocols, such as BACnet®, LONWORKS®, or Modbus®, in the 1990s. While the first decade of the new millennium brought significant progress with wide adoption of wireless technologies (ZigBee®, EnOcean®, Z-Wave®, Bluetooth®, etc.) that allowed individual devices and controls to communicate wirelessly, the follow-up trends evolved directly into the new era of the IoT that we experience today.

### *2.2.1 BAS Overview*

BASs deliver multiple functions, including the following:

- Control of the building's environment is primarily delivered through the automated control of the HVAC system and its individual components (air handlers, fan coils, fans, pumps, chillers, boilers, etc.). The other important aspects include the control of lighting and indoor air quality (IAQ).

- Energy management that aims to minimize overall energy costs in the building through the systematic monitoring and intelligent operation of HVAC systems with respect to occupancy, weather, and prices of electricity, gas or other energy sources.
- Monitoring of facility assets has the objective of detecting performance problems of HVAC equipment and addressing them early enough before they cause bigger issues.
- Security and access control help to minimize risks related to security breaches and improve situational awareness. Access control, video surveillance, and perimeter protection systems all together play important roles in mitigation of high-risk threats.
- Fire detection and life safety help to ensure that people and assets are protected from fires and other environmental risks. This is accomplished by the deployment of systems for fire and smoke detection, sprinkler supervision, and emergency communication.

All these functions enable building owners and facility managers to address a variety of operational goals, such as reducing energy consumption and maintenance and life-cycle costs, ensuring tenants' comfort and compliance with regulations (e.g., on the minimum required volume of fresh air in a given building), minimizing safety and security risks, and facilitating active participation in demand response or related energy trading schemes enabled by smart grid technologies [8].

The architectural complexity of today's building automation systems largely depends on the number of subsystems deployed. Figure 2.1 provides a complete view of one leading BAS, Honeywell Enterprise Building Integrator (EBI), with its modules for HVAC control (named Building Manager), energy management, life safety, security, and video surveillance. Each part can be installed independently of others but they together form a complete building management system.

### 2.2.2 HVAC Control Infrastructure

The infrastructure for monitoring and control of HVAC systems is the most commonly implemented part of any BAS, and it is also perhaps the largest and most complicated building system because of the variety of control devices involved and the multiple ways they can affect a building's operation. In Fig. 2.1, the HVAC control infrastructure is depicted in a simplified way inside the block labeled Building Manager. However, when physically deployed, it is usually structured into several logical layers of the traditional HVAC control architecture (see Fig. 2.2), complemented with the cloud environment that allows the implementation of additional functions:

- Field devices comprise the sensors, meters, variable speed drives, valves, and actuators that are used for monitoring or changing system variables, such as temper-

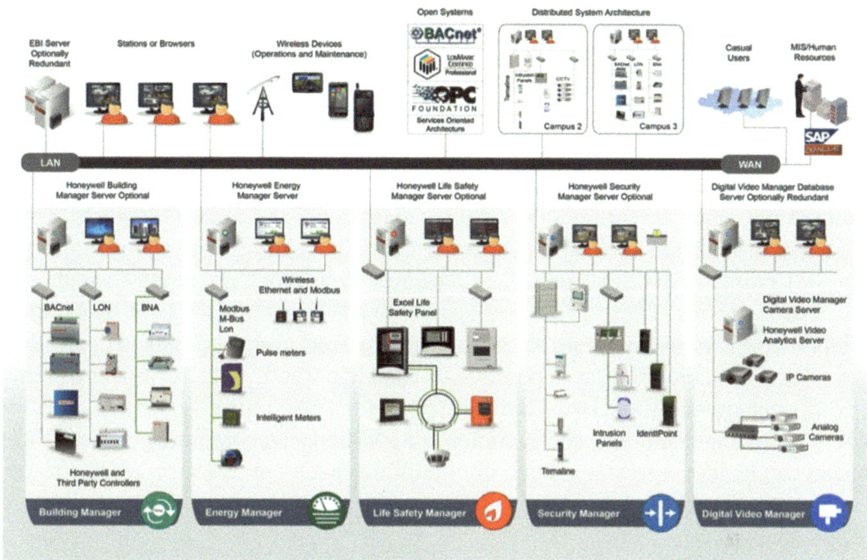


**Fig. 2.1** Architectural view of the BAS system [9]

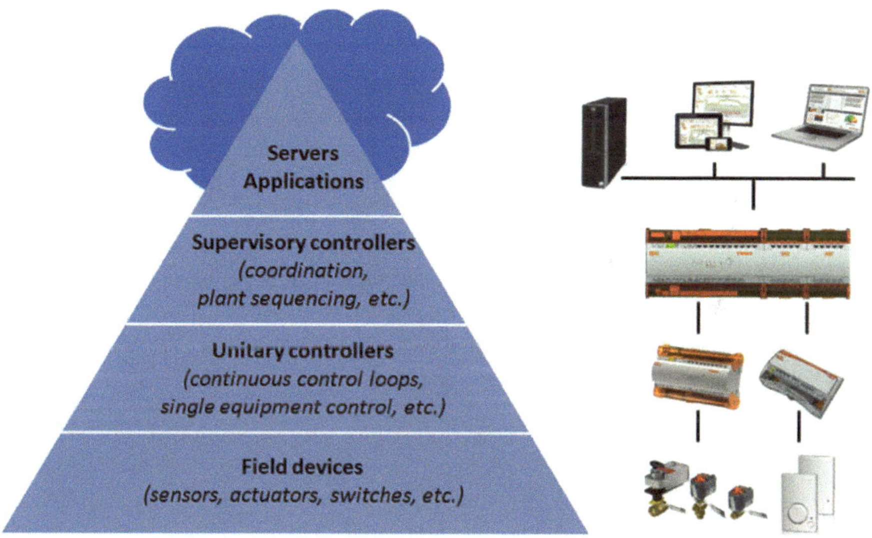


**Fig. 2.2** Hierarchical layers of HVAC control architecture

ature, humidity, flow rate, and pressure. Efficient HVAC control depends heavily on the quality of the field devices and the information these provide.

- Unitary controllers are stand-alone devices executing basic control algorithms and routines, designed for specific control applications, such as controlling a fan

coil unit or sequencing a pump. The scope of the controlled system is usually limited to a specific constrained environment, such as a single zone. The unitary controller receives signals from sensors and determines appropriate output signals for the actuators, based on the closed-loop control logic. Its operation can also be influenced by commands received from the respective supervisory controller.
- Supervisory controllers implement higher level control logic on the level of a plant, subsystem, floor, or entire building. This may include starting or stopping plants (boilers or chillers) according to correct or optimal timing, as well as adjusting the set points of unitary controllers based on conditions defined by weather or occupancy levels. These adjustments are typically implemented by a set of rules.
- Server applications integrate all useful data received from devices and controllers in a local data historian. It is possible to visualize trends, logs, alarms, and other events generated by the BAS. Server applications also typically include tools for system management and configuration. In addition, the computing power at this level can be leveraged for execution of building-wide optimization applications.

### *2.2.3 New and Emerging Trends*

BASs are continuously evolving to efficiently address new challenges and enable flawless and cost-effective operation of high-performance buildings. The whole ecosystem of technologies that are being deployed in buildings has a direct influence on BAS, their architectures and functions. The most important technology trends over the last few years relate to the increasing use of cloud technologies and data analytics [10], prevalence of the IoT paradigm [11], and the growing emphasis on user experience and comfort [12, 13].

The cloud and data analytics have made significant progress over the last years in many domains and they continue to create impact in building applications. The capability to collect data from multiple and potentially heterogeneous data sources and move them to a cloud repository allows the implementation of powerful applications that may provide insights into building operations [14]. Cloud connectivity and real-time processing will enable the data to become fluid versus static with vast new opportunities. The sophistication of new and more powerful building analytics will likely be increasing from visualization and reporting dashboards to fault detection and diagnostics too, ultimately, applications in predictive maintenance and holistic dynamic optimization of buildings [10]. Currently deployed building analytics can better inform facility managers about deviations from the expected energy consumption, likely HVAC equipment faults [15] and underperforming controllers [16].

IoT paradigms enable connecting building automation components to the IT network and generally improves the interoperability and connectivity of control devices. The IoT can help overcome the issue of isolated building systems and support creation of more cohesive environments. In this new context, building automation systems will potentially require fundamental changes in how they are designed and installed. New types of intelligent devices and systems will be required that collect and move data directly to cloud databases, where it can be used by specific analytics applications and exposed to end users. Also, new IoT technologies and applications will have a major impact on building occupants whose interactions with *smart things* could be monitored and leveraged for improved control and delivery of personalized comfort in interior spaces [17].

User experience aspects play an important role in the design of new applications for facility managers and building owners that take advantage of connected HVAC equipment, devices, and automation systems whose data can be shared in real time. New types of applications and user interfaces are delivered via smartphones and tablets that can provide multiple real-time functions such as secure monitoring of equipment operation, changing set points, viewing and acknowledging alarms, and adjusting schedules. The other categories of users are tenants and occupants where the main emphasis is on delivering healthy and comfortable indoor environments [18]. This goes beyond traditional thermal comfort [19]; for example, modern LED bulbs can make lighting conditions far more personalized than before.

*Collapse of control layers*—With the proliferation of the cloud, open architectures, and IoT technologies, we already observe tendencies to separate typical functions into only two levels: highly intelligent field devices and the cloud. Although this direction might be more typical for low-end installation of BAS, e.g., light commercial buildings [20]—it can imply that powerful building controllers may not be needed in some cases [21, 22]. The base level closed-loop control functionality will be implemented through a flat architecture of cooperating field devices, while supervisory functions will be pushed to the cloud environment. This concept can potentially be cheaper to deploy but the overall impact on the performance of such a control architecture still needs to be explored, primarily with respect to the potential issues with latency, jitter, or bandwidth in nondeterministic communication networks.

*Distributed optimization and analytics on the edge*—Economic optimization of building systems can be formulated at the whole-building level, integrating all important subsystems such as HVAC, lighting, onsite generation, and storage. However, the fundamental issue with this approach lies with building-wide optimization models, which will always be too complex and hampered by significant inaccuracy, uncertainty, and lack of data measurements. On the other hand, distributed optimization approaches could be more viable; these would focus on meaningful subsystems and their optimization according to their local objectives but not independently of others. The topic of distributed cooperative control has already been studied in the areas of renewable generation [23], power storage [24], and control of HVAC systems [25–27].

*Human-in-the-loop control*—Given the increased emphasis on user experience and occupants' health and productivity, the thermal comfort and other environmental aspects of buildings, such as lighting quality, should be maintained in a way that satisfies the maximum number of occupants. This can be achieved by allowing individuals to define their personal comfort preferences and providing immediate feedback on the current comfort conditions [28, 29]. Then new algorithms will be needed to aggregate and properly process all such inputs from occupants to determine new global set points in the most cost-effective way, or alternatively make localized adjustments that would respond directly to occupants' feedback. Beyond optimized comfort control, similar crowd-sourcing mechanisms have already proven to be useful for learning of occupancy patterns [30], one of the difficult-to-measure disturbance variables in HVAC control applications. Occupancy and behavior patterns can also be learned by processing of the data generated by indoor location tracking systems [17] under the assumption that the new family of smartphone-based context-aware applications would be properly integrated with BAS where they will be leveraged for improved control.

All these technological trends are expected to have an enormous impact on the architecture of BAS and the mechanisms they use for delivering control functions.

## 2.3 Adaptive HVAC Control: A Cloud-Based Solution

### *2.3.1 Rule-Based Methods*

Rule-based methods enable the translation of best practices, experience, and knowledge of HVAC control engineers into a set of rules, which are applied to manipulate key set points and schedules (optimal start/stop, pre-cooling, etc.) and ensure coordination between controllers. For instance, a rule-based control strategy for air handlers can involve the supply air temperature reset, night purge, CO2-based demand-controlled ventilation, and other concepts; see summaries in [31, 32].

Rule-based methods are popular because their implementation is intuitive and offers good opportunities to run HVAC more efficiently, under the assumption that rules are implemented properly. In practice, this approach has several limitations. With respect to the large variety of building types—and variety of HVAC systems used—application engineers have to configure customized solutions on a project-to-project basis, and the quality of rules may vary significantly with the knowledge and experience of the application engineer. Further, if the set of rules becomes too extensive, it is hard to ensure consistency within the rule set and the overall performance will deteriorate sooner or later. In other cases, the application engineers may have fairly limited time to tune the rules properly and keep the configuration regularly updated. Then, the natural tendency is to choose robust parameter settings for individual rule resets that will ensure occupant comfort for a wide range of conditions. However, this approach will control the HVAC system in a suboptimal way and with higher operating expenses.

### *2.3.2 Model Predictive Control*

An attractive possibility for addressing the above limitations is represented by the model predictive control (MPC) strategy, which can dynamically adjust all main

HVAC set points based on current/future conditions. Unlike a safely defined set of set point resets, which requires adjustments relatively rarely and thus operates the HVAC system in more or less constant regimes for long periods of time, MPC has the capability to adjust set points several times per hour, operate the HVAC system more efficiently and closer to its boundaries, and ensure it is responding dynamically to changes in outdoor or indoor conditions.

An MPC-based solution is usually formulated to address the primary goal of HVAC control: to maintain predefined comfort levels in zones while minimizing the overall operating costs, which are usually reduced to the costs of primary energy sources. The MPC controller then determines the optimal actions based on relationships among optimized variables, zone comfort, and energy cost. Mathematically, the optimal control problem is formulated over a finite future horizon:

$$\min_{u_i} \sum_{k=0}^{N-1} c_k(x_k, u_k, y_k) \tag{2.1}$$

subject to

$$\begin{aligned}
&x_0 = x \\
&x_{k+1} = f(x_k, u_k, d_k) \qquad y_k = g(x_k, u_k, d_k) \\
&y_{\min} \le y_k \le y_{\max} \\
&u_{\min} \le u_k \le u_{\max},
\end{aligned}$$

where $k$ is the discrete time step, $N$ is the prediction horizon, and $c_k$ is the cost function. $x$, $y$, $u$, and $d$ are vectors defined as follows:

- $x$ is a vector of system state variables that characterize conditions in zones (predicted thermal comfort, heating/cooling demand) or in the HVAC system (mode of operation).
- $y$ is a vector of system output variables, which are maintained as close as possible to their reference values. Output variables are the temperatures in zones and potentially other parameters such as humidity. Their reference values are given by the desired comfort conditions. The other part of output variables includes energy consumption of the HVAC system.
- $u$ is a vector of action variables or set points for supply air temperature, chilled water temperature, hot water temperature, pump speed, fan speed, and others.
- $d$ is a vector of disturbance variables, including usually weather conditions but potentially also occupancy, if available.

The cost function $c_k$ can be formulated as a tradeoff between the precision of tracking reference values (maintaining comfort) and energy costs, as discussed in detail in [33]. A numerical solver is typically used to minimize the function $c_k$ over the defined optimization horizon $N$ while keeping future comfort variables $y$ as close as possible to their reference values and all set points satisfying the box constraints $u_{\min}$ and $u_{\max}$ (e.g., pump speed between 60 and 100%).

### 2.3.3 *Implementation of Adaptive MPC*

A specific implementation of MPC was pursued by the Honeywell team with the objective of delivering a concept that would be easy to commercialize [16]. Taking into account the specifics of the building controls industry, two important design decisions were made:

- The MPC controller was implemented as a cloud-based solution, following the high-level architecture depicted in Fig. 2.3 where the existing BAS is connected with the cloud controller via a dedicated communication interface.
- The standard MPC scheme was extended by a module for regular adaptation of the predictive model for state variables. This need came from the fact that in many application projects there are not enough observations available to identify good models, while running step tests to get a dynamic response of building systems is impossible due to the costs involved. Model adaptation is also useful in the long run because built environments are always subject to change.

The execution engine (see Fig. 2.3) is initialized every 15 min by a timer and it runs through the following sequence of steps.

1. The engine receives new data from the local BAS, which includes the latest sensor and meter readings as well as all relevant control signals and other parameters. Some of the data points need to be processed; for instance, cumulative meter consumption is converted to interval consumption. Then all new points are inserted into the data storage.
2. Regular update of model parameters is initiated with respect to the recent data and applied to all predictive models, which are typically specific instances of a general multi-input multi-output autoregressive exogenous (ARX) class model. This includes the state model, disturbance model, and energy consumption model that

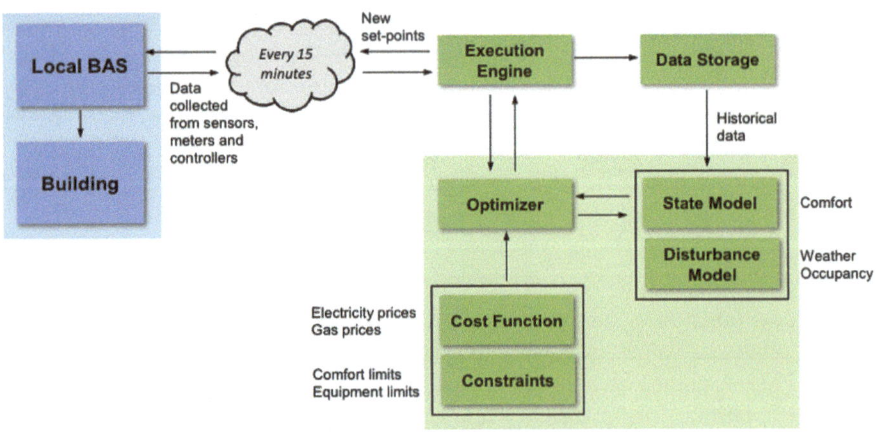


**Fig. 2.3** High-level architecture of the cloud-based MPC controller

is used for evaluation of the cost function. New model parameters are estimated in a robust way ensuring iterative removal of outliers.

3. The optimizer runs a multistep optimization algorithm that is applied over a configurable time horizon, at least four steps or 1 h ahead. The previously updated models are used to predict future disturbances, states, and outputs. For practical reasons, two types of limits—soft and hard–were introduced for comfort variables. The optimizer then searches for a combination of actions that will ensure all comfort variables are within soft limits, or at least do not violate hard limits. If a hard limit is reached in any of the building zones, the optimizer just firmly defines set points that correspond to the maximum (or minimum) heating or cooling in a given zone.
4. New set points are transmitted back to the local BAS system. Although the optimizer generates a 1-hour-ahead schedule, only the first step is distributed to the plant controllers, which is in-line with the principle of receding horizon control.

The adaptation of model parameters and identification of alternative model structures helps to accommodate various HVAC system changes that can occur relatively frequently over the solution life cycle. It is then ensured that all model structures and model parameters represent key relationships between optimized set points and estimated energy and comfort variables. This approach also reduces the need for the engagement of a control or optimization expert during the solution setup and maintenance.

### 2.3.4 *Validation*

The adaptive MPC controller was validated at several sites for an extended period of time. Typically, it was able to reduce the HVAC operating costs related to purchases of gas and electricity by 15–40%, which is in-line with savings achieved by MPC in similar applications [34, 35]. The initial set of six pilot buildings was intentionally selected in a way to ensure diversity of HVAC systems ranging from the relatively simple (one boiler, chiller, and air handler, up to five zones) to the rather complex (several boilers and chillers, at least five air handlers, at least 15–20 zones). However, we realized that the levels of savings, and thus the commercial success, are influenced by many other aspects:

- The performance of the baseline solution, i.e., the control strategy that is currently in use, has a significant impact because in the case of poorly operated systems it is easy to achieve savings, while in other cases the bar is higher.
- The level of instrumentation in some building is insufficient for the implementation of the advanced control solution and the need to install new sensors or meters makes the ROI less attractive.
- Legacy control systems may prevent the manipulation of some set points, e.g., chilled water temperature, which means the new concept cannot be utilized to its full potential.

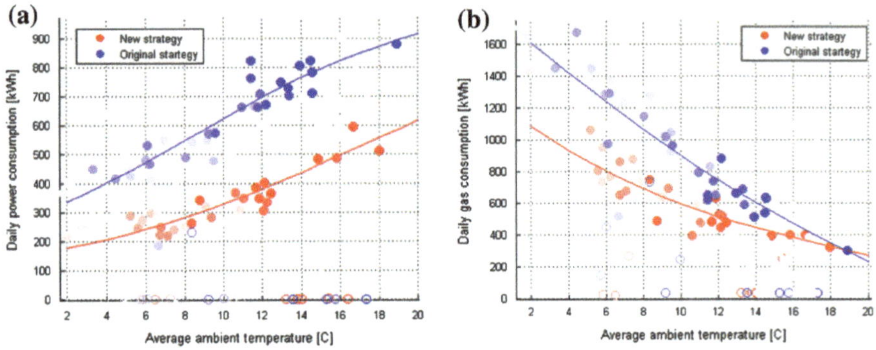


**Fig. 2.4** Daily energy consumptions in power and gas conditioned by average ambient temperature

- In the case of a cloud-based solution, the quality of control can be influenced by such simple things as is the reliability of the Internet connection. Efficient local backup must be ready for the case when this connection drops.

As a part of the validation process, it was necessary to quantitatively assess performance improvements of the new solution in comparison with the original solutions. Despite differences in specifics, these legacy strategies primarily differed in how they manipulated important HVAC set points. Given that the operational patterns of any building follow regular daily cycles, the most suitable validation scenario was to switch between the two solutions on a day-by-day basis. Consequently, the results were assessed over the respective 24 h intervals. But when two different control solutions are running on different days, it is important to consider the different operating conditions, which are primarily characterized by the occupancy patterns and weather conditions. In some cases, it might be reasonable to omit the information about the occupancy. For instance, in the case of an administrative building with stable occupancy patterns during the working days when the individual control strategies are validated just on these days. But in general, the occupancy is an important parameter for the comfort control and should be used wherever it is possible to quantify it. Regarding the weather conditions, the most important influencing factor for the energy consumption is the ambient temperature. Figure 2.4 illustrates results of comparison of the original (baseline) and new control strategy, which were conditioned by the ambient temperature only. Despite some variation, it illustrates the systematic reduction in both power and gas consumptions achieved on our pilot sites.

## 2.4 Central Plant Optimization: Concepts and Prototype

In campuses such as universities and government facilities, large central plants deliver cooling or heating to individual buildings. In large buildings as well, a central plant is usually the primary source of cooling or heating, delivering thermal energy as chilled or hot water to the forced air HVAC system, or to radiators and other terminal units. Improving the efficiency of the generation and distribution of thermal energy reduces energy wastage at the source. A small percentage improvement can produce large overall savings because of the aggregation of energy production and distribution. There is potential for savings because central plants are currently operated to meet all demands reliably and not necessarily for fuel economy or energy efficiency. Plant operators run the equipment according to a preset, fixed strategy. However, plant equipment efficiencies vary with load and external conditions such as ambient temperature. In addition, central plants have multiple chillers, boilers, and power generators, which may differ from each other in capacities and performance. The ability to select equipment and operate it at optimized points to minimize the total energy cost of the plant is not intuitive to plant operators and has the potential to offer great benefits. Modeling the load dynamics offers the additional benefits of predictive optimization, for not just instantaneous energy savings but over future time horizons so that time-varying energy prices or weather-dependent equipment efficiencies can be considered in operating the entire system.

Optimization of HVAC systems and chiller plants has been a topic of research for many years. In [36], we find a comprehensive review of supervisory and optimal control of HVAC systems. An example of a more recent model predictive control of a chiller plant is in [37]. There are several other research implementations and some commercial products. However, many of these tackle either one-off implementations for research, rather than a general advanced optimization software product, or they tackle optimizing only part of the plant. We describe in brief a prototype central plant optimization system that has shown promise from the energy savings achieved in several pilots. The optimization system was recently implemented at a DoD central plant and this implementation is described next [38]. Our objective in this section is not to present another technique for optimization and its benefits, but rather to present the considerations involved in the practice of translating advanced control and optimization solutions developed in a research lab to a profitable commercial offering. In this context, see also [35], where the challenges in implementing model predictive control in buildings are described. Despite a mature prototype, field implementation remains difficult in advanced control applications and we explore the process, architecture, and standardization needed for easy diffusion in the marketplace.

### *2.4.1 Overview of Supervisory Optimization*

The central plant optimization solution provides optimal schedules and operating points for all equipment in the plant. It relies on equipment performance models, forecasted load, a building load model, and energy price information. The equipment and building models are set up based on historical data and updated as new data becomes available. The optimization is based on minimizing energy costs and uses an evolutionary algorithm. The solution concept is illustrated in Fig. 2.5.

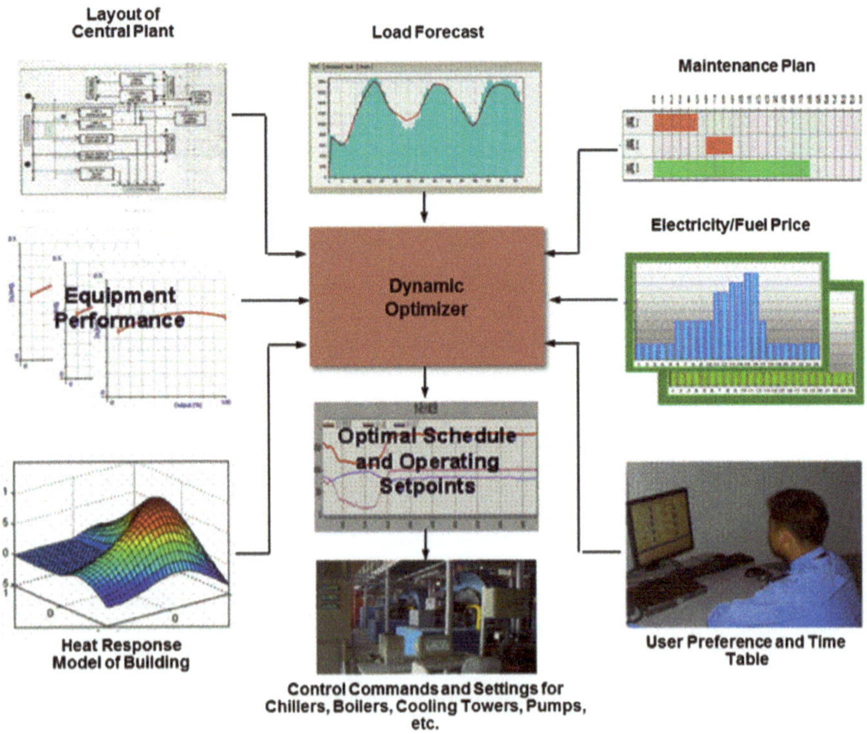


**Fig. 2.5** Technology overview

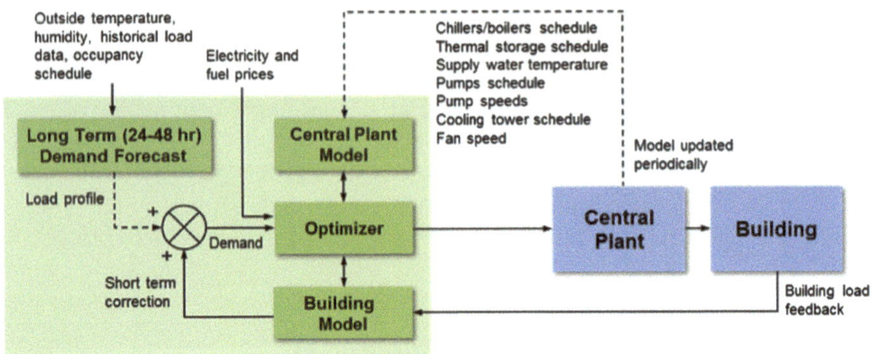


**Fig. 2.6** Optimization solution

The online information flow is conceptualized in Fig. 2.6. A demand forecaster predicts loads for the next 24 h period of optimization based on the current weather, load history data, and occupancy criteria. The central plant model is configured from a library containing models of chillers, boilers, cooling towers, and pumps. A dynamic load model represents the building response to changes in energy supplied. Based on the inputs of upcoming demand loads, central plant performance, and building response, the optimizer solves the schedules and operating commands for the major equipment in the supply and distribution of chilled and hot water. Feedback from the buildings provides corrections to the long-term forecast load that is used to adjust the energy supplied.

The model library is an integral part of the optimization solution. Models are developed using historical data and are periodically updated with newly arrived data. The optimizer models are continuously updated and do not lose their efficacy when the equipment deteriorates.

The central plant optimizer had been piloted previously and has shown promising energy savings ranging from 9% to more than 40%. The energy savings depend on the extent of automation, existing controls, and processes in the baseline operation. The implementation described next is at a site where a cloud-based solution or remote monitoring and support was not possible.

## 2.4.2 Prototype Implementation

The control implementation architecture for a prototype version of the real-time supervisory optimizer implemented at a Department of Defense (DoD) facility (Fort Bragg, NC) [38] is shown in Fig. 2.7. The chiller plant serves about 80 buildings and consists of four chillers (total of 6300 tons), a free cooling heat exchanger, a 2

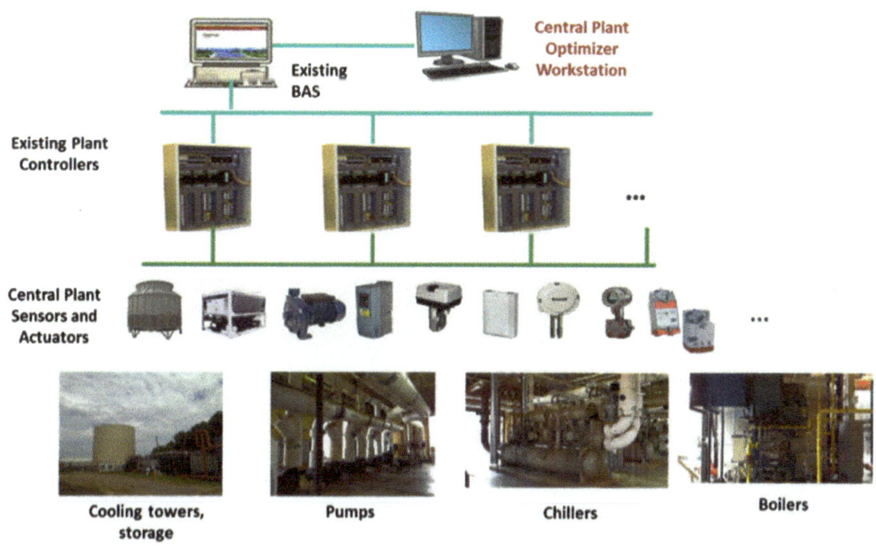


**Fig. 2.7** Implementation architecture

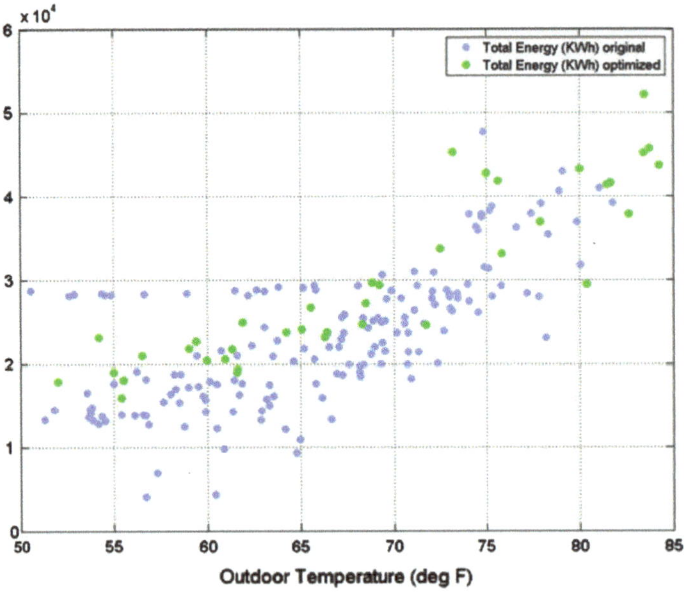


**Fig. 2.8** Daily energy usage comparison

million gallon chilled water storage tank, and several primary and secondary pumps and cooling towers. For the prototype version of the optimizer, the control interface was through the Honeywell EBI building automation system.

The optimizer was brought online by following a systematic and thorough testing and commissioning process. It was handed over to the facility staff after training, but without remote access to the optimizer for observation or troubleshooting. The optimizer software was connected at the chiller plant for about a year, but controlled the plant only when enabled. The prototype included a mechanism for switching between the existing fixed control rules and optimized control to gather data for both systems. The data shows that the optimizer was enabled to operate the plant for 39 days in several continuous periods. During the same period, the data shows 164 periods of original control days.

The original control days data was normalized for weather—we considered several factors and used the best fit model after evaluating combinations of factors and regression model algorithms. The final analysis comparing daily overall energy usage between normalized baseline control and optimized control showed similar energy usage, within one standard deviation in most cases (see Fig. 2.8).

Comparing this overall result with an example of previous promising savings (Figs. 2.9 and 2.10) illustrates the inconsistency in the range of savings. Not only are baselines different for different implementations, but also conditions during operation can be tricky to monitor: Was the optimizer system operating ideally with the correct inputs? Were there other mitigating circumstances? We discuss the issue of

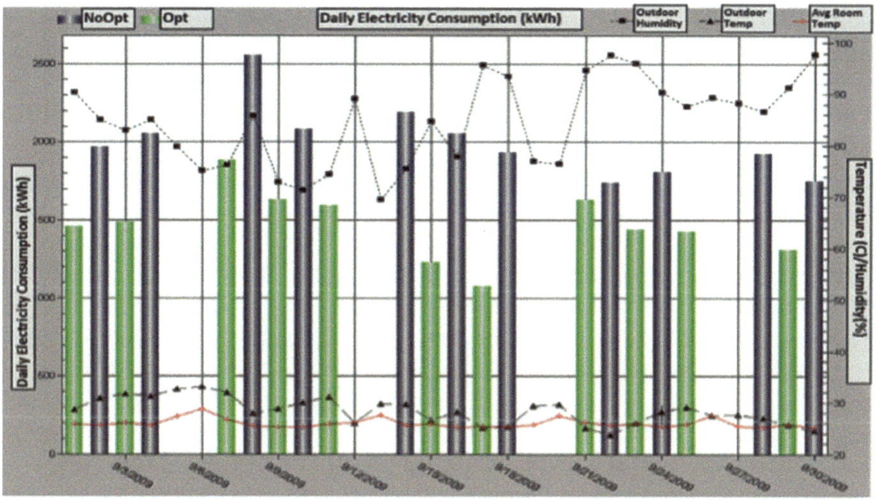


**Fig. 2.9** Daily electricity consumption from the previous pilot with alternate day testing

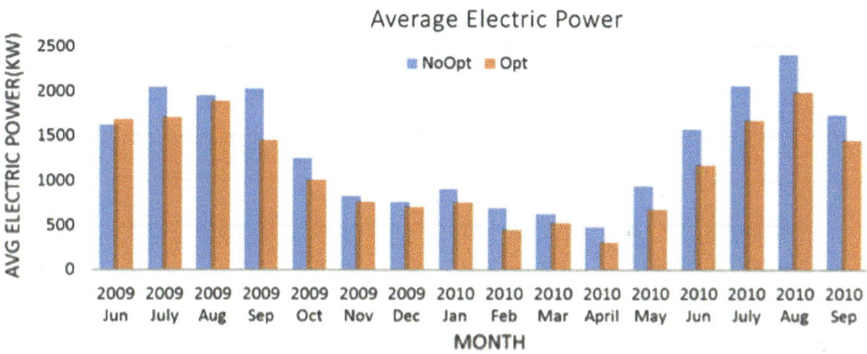


**Fig. 2.10** Average electric power from the previous pilot

data quality in the next section, and further expand on the considerations involved in transferring advanced solutions from concept to commercialization later in this chapter.

### 2.4.3 *Role of Data Quality*

The prototype was the first to be implemented at a site with no possibility of remote observation of the system or ready access by the developers. The post-analysis showed several discrepancies in the data used as optimizer inputs. These included condenser, primary and secondary pump power, cooling tower power, and indoor and outdoor temperatures. Nonoptimal solutions were possibly provided to the plant based on wrong data, although the safety limits and layers in the optimizer prevented incorrect or unsafe operation of the plant. The safety layers also prevented knowledge of non-optimized operation when the optimizer was "enabled". On being apprised of a huge power spike in a condenser pump, the site staff immediately said that was probably why the chiller connected to it was never being switched on: "the optimizer hated chiller #4." The site operations staff are qualified to ensure that the plant operates correctly; however, it needed research labs staff to ensure the optimizer operates with the correct data inputs. In conclusion, as noted above, we were unable to validate the effectiveness of the optimization solution implemented, in large part, we believe, because of data quality issues. The root cause of bad data might have been communication issues or bad sensors. A supervisory optimizer requires data with a quality, resolution, and frequency that are not very common in current BASs. A reliable data infrastructure and data fault detection and adaptation system are essential for providing a supervisory advanced optimization solution. This experience highlights the importance of implementation-related issues that algorithm developers are often unaware of and that are often not resolved in time when research teams implement their prototype solutions. The challenges in transferring technology to the market are not restricted to data infrastructure only, but a combination of several factors that may suppress or augment technological weaknesses. We further discuss the considerations of transferring technology from R&D groups to commercial operation in a later section.

## 2.5 Automated Demand Response: Smart Buildings Meet Smart Grids

In many locations, building owners and operators have an opportunity to reduce their electric energy cost by participating in demand response programs offered by electric utilities or grid operators. These programs provide an economic reward for partially reducing or time-shifting electric demand during peak periods or other times when the electric grid is under stress. The rich sets of operating data available in intelligent buildings are a key enabler to make the best use of these demand response programs.

### *2.5.1 Background*

Until relatively recently, electric utilities and grid operators in the United States generally had an abundance of generating capacity (aside from periods of equipment outages or extreme weather conditions) and had few constraints in using that capacity to satisfy electric demand. As the industry began to retire older less-efficient generating resources, they began to seek ways to make the grid more energy and cost efficient as well as more environmentally friendly. Studies found that in many cases, building owners and other electric customers' demand profiles could be adjusted slightly in order to assist with the balance between supply and demand on the electric grid. Experience has shown that some electric customers are willing to occasionally reduce their demand in return for some form of economic benefit, through demand response programs offered by their electric utility provider. These demand-side reductions can be either directly controllable by the electric utility (e.g., residential HVAC programs) or indirectly controllable at the option of the customer (e.g., for commercial building HVAC, lighting, etc.). Industrial customers have also been able to identify similar demand response (DR) opportunities in their operations. These reductions in electric demand (or "negawatts") are utilized by the utility or grid operator to fill imbalances between supply and demand, with the objective of ensuring grid reliability.

The U.S. Department of Energy (DOE) defines demand response as *a tariff or program established to motivate changes in electric use by end-use customers in response to changes in the price of electricity over time, or to give incentive payments designed to induce lower electricity use at times of high market prices or when grid reliability is jeopardized* [39]. The U.S. Federal Energy Regulatory Commission (FERC) defines a demand response event as *a period of time identified by the demand response program sponsor when it is seeking reduced energy consumption and/or load from customers participating in the program. Depending on the type of program and event (economic or emergency), customers are expected to respond or decide whether to respond to the call for reduced load and energy usage. The program sponsor generally will notify the customer of the demand response event before the event begins, and when the event ends* [40].

## 2.5.2 Participation in Retail and Wholesale DR Programs

Retail DR programs offered by electric utilities are often associated with the customer's electric utility rate tariff. In areas where organized wholesale energy markets exist, the procurement of electricity is typically orchestrated by the grid operator's wholesale market, which can include both supply-side and demand-side energy providers (which can include building owners).

A recent U.S. Department of Energy technical report outlines the utilization of various demand response resources in the planning and operation of the electric grid [39]. DR opportunities for building owners can include both price-based DR and incentive-based DR programs. Examples of price-based DR programs include the following:

- Time-of-use (TOU): an electricity rate having different unit prices for different blocks of time, typically defined across a 24 h day.
- Real-time pricing (RTP): an electricity rate in which the price fluctuates hourly reflecting changes in the wholesale price of electricity. Customers are notified of these prices on a day-ahead or hour-ahead basis.

- Critical Peak Pricing (CPP): a hybrid TOU structure having a provision for replacing the normal peak price with a much higher CPP event price under specified trigger conditions (e.g., when grid reliability is under stress or wholesale prices are very high).

Examples of incentive-based DR programs include the following:

- Demand Bidding/Buyback Program: a retail tariff with a DR option that enables the customer to offer bids to curtail electric demand (typically driven by high wholesale electricity prices).
- Emergency Demand Response Program: an option that provides utility bill credits for load reductions during specified periods (e.g., when there is a shortfall in electricity supply reserves).
- Interruptible/Curtailable Service: a retail electricity tariff with a DR option that provides a rate discount or bill credit for reducing electric load during utility or grid-level contingency periods.
- Ancillary Services Market Program: a program in which customers can bid load curtailments into the wholesale electricity market. If their bids are accepted, customers are paid the market price for committing to be on standby to reduce load. If their load curtailments are needed, they are notified by the utility or grid operator, and are typically paid the wholesale spot market energy price.
- Capacity Market Program: a program in which customers can offer load curtailments to serve as additional grid system capacity, to augment conventional generation resources.

Each of the above DR programs functions in a different way to contribute to grid reliability. These actions are applied at different points along the time continuum of utility and grid operations, from system planning (across months or weeks), down to timescales of minutes and less. This set of coordinated services is carefully managed by utility and grid operators to ensure reliable power delivery to electric customers. Intelligent buildings can play an important part in this complex and highly interactive system.

Recent rulings by utility regulators and policy changes at independent system operators and regional transmission organizations are creating new opportunities for building owners in wholesale electricity markets [41]. In the past, these grid balancing services were provided only by conventional electric power generation sources.

Demand response services in the wholesale market are delivered by qualified providers (or through qualified intermediaries) to the electric grid operator. Except for very large electric customers, building owners will typically participate in these markets through a qualified intermediary, either by contracting with their electric utility for the appropriate electric tariff or by contracting with a qualified demand response aggregator. In these arrangements, the electric utility or the DR aggregator participates in the electric grid market on the behalf of the building owner. Intelligent buildings are well positioned to participate in this process.

### 2.5.3 Automated Demand Response and OpenADR

Early experience with DR programs revealed that manual communication methods (i.e., telephone and fax notification of pending DR events) and manual control of equipment (i.e., manually shutting off power to equipment) were less reliable or predictable than desired. For this reason, work began on ways to automate the demand response. Over the past 20+ years, automated demand response (AutoDR) has progressed to an advanced state that now includes a broad range of HVAC DR control strategies.

AutoDR and intelligent buildings' participation in the smart grid vision requires an open, interoperable, and secure automation and communication method to facilitate reliable and cost-effective communication of electricity price and electric grid reliability signals. The Open Automated Demand Response (OpenADR) standard was developed beginning in 2003 to provide this capability and enable automated interactions between buildings and their electric utility and grid operator partners. OpenADR is being successfully applied by numerous utilities and grid operators.

To extend and enhance the OpenADR standard, an international group of smart grid and buildings stakeholders have formed a working group in the Power Systems Management technical committee of the International Electrotechnical Commission (IEC) [42]. This activity will result in the development of an international standard for automated communications between intelligent buildings and the smart grid, which can be applied worldwide.

### 2.5.4 AutoDR Control Methods for HVAC Applications

Building owners and facility managers can benefit by investigating utility and grid operator DR programs to identify ways to take advantage of operating cost benefits resulting from AutoDR. Assistance from utility customer service representatives is helpful in determining the best courses of action. AutoDR control strategies are typically implemented in the electric customer's building automation system, including the necessary digital communications link to the electric utility or grid operator. These strategies can be selected and configured as a collaborative effort of the building facility manager, operating staff, BAS provider, and outside consultants as appropriate. The implementation of AutoDR control strategies using OpenADR communications and the role of the BAS are shown in Fig. 2.11.

The energy R&D community has devoted significant effort to identify and implement DR applications and control strategies for HVAC systems. Examples of proven DR applications can be found in the published literature [8]. These applications include DR control strategies which apply to most or all of the various types of HVAC equipment. The timing of DR events often coincides with peak demand periods, which are typically driven by cooling-related energy use. DR events can also be driven by peaks in heating-related energy use, depending on the locale.

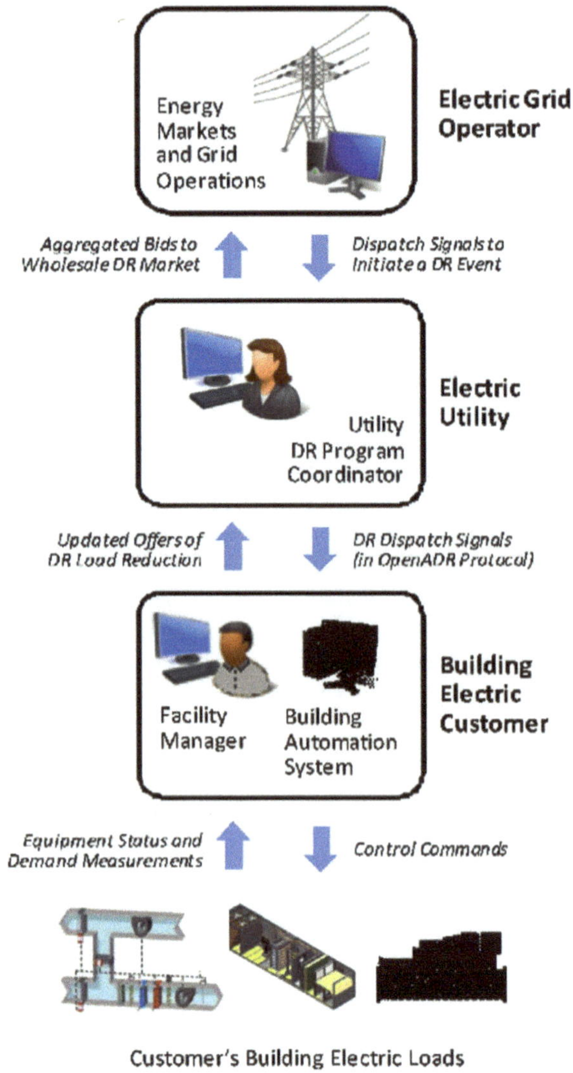


**Fig. 2.11** AutoDR control and OpenADR communications

Example DR applications for HVAC include control adjustments to [43]

- Space temperature set points (e.g., variable air volume systems),
- HVAC supply air temperature,
- Chilled water temperature,
- Duct static pressure,
- Motor speed control (e.g., fans, pumps, etc.),
- Demand limiting for major equipment (e.g., for chillers).

These HVAC DR strategies can be scheduled to begin at the start of a DR event. The strategies can be employed with price-based DR as well as incentive-based DR programs. Appropriate configuration of the DR strategies can limit associated indoor comfort impacts that could occur during and after the DR event. Other non-HVAC-related DR opportunities also exist in most buildings. Examples include indoor lighting, miscellaneous equipment and appliances, etc.

Some of these DR strategies can also be set to occur prior to a DR event to time-shift the building's demand profile, thereby providing a demand reduction during the period of the DR event. Example time-shifting strategies include pre-cooling of the occupied space. With careful design and configuration, these time-shifting strategies have shown good results.

### *2.5.5 Role of Intelligent Buildings*

Initial efforts in developing DR control strategies generally relied on the rather limited amount and quality of measured data that was available from typical BAS. As BAS technology has advanced in recent years, more sophisticated DR control strategies are now practical due to the greater amount of data that is made available in today's BAS.

The rich building HVAC operating data which is present in intelligent buildings can be utilized to develop improved AutoDR control strategies. These strategies can provide greater economic benefits to building owners and improved DR performance desired by utility and grid operators. These improved AutoDR control strategies can be tailored to make best use of the various DR programs which are available to each specific building and locale. An intelligent building is also an important enabler for emerging advancements in AutoDR. Examples include data-driven control strategy development and improved building models for demand response [44–46].

Taking a holistic view, intelligent buildings can integrate their demand response strategies with related energy systems in complementary ways to drive greater cost and energy benefits. Examples include renewable energy, energy storage, microgrids, transactive energy, energy efficiency initiatives, and onsite power generation.

In [7], four international implementations of automated demand response, all relying on the OpenADR standard noted above, are discussed:

- China's first automated demand response pilot project, implemented in the Tianjin Economic Technological Development Area (TEDA). Additional projects are underway as a result of this project.
- A microgrid for the U.S. Food and Drug Administration's White Oak campus. The microgrid can operate in both grid-connected and islanded modes. With its onsite generation capacity the microgrid supplies more electricity to the grid than it purchases from it.
- An ancillary services project at the Los Angeles Air Force Base, which relies on electric vehicle batteries to provide frequency regulation to the grid. Fleet vehicles

at the base have been replaced with electric vehicles and charging stations. 700 kW of power will be available for the grid when the project is complete.

- The Thames Valley Vision Project in the U.K. As of June 2014, 20 buildings were participating in the project and over 100 load-shed events had been conducted with an average participation rate of 98%.

## 2.6 From Research to Practice—Considerations and Challenges

### *2.6.1 Understanding the Context for Applied Research*

Technology transfer from R&D to commercial products has several challenges. There are gaps in translating the technology to practice, developing tools for smooth deployment, and mobilizing a trained team for installing, commissioning, and troubleshooting. Adding to that, advanced automated optimization and control principles create black-box-like paradigms for the operation of a plant or building. These do not always translate to operator familiarity and comfort with the system. Hence, user experience principles must be baked into the commercial offering.

In academia and industrial research teams, the focus for advanced control researchers is on proving a concept and in developing a working prototype. The engineering problem to be solved is abstracted up to a mathematical or statistical problem. Research involves developing and testing computation-aided solutions which are tested in simulations, which tests the technological feasibility and performance, but not the other aspects of field implementation such as configuration or end-user experience. Assumptions are made during research to contain the scope and solve the core complex mathematical problem, without getting bogged down by varied field control systems and equipment. This leads to several valuable theoretical advances and evolves the state of the art. However, the commercialization plan is about making a profitable offering that solves a real-world problem. There are many considerations in addition to the advanced solution which may determine the technology's viability and success in the marketplace.

The following questions are important when fielding an advanced optimization and control product:

- What is the current state of a typical building or plant? For example, is it operated manually, and if automated, how often does the operator revert to manual operation?
- How well is the building or plant instrumented? Is enough data being obtained at the resolution and sampling time the advanced algorithm requires?
- Is the typical building automation and communication system robust? Will there be periods of no communication or delayed communication?

- What is the typical architecture of the control and automation system? Do we need to bring information from different systems together, or pipe in several different sensor or control inputs?
- What is the typical information about a plant or building needed to customize or configure the advanced solution? How will this information be readily available in a typical facility?
- Are the right people in place to be first adopters (wanting to solve a problem, save energy, etc.) and dedicated to transitioning to a new system?

The answers to these may be more important than the advanced control solution itself and have the potential to derail a promising solution from moving forward as a commercial offering.

Unlike other domains such as aircraft systems and refineries, buildings, and their control systems are very diverse. There are numerous ways an HVAC system may be configured, and there are always exceptions to established practices in the field. Buildings change over time where additions are built or modified and occupancy changes, and in addition the end user has some control over the environment. These often result in ad hoc control changes to facilitate the addition of new equipment or changes in zone functions. Since these could be undocumented, it usually requires the knowledge of a facility operator to unravel the HVAC infrastructure and control system changes and to understand the full picture. This full picture is needed if a supervisory level optimized control is to be implemented.

The end user experience cannot be over-emphasized and it is important for applied researchers to understand the work environment. The workflow in large facilities is very cost efficient. One or two operators may manage several plants (roving operators) and are trained to know only what is needed to operate the plants and buildings correctly; special displays are created so that the operators may focus on only the most important information at that time. Control technicians program controllers and have knowledge of the logic implemented in the controllers for different systems and equipments. The building automation programmers connect all pieces together including controllers, sensors, and meters. The operations staff are not motivated to try new controls and optimization software, because operational reliability is their number one goal. In many cases, control expertise or knowledge of the specific system does not reside with facilities management of a building, but with control and building automation services provided by automation companies with their own technical staff.

### 2.6.2 *Role of Architecture*

Typically, feasibility is established when the advanced algorithm produces correct or optimal results for test cases. The research engineers usually work with building experts to develop a prototype that works in the field. However, this is also the phase when the control and software architecture and user interfaces need to be designed because the advanced solution must create the least disruption for the facility operators, even at the prototype stage, to be accepted, and this can be critical in the introduction of new operational software. This does not mean that the prototype must be the final product. However, the main structure and functional delineations of the control decisions should have been defined, because of their critical role in proliferating technological solutions smoothly. A typical architecture for supervisory optimizers is shown in Figs. 2.2 and 2.7. The advanced optimization at the supervisory level provides top-level set points and on-off commands to major equipment. Lower level feedback control loops such as those for pump flow and cooling water temperature run on local controllers. The local controllers also encompass individual equipment control from the manufacturer, such as chiller control.

Let us illustrate these remarks using the prototype chiller plant optimization system described previously. A chiller plant with varying chiller sizes and types, and not undersized, is generally a good candidate for optimization, because the loading and operating set points for maximum efficiency of the plant as a whole are not intuitive to the operator. However, the chiller plant is also complex and optimizing the entire chiller plant including the pumps and fans makes delineation of the control hierarchy difficult. The schematic of one of the example chiller plants is shown in Fig. 2.12.

Commands such as chiller on/off switching or supply temperature set points may be provided by a supervisory level optimizer. Pumps are part of the lower level functionality in the chiller plant: they deliver the required flow to the chillers, the cooling towers, and the building loads. Optimizing pump speeds and switching on/off schedules intertwines the supervisory control with the lower level control. Although

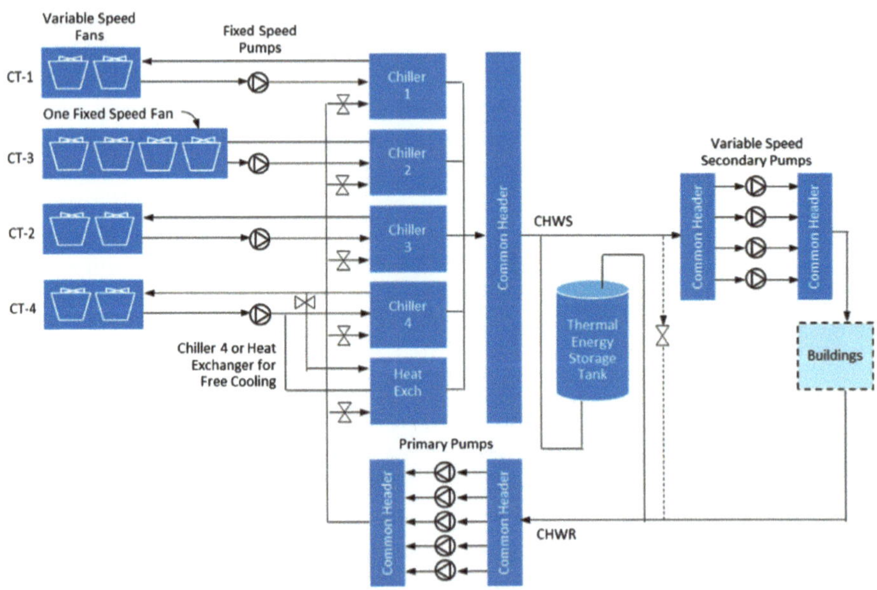


**Fig. 2.12** Chiller plant layout

chillers are the main energy consuming equipment in the plant, the pumps and fan motors in a large plant can add up to a high energy cost. If our objective is to optimize the operation of all equipment, including the primary and secondary pumps and cooling tower fans, which are operated by lower level closed-loop control, then the clean delineation requires thought and careful design. For example, a chiller on/off command involves a combination of primary pumps (shared among several chillers that may be on or off) to be switched on or commanded to increase flow within several constraints. The controller is programmed with preset rules for pump combinations and speeds to satisfy the chiller flow requirements. In the case of secondary pumps, the lower level controllers use a differential pressure set point across the pumps to control their speeds.

The initial control architecture of the prototype optimizer with black-box models generated a hybrid set of outputs: set points and on/off commands for the chillers and the control actions for low-level equipment such as pumps and cooling tower fans. This may work temporarily in the prototype test scenario, but may be accepted with reluctance by the facility operators and managers. The more effective process is to construct the architecture for the advanced optimizer to specify the set points (within operating limits) for all lower level controllers, but this is not trivial in some cases. Taking the case of the secondary pumps, specifying the differential pressure set points to lower controllers involves modeling the flow loop and the pump curves (see Fig. 2.13). This is not a trivial task and to be able to generalize to a software product software tools will need to be built for configuring such models; the tools would bridge the gaps between the supervisory and local control, and the optimization developer and the application engineer. Without such translation from advanced solutions to field practice, the operational costs of advanced modeling and optimization skills needed in implementation will be high, and productization cannot be sustained.

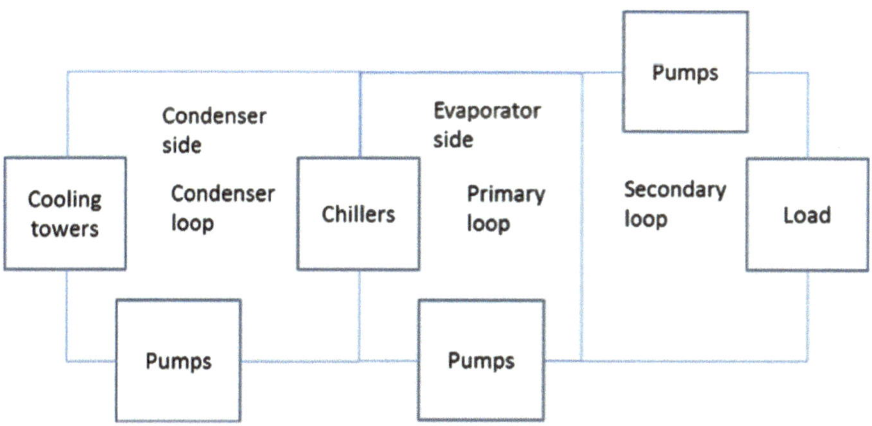


**Fig. 2.13** Example layout for flow modeling

### 2.6.3 "Supply Chain" for Technology Transfer

First, let us look at what happens when an advanced control solution is sold and implemented at typical large facilities (Fig. 2.14). A typical implementation progresses from initial customer contact by a sales team to a site audit and putting together the solution to meet the customer needs by a solution assembly team. This team would typically assemble the full solution (that includes upgrades other than the advanced control) for the customer by selecting from a set of equipment, control, and automation solutions. Once approved by the customer, application engineers would configure the advanced control solution for the site (setting up the generic software for the site: number of chillers and their interconnections, specifications of equipment, and so on). The solution would also be set up with all I/O points for the specific plant to communicate with the building automation system or controllers during installation and commissioning, and handed over to facility operators after training.

Next, who should be in the "supply chain" of technical professionals who develop the solution? The advanced control solution is only one part of several functions that need to be designed for a cost-effective product that fulfills the objective, be it energy savings, comfort, or staff productivity. Figure 2.15 shows the different functions and skillsets that should be involved; this is not meant to convey that each team is separate, or that there need to be separate individuals for each role, but that the essential role of the skills and functions should be acknowledged and understood early, so that the advanced solution may be architected for ease of deployment and maintenance. The advanced control developer is the control theory and optimization expert who researches and develops the computational algorithms. The building domain expert has general knowledge of all aspects of a typical building: the equipment, the building

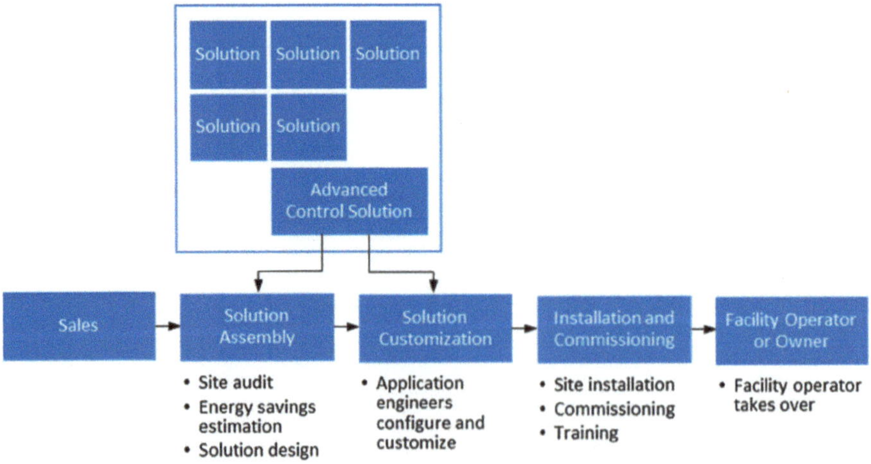


**Fig. 2.14** Organizational chain in advanced control application

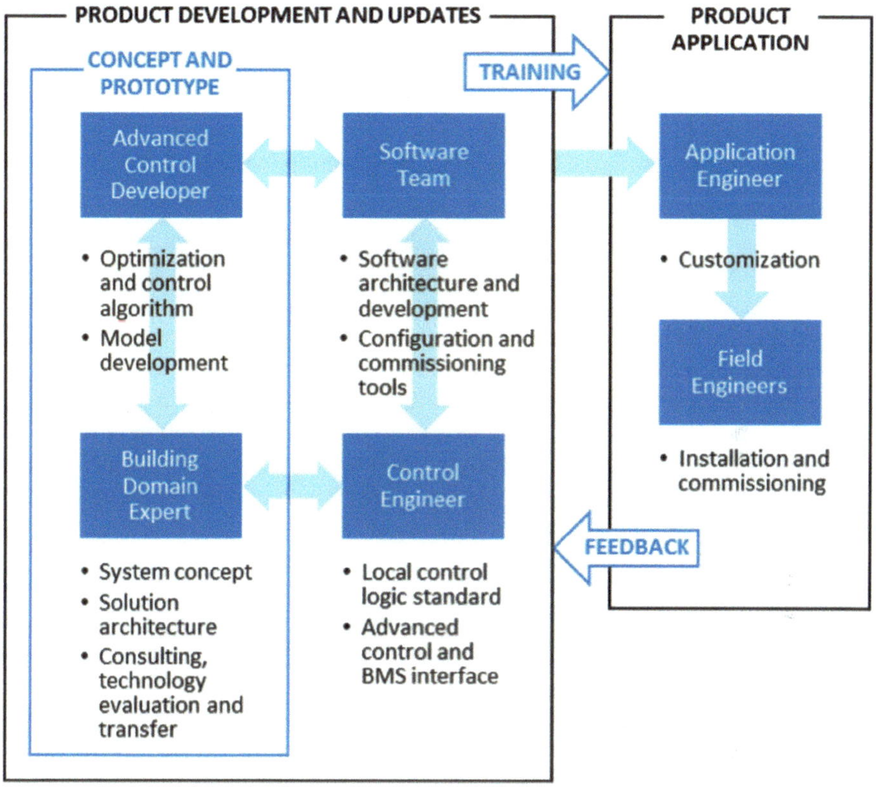


**Fig. 2.15** Supply chain of professionals in tech development and transfer

and plant controls, HVAC distribution, local control, and BAS. The control engineer has practical knowledge and experience of control programming and implementation with several low-level controller brands. He/she works with the advanced control researcher and building domain expert to architect the control functional layers, and provides the local control knowledge to standardize on them for use by the advanced solution. The software team collaborates with the other teams to design the right software architecture for the computation, storage, data access, and cycle times required. These separate entities bring their expertise so that together the constraints and the best practices from each may be incorporated in the final product.

## 2.7 Conclusion

Our experience with implementation of mature prototypes, their deployment and validation in real pilots, and their transfer to commercial teams, as well as others' similar experiences, prompt a few thoughts and recommendations.

- Advanced control and optimization bring real benefits to building energy management and bring us much closer to reducing the building sector's energy use and emissions. It is vital that control and optimization experts also become the champions in translating advanced concepts to widespread commercial use.
- The challenge to bring to market innovative energy-saving technologies is exacerbated by the low ROI compared to other investment options, as noted before. Indeed, an unfavorable cost-to-benefit ratio continues to be one of the major limitations for a wider adoption of advanced control and optimization applications in buildings. The costs considered over the entire solution life cycle include primarily installation costs and maintenance costs, which creates pressures on short and easy configuration, discourages inclusion of additional hardware, and favors keeping the effort needed for supervision and re-tuning to minimal levels.
- Energy projects shift from purely energy-savings-driven to automation-upgrade-driven, depending on market conditions. It is important for practitioners to understand the market driver, and develop the tools and mobilize the trained workforce that makes a commercially viable product or service.
- The energy savings benefit provokes the chicken-and-egg game of demonstrating savings in order to find the investment to mature the technology. A prototype can only show the benefit if well deployed and accepted, which needs additional investment. Therefore, control architecture and operator considerations are of top importance, even at the prototype stage.
- Standard implementation tools must also be developed to quickly and reliably configure the advanced software and connect it to the local control on site. These includes general tools such as ontology-driven data modeling libraries [47] that facilitate all data-intensive applications, as well as specific tools for pre-configuring of software and plug-and-play type implementation. The new solution should not require a skilled advanced control expert as is the norm in the industrial domain. HVAC field engineers are heavily time constrained and that is why they prefer an intuitive plug-and-play configuration, which ideally does not require any additional sensors, actuators, or meters.
- In those situations where building operators use new tools as advisory systems, their experience and acceptance should be improved by providing explanations of major actions by the supervisory optimizer, such as "turning off chiller 4 and turning on chiller 2 at higher load operates the plant at 4% higher efficiency."

For complex solutions that require advanced knowledge and skillsets, considerations of control and software architecture force the question of marketplace viability early in the development process. Appropriate decisions at that stage can then guide the developers in formulating and solving the most commercially viable problem, and not just pushing through a mathematically complex or convenient problem. This also informs where additional development funding needs to be focused, for smooth technology adoption and diffusion in the market.

**Acknowledgements** Part of this material is based upon work supported by the U.S. Army Corps of Engineers, Humphreys Engineer Center Support Activity under Contract No. W912HQ-13-C-0058. Any opinions, findings, and conclusions or recommendations expressed in this material are those of the author(s), and do not necessarily reflect the views of the U.S. Army Corps of Engineers, Humphreys Engineer Center Support Activity, and should not be construed as an official Department of Defense position or decision unless so designated by other official documentation.

**Note**: BACnet is a registered trademark of ASHRAE. LonWorks and LON are registered trademarks of Echelon Corporation. Modbus is a registered trademark of Schneider Automation Inc. ZigBee is a registered trademark of ZigBee Alliance, Inc. EnOcean is a registered trademark of EnOcean GmbH. Z-Wave is a registered trademark of Zensys Inc. Bluetooth is a registered trademark of Bluetooth SIG, Inc. Honeywell Enterprise Building Integrator (EBI) is a trademark of Honeywell International, Inc. OPC is a registered trademark of OPC Foundation.

## References

1. U.S. Department of Energy, Buildings energy data book, building technologies program, U.S. Department of Energy, Technical report (2012). http://buildingsdatabook.eren.doe.gov/. Accessed 5 September 2016
2. Environmental and Energy Study Institute, Buildings and climate change, Environmental and Energy Study Institute, Technical report (2016). http://www.eesi.org/files/climate.pdf. Accessed 5 September 2016
3. Brambley MR et al (2016) Advanced sensors and controls for building applications: Market assessment and potential r&d pathways, Pacific Northwest National Laboratory, PNNL-15149 (2005). http://apps1.eere.energy.gov/buildings/publications/pdfs/corporate/pnnl-15149_market_assessment.pdf. Accessed 14 September 2016
4. Kelly H (2016) The potential of more efficient buildings. Issues Sci Technol 32(4):21
5. Markets and Markets (2016) Building automation system market worth 100.60 billion usd by 2022. http://www.marketsandmarkets.com/PressReleases/building-automation.asp. Accessed 5 September 2016
6. Honeywell Building Solutions (HBS), HBS retail complex diagram (2016). https://buildingsolutions.honeywell.com/en-US/newsevents/resources/Publications/honeywell-hbs-retail-complex-diagram.jpg. Accessed 6 September 2016
7. Sauter T, Soucek S, Kastner W, Dietrich D (2011) The evolution of factory and building automation. IEEE Ind Electron Mag 5(3):35–48
8. Samad T, Koch E, Stluka P (2016) Automated demand response for smart buildings and microgrids: the state of the practice and research challenges. Proc IEEE 104(4):726–744
9. Enterprise Building Integrator (EBI), Building integration, Honeywell, Technical report. https://www.ebi.honeywell.com/PublishingImages/Integration.jpg. Accessed 5 September 2016
10. Association CABA (2015) Report on intelligent buildings and big data. Technical report, CABA Association
11. Friess, P (2013) Internet of things: converging technologies for smart environments and integrated ecosystems. River Publishers
12. Clements-Croome, DJ (2014) Sustainable intelligent buildings for better health comfort and well-being, Report for DENZERO project

13. World Green Building Council (2015) Report on health, wellbeing and productivity in offices, the next chapter for green building, World Green Building Council, Technical report
14. Derguech W, Bruke E, Curry E (2014) An autonomic approach to real-time predictive analytics using open data and internet of things. In 2014 IEEE 11th international conference on ubiquitous intelligence and computing and 2014 IEEE 11th international conference on and autonomic and trusted computing and 2014 IEEE 14th international conference on scalable computing and communications and its associated workshops (UTC-ATC-ScalCom). IEEE, pp 204–211
15. Capehart BL, Brambley MR (2014) Automated diagnostics and analytics for buildings. Fairmont Press, Inc
16. Stluka P, Mařik K, Endel P (2014) Advanced control solutions for building systems. In IFAC, vol 47, no 3. Elsevier, pp 606–611
17. Akkaya K, Guvenc I, Aygun R, Pala N, Kadri A (2015) Iot-based occupancy monitoring techniques for energy-efficient smart buildings. In Wireless communications and networking conference workshops (WCNCW), 2015 IEEE. IEEE, pp 58–63
18. Bluyssen PM (2009) The indoor environment handbook: how to make buildings healthy and comfortable. Routledge
19. Zhao Y, Zhao Q, Xia L, Cheng Z, Wang F, Song F (2013) A unified control framework of HVAC system for thermal and acoustic comforts in office building. In 2013 IEEE international conference on automation science and engineering (CASE). IEEE, pp 416–421
20. Samad T (2016) Control systems and the Internet of Things. IEEE Control Syst 36(1):13–16
21. Li F, Vögler M, Claeßens M, Dustdar S (2013) Efficient and scalable IoT service delivery on cloud. In IEEE sixth international conference on cloud computing (CLOUD). IEEE, pp 740–747
22. Vögler M, Li F, Claeßens M, Schleicher JM, Sehic S, Nastic S, Dustdar S (2015) COLT collaborative delivery of lightweight iot applications. In Internet of things. User-centric IoT. Springer, pp 265–272
23. Spudić V, Conte C, Baotić M, Morari M (2015) Cooperative distributed model predictive control for wind farms. Optimal Control Appl Methods 36(3):333–352
24. Morstyn T, Hredzak B, Agelidis VG (2015) Distributed cooperative control of microgrid storage. IEEE Trans Power Syst 30(5):2780–2789
25. Putta V, Donghun K, Cai J, Hu J, Braun J (2014) Distributed model predictive control for building HVAC systems: a case study. In International high performance buildings conference, West Lafayette, IN
26. Joe J, Karava P, Hou X, Hu J (2016) Model predictive control of a radiant floor cooling system in an office space. In 4th international high performance buildings conference, West Lafayette, IN
27. Joe J, Karava P (2017) Agent-based system identification for control-oriented building models. J Building Perform Simul 10(2):183–204
28. Purdon S, Kusy B, Jurdak R, Challen G (2013) Model-free HVAC control using occupant feedback. In IEEE 38th conference on local computer networks workshops (LCN workshops). IEEE, pp 84–92
29. Gupta S, Kar K, Mishra S, Wen J (2014) Smart temperature control with active building occupant feedback. In 19th world congress of the international federation of automatic control
30. Maaijen R, Zeiler W, Boxem G, Maassen W (2012) Human centered energy control: taking the occupancy in the control loop of building systems. REHVA J
31. Mařik K, Rojiček J, Stluka P, Vass J (2011) Advanced HVAC control: theory versus reality. IFAC 44(1):3108–3113
32. Gwerder M, Gyalistras D, Oldewurtel F, Lehmann B, Wirth K, Stauch V, Tödtli J (2010) Potential assessment of rule-based control for integrated room automation. In 10th REHVA world congress clima, pp 9–12
33. Cigler J, Siroky J, Korda M, Jones C (2013) On the selection of the most appropriate mpc problem formulation for buildings. In 11th REHVA world congress CLIMA 2013, no. EPFL-CONF-183457

34. Široký J, Oldewurtel F, Cigler J, Privara S (2011) Experimental analysis of model predictive control for an energy efficient building heating system. Appl Energy 88(9):3079–3087
35. Cigler J, Gyalistras D, Široky J, Tiet V, Ferkl L (2013) Beyond theory: the challenge of implementing model predictive control in buildings. In Proceedings of 11th Rehva world congress, Clima, vol 250
36. Wang S, Ma Z (2008) Supervisory and optimal control of building HVAC systems: a review. HVAC&R Res 14(1):3–32
37. Ma Y, Borrelli F, Hencey B, Coffey B, Bengea S, Haves P (2012) Model predictive control for the operation of building cooling systems. IEEE Trans Control Syst Technol 20(3):796–803
38. Parthasarathy G (2013) Central plant optimization for waste energy reduction (cpower): final report (draft), ESTCP Program, EW-201349
39. U. D. of Energy (2006) Benefits of demand response in electricity markets and recommendations for achieving them: a report to the united states congress pursuant to section 1252 of the energy policy act of 2005, February 2006. http://energy.gov/oe/downloads/benefits-demand-response-electricity-markets-and-recommendations-achieving-them-report
40. F. E. R. Commission (2007) Assessment of demand response and advanced metering - staff report, September 2007. http://www.ferc.gov/industries/electric/indus-act/demand-response/dem-res-adv-metering.asp
41. Rahimi F, Ipakchi A (2010) Demand response as a market resource under the smart grid paradigm. IEEE Trans Smart Grid 1(1):82–88
42. International Electrotechnical Commission, WG 21, Interfaces and protocol profiles relevant to systems connected to the electrical grid. http://www.iec.ch/dyn/www/f?p=103:14:0::::FSP_ORG_ID,FSP_LANG_ID:8572,25
43. Motegi N, Piette M, Watson D, Kiliccote S, Xu P (2007) Introduction to commercial building control strategies and techniques for demand response, Lawrence Berkeley National Laboratory, LBNL-59975, May 2007. http://drrc.lbl.gov/publications/introduction-commercial-building
44. Lee K, Braun J, A data-driven method for determining zone temperature trajectories that minimize peak electrical demand, California energy commission, public interest energy research program. http://drrc.lbl.gov/sites/all/files/lee-braun-data-driven.pdf
45. Piette M, Granderson J, Wetter M, Kiliccote S (2012) Intelligent building energy information and control systems for low-energy operations and optimal demand response, Lawrence Berkeley National Laboratory, LBNL-5894E, May 2012. http://drrc.lbl.gov/publications/intelligent-building-energy
46. Rongxin Y, Kiliccote S, Piette M (2014) Linking measurements and models in commercial buildings: a case study for model calibration and demand response strategy evaluation, Lawrence Berkeley National Laboratory, LBNL-7006e, December 2014. http://drrc.lbl.gov/publications/linking-measurements-and-models
47. Prairie D, Samy MKG, Petze J, Petock M (2016) Project haystack, a caba white paper. http://project-haystack.org/file/22/CABA-White-Paper-on-Project-Haystack.pdf

# Chapter 3
# Operating Systems for Small/Medium Commercial Buildings

**Marco Pritoni and David M. Auslander**

## 3.1 Introduction and Motivation

Small and medium commercial buildings constitute 90% of the commercial buildings in the US (by number) and use more than 50% of the energy of the commercial sector. Space heating, cooling, and ventilation (HVAC) together represent the largest portion (44%) of this energy use, followed by miscellanea (13%) and lighting (10%) [1]. A recent scoping study has shown that a large fraction of the energy use and energy cost in small-medium commercial buildings is wasted due to poor control strategies [2, 3]. Further, energy cost is likely to increase in the next future. In fact, the advent of the smart grid will enable utilities to adopt time-varying prices, but existing controls cannot optimize for these dynamic changes in the price of electricity, with the result of even higher costs for building owners.

Typically , in small commercial buildings, lighting and plugs are controlled manually, and HVAC controls lack the sophistication of large building automation systems (BAS) described in Chap. 2. Small buildings often utilize multiple rooftop forced-air packaged units (RTUs) for HVAC, providing heating, cooling, and ventilation to the building through ductwork. RTUs are used because they are relatively inexpensive, easy to install and maintain, and reliable. In buildings with multiple tenants, they also allow separate metering and flexible reconfiguration of the spaces, without significant work on the HVAC infrastructure. However, RTUs are also relatively inefficient devices as compared to large building chiller and boiler systems. Most RTUs have one single-speed compressor, single-speed fans, fixed damper for outdoor air intake, and are controlled by simple analog thermostats [4], very similar to those used in residential applications. Each thermostat independently controls a single RTU. While the thermostats often are programmable (i.e., they contain a user-modifiable schedule of set points), there is usually no central control system as is found in buildings that

M. Pritoni (✉)
Lawrence Berkeley National Lab, California, USA
e-mail: MPritoni@lbl.gov; marco.pritoni@gmail.com

D.M. Auslander
University of California, California, USA
e-mail: dma@me.berkeley.edu



J.T. Wen and S. Mishra (eds.), *Intelligent Building Control Systems*,
Advances in Industrial Control, https://doi.org/10.1007/978-3-319-68462-8_3

use central HVAC facilities [2]. Replacing or retrofitting RTUs with more efficient HVAC systems to reduce energy cost is economically unfeasible unless the units are at the end of their useful lives (typically 15 years or more) [5]. At the same time, upgrading to a full-scale (traditional), BAS is also prohibitively expensive.

In the last few years, government, industry, and academics have been developing and proposing control systems to reduce energy demand in small-medium commercial buildings. The US department of energy (DOE) has engaged industry promoting and testing advanced rooftop control (ARC) retrofits. These retrofit kits have significant potential to save energy, but require hardware upgrades (e.g., addition of variable speed drives to motors and compressors), and may be economically feasible only on large RTUs [5–7]. While traditional manufacturers of control systems have been slowly introducing new integrated products, capable of controlling lighting, plugs, and HVAC (for instance: [8, 9]), most of the innovation comes from new players.

Technologies created by information technology companies for consumer electronics (e.g., mobile phones, wearables, and other Internet of Things (IoT) devices) are rapidly finding applications in building controls. For instance, there has been an increase in the number of connected devices, such as smart thermostats, lights, and plugs that can introduce intelligence into previously unconnected subsystems [10–13]. These devices offer remote access, improved schedules, or better control algorithms, and require minimal hardware. Most of these innovative products are marketed for residential applications. A recent review of the home automation market found more than 350 energy management products available in the US [14]. While many products exist, these solutions frequently use proprietary hardware and software stacks. That is, software running on the device and the cloud platform are not open, and it is difficult to integrate them in a larger software environment. To connect these devices together, integrators need gateways and hubs, due to the lack of standardization in the network layer (e.g., Wi-Fi, ZigBee®, Z-Wave®, Bluetooth®, and other proprietary protocols are used) [14]. In addition, multiple competing software platforms are now emerging, trying to gain market traction and to become the dominant aggregator of devices (e.g., *works with Nest* ® [15], *works with SmartThings*® [16], *Lowe's Iris* ® *platform* [17], and *Apple HomeKit*® *platform* [18]). In parallel to commercial software, several non-academic open-source home automation projects have emerged for consumers that want to build their own system and have some programming skills [19–26] ®. Currently, these platforms target residential customers and lack features required in small commercial buildings, such as multi-user authorization profiles, and management of multiple zones and more complex HVAC systems.

Over a decade ago, Tridium introduced in the commercial building market, the first Java-based platform, called Niagara Framework [27], aimed at bridging the interoperability problem between BAS vendors. The product now offers an open API enabling software extensions, and uses well-known programming languages (i.e., Java and HTML 5), but mostly targets large commercial buildings with direct digital control (DDC) systems. There also exist commercial products (e.g., [28]) as well as a few open-source projects (e.g., [22, 29]) targeting small commercial buildings, but so far they had limited success compared to the products sold in the residential market. Looking forward, the control software could also be built on top of emerging IoT platforms [30–35], which target a wider spectrum of applications, including integration with the grid and renewables. This rapidly changing landscape and the wide variety of data models, application programming interfaces (APIs), and network protocols limit the development of cross-platform analytics and control applications for buildings [36, 37].

Academics have taken advantage of added connectivity offered by connected devices and advances in information and communication technology to research two main areas: the design of software platforms, frequently called building operating system (BOS), and the development of controls and analytics applications to improve building operations. In this chapter, we review the existing literature and present our contribution in the attempt of increase portability of advanced control sequences.

## 3.2 Previous Work: Building Operating System (BOS)

The wireless sensor network community first defined the role and characteristics of the middleware called BOS, and identified the following requirements [38]. The BOS should

1. accommodate new types of sensor (new occupancy sensors, wearable, virtual sensors) and higher level third-party applications;
2. allow occupants to participate in setting the control policies for their buildings, without compromising security and safety;
3. interface and exchange information with systems external to the building, such as the electrical grid, the transportation network, or other communication networks.

Initial work in this area focused on integrating different sources of physical information (e.g., devices connected to different networks, data from websites) under a single data plane, to make them accessible in a unified way by higher level applications [39–41]. Although early systems only allowed reading from sensors [42–44], more modern software enables bidirectional communication and allows actuating devices [43, 44]. These studies did not directly target controls in small commercial buildings, but they provided a general framework that later studies adopted extensively. The BOS software is developed in tiers, and it abstracts common functions with the intent of enabling programmers to write "portable"[1] applications, which can automatically adapt to different buildings [45]. Different architectures were proposed, but the most general consists of three tiers [36, 45–48] depicted in Fig. 3.1.

[1]"Write once run everywhere" [45]. This means that it is possible to run the same code on multiple buildings with little or no customization.

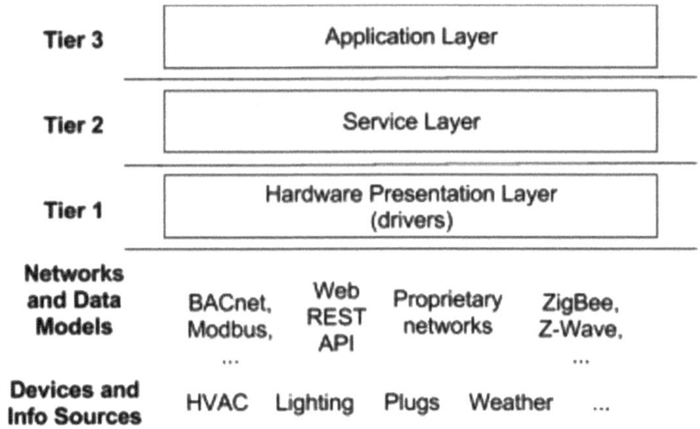


**Fig. 3.1** Three Tiers of a BOS

### *3.2.1 Tier 1 : Hardware Presentation Layer*

The lower BOS layer deals with connectivity and uses drivers (also called adapters or connectors) to interconnect heterogeneous data sources to the higher layers. Network protocols are translated and contained messages are converted to a uniform API. Dawson-Haggerty et al., for instance, used small RESTful[2] web services to "decode" the information from a sensor and transmit it in JSON format over HTTP protocol using a common interface [41]. Rowe et al., instead, used adapters developed in a library over XMPP protocol, for their Sensor Andrews [40]. Weng et al. used connectors that communicate over HTTPS or use RPCs over RabbitMQ [37]. Some researchers go one step further and attempt to abstract generalized device-type functionalities instead of just developing interfaces for each single device [47, 49]. For instance, a generic thermostat interface would allow developers to ignore implementation details specific to a particular brand or model, focusing on shared concepts such as state, mode, set points, and temperature [49]. We will call this layer the Hardware Presentation Layer (as in [36, 45, 46, 49]).

### *3.2.2 Tier 2: Service Layer*

The intermediate tier provides services required by most applications including data storage, authentication and authorization, information delivery, and device discovery.

[2]http://www.ics.uci.edu/~fielding/pubs/dissertation/rest_arch_style.htm.

#### 3.2.2.1 Automatic Discovery

Ideally, when a new device is connected to the network, we would like to discover its presence, its functionalities, and automatically configure it, similar to what happens when we connect a new printer to a Wi-Fi network [50, 51]. Most BOSs provide an automated or semi-automated discovery function. However, there are limitations in the information that can be discovered, since most devices do not natively support this feature or share a limited set of information.

#### 3.2.2.2 Data Storage and Retrieval

In contrast to traditional control systems for small commercial buildings [8], BOSs save and give access to historical data, thus enabling developers to write richer applications. However, storing a large volume of time series data in traditional relational databases is not efficient, and data retrieval can become very slow. For this reason, the majority of the BOSs use specialized time series databases or other non-relational databases (sometimes called "historians") that exhibit dramatically better performances in saving and recovering data. Both open-source [52–57] and commercial products [58–60] are now available. To store the information about all the data points and their relationships, these databases need to be coupled with metadata (descriptors about the data) repositories. Typical metadata include sensor name, function, location, and relationships with other systems or subsystems. In traditional building automation systems (BAS), the metadata is encoded in sensor names (e.g., a temperature sensor for zone 11_3 could be called "GIEDT.AHU1.RM11_3.RMTMP") or in the graphical interface. The sensor names are frequently limited in number of characters. Meant to be interpreted by a facility manager, who understands its context, the information can be very cryptic and difficult to decypher for a machine [61]. Figure 3.2 shows an example of a legacy BAS interface with point names. Metadata databases offer the opportunity to define these name tags in a more clear and uniform way, using e.g., "project haystack" tagging convention [62].

#### 3.2.2.3 Queries, Templates, and Configuration Files

One of the most innovative features of BOSs is the ability to write control sequences that do not hard-code-specific sensor names (as is practice in the BAS industry), but use variables, templates, and objects that can be reused on other buildings. Taking advantage of modern object-oriented high-level programming languages like Python, developers can now write classes of control sequences that can be dynamically instantiated for a specific building. This is possible because developers can query the metadata database and dynamically retrieve the specific reference to the sensor (e.g., the occupancy sensor and the temperature sensor in zone 123)

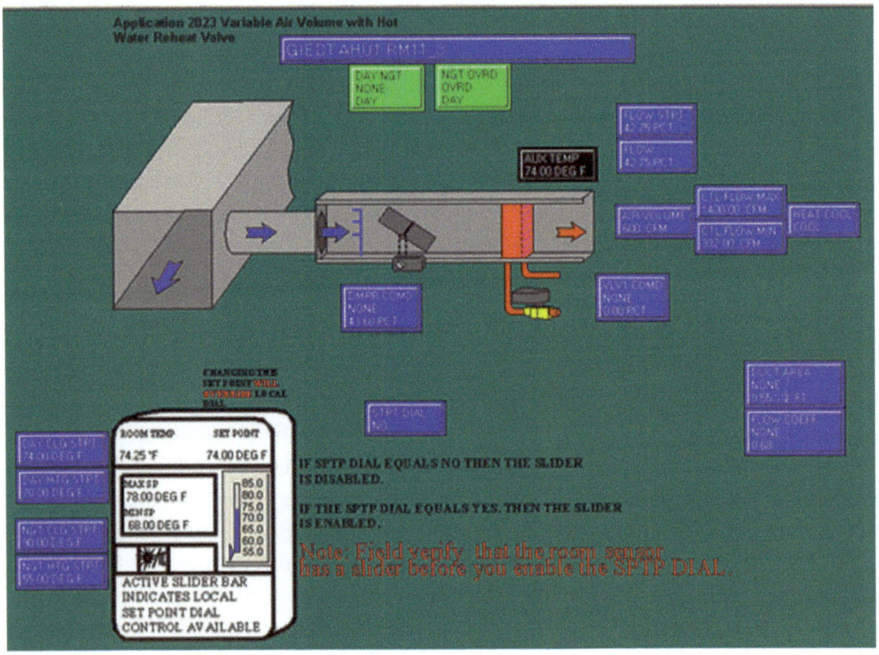


**Fig. 3.2** Interface of a Traditional BAS with Sensor Names

[36, 37, 41, 45, 46, 63]. Configuration files describing the relationships between points and system templates are developed either manually or as a result of the automatic discovery process [37, 49].

#### 3.2.2.4 Efficient Information Delivery

Control systems may need to poll sensor data frequently, and repeated requests for updates may increase the traffic on the network and overload the controller. Most BOSs borrow from the information technology (IT) industry and use a publish-subscribe information architecture. Similar to web feeds (e.g., RRS, Atom [64–67]), the sensors provide real-time notification of changes, to subscribed clients (the controller, in this case), only when a change of value happens. When several clients need the same information (e.g., outdoor air temperature), or when the variable does not change very often (e.g., occupancy or scheduled set points), this delivery method is very efficient. Certain control strategies, such as event-driven controls (e.g., occupancy-driven lighting), can benefit significantly from this efficient "information exchange bus" [47, 67].

#### 3.2.2.5 Authentication and Authorization

Cybersecurity has become a very important issue for any IT network, and control systems are not an exception [68]. For instance, hundreds of thousands of IoT devices were recently used to conduct a distributed denial-of-service attack to a major DNS provider rendering many websites unavailable for an entire day [69]. Academic projects developing BOSs implement some security features, such as authentication and authorization of clients and applications [38, 40, 49], but more work needs to be done to analyze potential threats in each layer of the architecture, starting from the physical devices.

#### 3.2.2.6 Other Features

In addition to the core services described above, BOSs provide an execution environment, where applications run. Some BOSs distribute applications in the form of agents [40, 47, 63, 70], while others leave the installation and management of new applications to the developers. The operating systems can be completely local to the building (Fig. 3.3a), completely in the cloud (Fig. 3.3b), or a mix of the two. The software can also run on separate hardware components distributed in the network. Figure 3.3a shows two drivers and the higher software layers run on a small PC and a third driver run on a single-board computer. The two components are networked together and form the BOS.

New BOSs promise to be "extensible", i.e., to have the ability to seamlessly incorporate new functionalities to adapt to new sensors, new controllers, or new applications. This philosophy, often ignored in traditional BASs, is particularly important in this period of rapid change of the energy landscape. While in the past BASs were isolated systems and were charged with limited tasks (e.g., keep the rooms within comfort bands), now control systems are asked to interact with the grid, on site renewables, electric vehicles, and other systems. To promote a better building-grid interaction, the US DOE recently developed an open-source, distributed, control system called VOLTTRON supporting the vision of a "transactive" (i.e., based on energy transactions) energy system. As of 2016, universities and national laboratories developed applications for this platform and some industry partners seem interested in adopting it to develop commercial products.

In addition to the grid, BOSs should be able to communicate with the growing ecosystem of IoT devices, such as phones, computers, wearables, appliances, and even furniture. An interesting example is personal comfort systems, such as heated and cooled chairs, foot warmers, and fans, proposed by some researchers to improve occupant comfort [71]. These systems can also save energy if they can coordinate with the HVAC system. For example, if occupants can adjust their local thermal preferences, the set point in a room can be relaxed [72, 73]. Wearable devices can also provide invaluable information about occupant comfort, as well as provide information about room occupancy. Integration of these new devices calls for an extensible design of the control system.

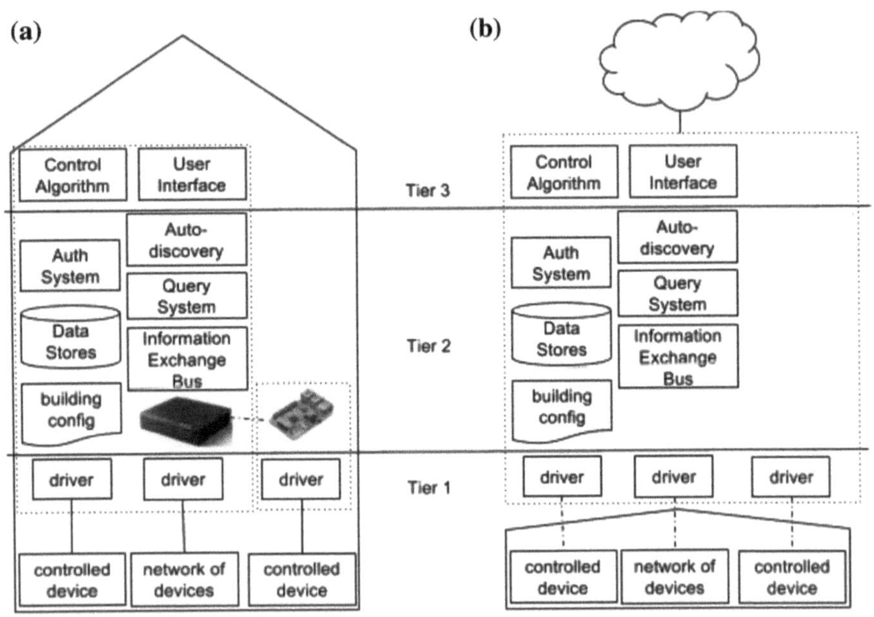


**Fig. 3.3** Two deployment options for a BOS. **a** All the components run on-premises in a small PC and a lightweight single-board computer. **b** All the components, including the drivers, run on the cloud

### *3.2.3 Tier 3: Application Layer*

The top tier of a BOS is composed of applications (Fig. 3.1). Some examples include user interfaces (e.g., web displays, phone apps), occupancy prediction and modeling, energy allocation and apportionment (e.g., nonintrusive energy disaggregation), occupant feedback (e.g., participatory comfort systems), fault detection and diagnostics, demand response, and advanced controls [74]. In the last 15 years, academics in the fields of building science, energy modeling, and control systems have proposed and developed innovative control algorithms for buildings. Primarily targeting HVAC systems, researchers proposed fuzzy , adaptive , and optimal controls trying to improve energy efficiency, comfort, and health (e.g., air quality) [75]. Two recent reviews identified more than 120 papers written on the subject in the last decade [75, 76], to which we need to add the more recent contribution of the wireless sensor networks community [74]. While fuzzy logic was a popular approach in the early studies, most of the new research uses model-based algorithms, such as model predictive control (MPC) or model-free learning algorithms, based on machine learning. Multi-agent approaches have also gained some popularity.

While the lighting industry has integrated some of the advanced control algorithms in actual products (e.g., daylighting controls), very little of this research has been adopted by the HVAC industry. This is due to a mix of limited incentives in the market and complexity of the actual algorithms. Recently, a few automation startups have begun offering advanced controls such as MPC [77–79], mostly targeting large commercial buildings.

Despite the abundance of studies on this topic, most papers test these innovative control schemes using computer simulations, without field deployment. The few implementations in real buildings are designed as "ad hoc" experiments, where researchers try to test the algorithms but do not provide a general solution to the problem. As a result, the strategies are difficult to replicate in other buildings, without complete redesign of the controller. This problem of low "portability" of the solutions is common also in industry, as most buildings and HVAC systems are custom-designed and control engineer hard-code control sequences, referencing specific component names and addresses, or using simple lookup tables [45]. While this problem is somewhat simpler in small commercial buildings, there is still a need for a software that can be rapidly configured for different deployments.

The literature of advanced controls for small commercial buildings is sparse. Model-based controls are the most common strategies researched. Early studies propose methods to create gray-box energy models that can be used to control single zone or simple multi-zone buildings. Different models have been created for specific applications, such as retrofit verification, commissioning, performance evaluation, fault detection, and demand response, and very few have general applicability [80–85]. Setup of advanced controls takes considerable engineering time, because models have to be created and validated by engineers. Recent studies try to simplify the deployment of these control algorithm investigating "plug-and-play" solutions [86]. Still, lack of a general framework for creating extensible energy models prevents wide adoption of these control strategies.

## 3.3 Software Design of a New Operating System for Small–Medium Commercial Buildings

This section describes the key criteria that guided the design of our BOS system and its features. While significant progress has been made by the research community in developing different BOS platforms, its use to support advanced controls applications in small–medium commercial buildings has not been extensively explored. We sought to develop software that could work with different types of devices and sensors, interface with other sources of information, and incorporate occupants' inputs. By targeting buildings under 50,000 ft2, the solution aimed at being cost-effective, reliable, and easy to install and manage. In the US, the small commercial segment is very cost-sensitive, as building services (e.g., HVAC and lighting) are considered commodities, and access to capital is limited, especially in case of split incentive between tenants and owners. For this reason, in our implementation, we use inexpensive wireless connected devices and analyze the deployment process that can represent a substantial part of the cost.

While a BAS for large commercial buildings is typically custom-built and custom-programmed on-site by a control technician, the same process would be too expensive for a small building. The solution so far has been to use very simple pre-programmed controllers (thermostats, timers, or manual switches). Small commercial buildings are also traditionally under-sensed (very few sensors are deployed) and under-actuated (multiple rooms are conditioned using ducts that miss local control). In addition, current HVAC controllers often

- use fixed and conservative set points (for comfort, not energy savings),
- do not coordinate their set points and the runtime of the units they control,
- do not measure occupancy and use conservative schedules (starting too early and ending too late compared to actual occupancy),
- may be heating and cooling the same space at the same time, and
- lockout occupants who cannot adjust them.

To address current limitations, our controller aims to

- offer a simple way of coordinating multiple schedules,
- offer more flexible and powerful control strategies,
- provide a platform to write and test different supervisory control algorithms that can be easily used in other buildings,
- incorporate additional sensors that are becoming less expensive to buy and install (with wireless communication, building can now be well sensed but still under-actuated), and
- minimize the installation and operational cost.

To reduce costs, and offer flexibility in the control sequences, we set out to pre-program the BOS with drivers and control algorithms that can be configured (without additional coding) for the specific installation (Fig. 3.4). Once new devices are connected (a), the BOS discovers them (b), loads the drivers and updates the building configuration (c), determines whether new algorithms are available and can be used with the devices (d), configures the algorithms (e.g., it sets parameters or creates models) (e), and runs them (f). Additional control sequences or drivers can also be added to the library and made available through software updates.

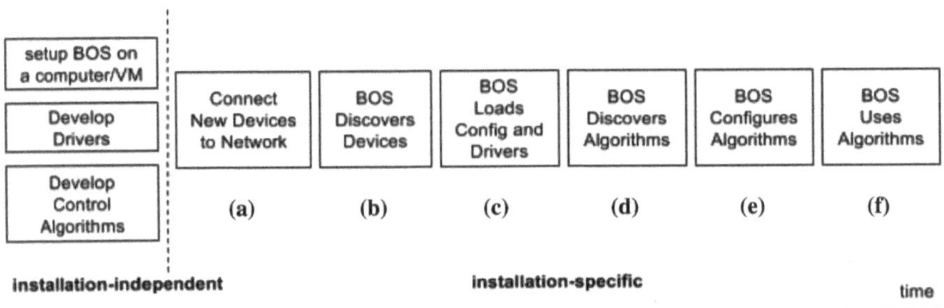


**Fig. 3.4** Pre-programmed tasks (installation-independent) and configuration (installation-specific) steps in the deployment of our BOS

We chose to build our software over sMAP (Simple Measuring and Actuation Profile) [41]. sMAP is a RESTful web service that allows instruments, sensors, and other producers of physical information to directly publish their data. More than 50 drivers for different devices and services were developed in the last few years, allowing bidirectional communication with connected thermostats, lights, and plugs. The data is stored in a time series database, coupled with a metadata repository, and can be accessed through an API. Simple queries against the metadata can be used to dynamically reference sensor and actuation points, to retrieve data, and to send commands [41]. sMAP provided the first layer and part of the second layer (i.e., storage, queries, information bus) of the three-layer architecture shown in Fig. 3.1. We also used an extension of sMAP, developed during a previous project in collaboration with UC Berkeley, UC Davis, and Lawrence Berkeley National Laboratory [87]. Using this add-on, we were able to automate the discovery and configuration of the devices. Defining and changing schedules for all the devices in the building became a trivial task using these two layers of the software platform. Further, one device could be easily set to "listen to" a sensor, reacting to events such as changes in occupancy or temperature [49].

The control library operates on top of the sMAP BOS stack and allows for testing and comparing of multiple control strategies. The implementation uses object-oriented programming (OOP) in Python (https://www.python.org/) with classes that can be simply instantiated in different buildings without rewriting them from scratch. The control library can work with different sets of sensors and devices and it is easily extensible to incorporate other configurations, algorithms, or sources of information. Table 3.1 shows several examples of control strategies that can be implemented with the library. The column named "system interaction" specifies whether the control strategy has impact on other systems (e.g., two thermal zones in the same room) and the effect is explicitly considered. The column "model" indicates whether a model is used to calculate control parameters (e.g., learning thermal response of a zone) or to predict future responses of the system (e.g., in MPC). The examples presented focus on simple RTU-thermostatic controls that are the most common HVAC control systems in small–medium commercial buildings, but the library can be used to model more complex systems (e.g., VAV-box systems in large commercial buildings).

In the literature, there are examples of how to write compact and simple control code in case of supervisory, stateless, independent, and model-free control sequences [36]. For instance, an event-based algorithm such as "if occupancy is detected then turn the light on" (example #2 in the table above) can now be implemented with one click using "If This Then That" (IFTTT), a free online web service that connects APIs of two networked devices [88]. However, the generalized implementation of more complex algorithms has not been explored thoroughly. For instance, a model-based control strategy requires model parameters to be identified before using it. This step is typically performed by an engineer, adding significant cost to the deployment.

In our implementation, when new algorithms or new devices are added, the BOS scans the available control strategies against the building configuration and discovers what sequences can be used. Each control module defines the inputs and outputs

**Table 3.1** Examples of control strategies implemented

| # | Control strategy | Description | System interaction | Model | Equipment controlled |
|---|---|---|---|---|---|
| 1 | Building-level coordinated schedules | Simple coordination of schedules in a building | Independent Systems | No | All Systems |
| 2 | Response to Occupancy | Integration of schedules with response to occupancy recorded by additional sensors | Independent Systems | No | All Systems |
| 3 | Response to occupancy with sensor fusion and prediction | Integration of schedules with response to virtual sensors estimating and predicting occupancy | Independent Systems (HVAC zones) | Yes for prediction | HVAC (zone-level) |
| 4 | Response to Thermal Feedback (Votes) | Integration of schedules with user overrides using thermal feedback | Independent Systems (HVAC zones) | No | HVAC (zone-level) |
| 5 | Coordinated Operation: avoid simultaneous heat and cool | Coordination between multiple HVAC units serving the same space to avoid heat and cool at the same time | Multiple zones (HVAC) in same space (room) | No | HVAC (zone-level) |
| 6 | Coordinated Operation: Limit Demand | Time coordination between multiple HVAC units to limit simultaneous operation | Multiple HVAC units in building | Yes | HVAC (all levels) |
| 7 | Coordinated Operation: Select Stages | Best use of stages from multiple units to reduce energy consumption or peak power | Multiple HVAC units in building | Yes | HVAC (all levels) |
| 8 | Learning: learn thermal response of building/HVAC | Learn thermal response to calculate time-to temperature and adapt set points | Independent Systems (HVAC zones) | Yes | HVAC (zone-level) |
| 9 | Learning: learn thermal preferences | Learn comfort temperatures to adapt set points | Independent Systems (HVAC zones) | Yes | HVAC (zone-level) |
| 10 | Model-Based Optimization (MPC) | MPC: optimize energy use, cost or peak | Independent Systems (HVAC zones) | Yes | HVAC (zone-level) |
| 11 | Subzoning: add sensors in subzones | Add sensors such as additional temperature sensors, creating subzone | Subzones + Zone interaction | No | HVAC (subzone) |
| 12 | Subzoning: add actuation | Add subsystems capable of providing services to subzones (e.g., personal comfort or task lighting). Coordinate with zone-level system | Subzones + Zone interaction | No | HVAC and lighting (subzone) |

needed in terms of queries to the metadata database. For instance to understand if a simple occupancy-responsive control strategy (strategy #2 in Table 3.1) is available, the BOS performs the following queries:

1. List all the thermal zones
2. For each thermal zone, list the zone temperature streams and occupancy streams (sensor inputs)
3. For each thermal zone, list the set point streams (actuation outputs)

For each zone with complete sensor and actuator sets, the BOS reports that the selected control sequence is available. It also details what other objects need to be created and instantiated before the sequence is ready to run. For instance, Fig. 3.5 shows that algorithm #3 in Table 3.1 requires the definition of a sensor fusion rule, while algorithm #10 needs a set of objects: an optimizer, a cost function, constraints, a model with identified parameters, and new data streams forecasting future values of input variables. There are two categories of objects: static objects containing variables, equations, algorithms, and objects that generate additional time streams (e.g., occupancy forecast). When the control sequence is instantiated, static objects are automatically created and populated, when possible, with preset values. During the initialization, new time series are also created as persistent processes that report values to the time series database. In case of forecasts, a prediction algorithm needs to be specified. When the control strategy needs a model, which requires training, the BOS can train the model with past data using a system identification algorithm. The identification process provides error metrics and can run different algorithms

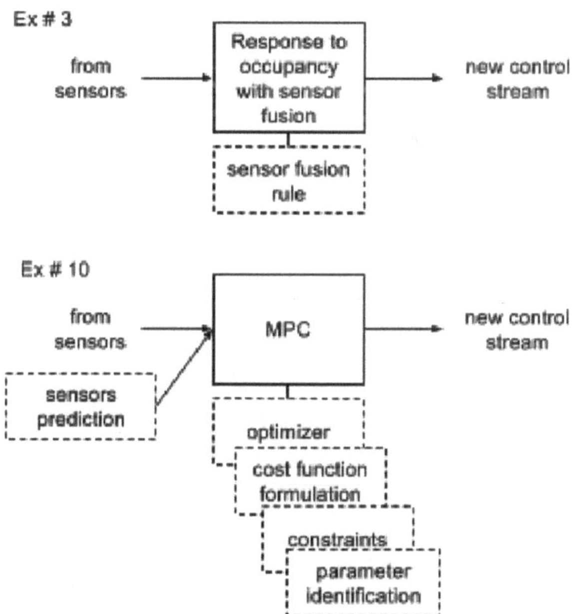


**Fig. 3.5** This graph shows what additional objects (dashed boxes) are necessary before starting the control sequence

iteratively to find the one that minimizes the error. When the training is completed, the model parameters are saved and the model can be used to predict new output values given the inputs.

Once a control algorithm is instantiated, it starts reporting control signals to the sMAP archiver. Alternative algorithms can run in parallel and the controller (e.g., a thermostat) can subscribe to the output of one of them at the time.

## 3.4 Implementation and Results

We tested this software in a small commercial building located in University of California Davis campus, in Davis, CA. The building was built in 1955 and renovated in 1999 (Fig. 3.6 and Table 3.2). The total building floor area is about 3,160 square feet and each zone has one or two large walls exposed to the outside (i.e., there are no "core" zones). Walls are wood framed with insulation; offices have ceilings but

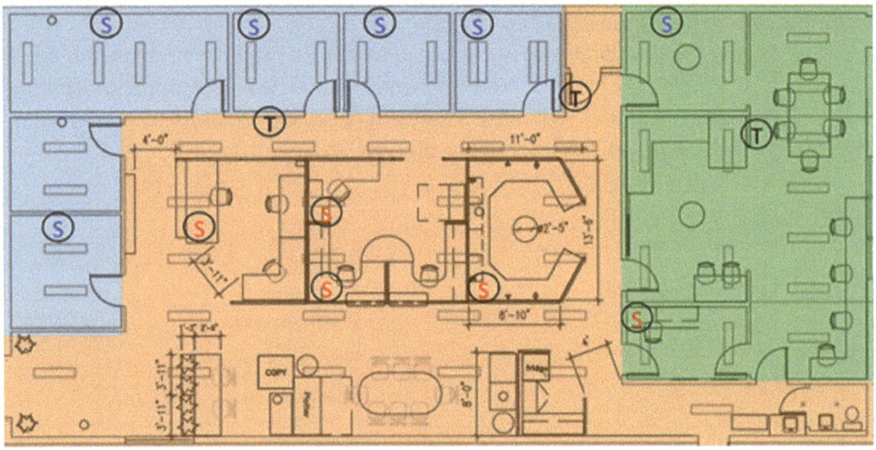


**Fig. 3.6** Plant of the test bed with three different colors for each thermal zone. The position of the thermostats is indicated with "T" and the position of additional occupancy/temperature sensors with "S"

**Table 3.2** Characteristics of the test bed building and the three thermal zones

| | Zone 1 (East end) | Zone 2 (Openspace) | Zone 3 (Offices) | Whole building |
|---|---|---|---|---|
| **Area (square feet)** | 790 | 1,740 | 630 | 3,160 |
| **RTU cooling capacity (ton)** | 4 | 5 | 5 | 14 |
| **RTU heating capacity (kBtu/h)** | 60,000 | 72,000 | 72,000 | 204,000 |
| **Electricity and Gas Meter** | N | N | N | Y |

the roof is not insulated. There are two sets of non-operable windows, one set on the south wall and one on the east wall. Figure 3.6 shows the three thermal zones in different colors, the positions of the thermostats (T) and the positions of additional temperature and occupancy sensors (S). Each zone is conditioned by an RTU that has a single-speed single-stage direct expansion cooling system and uses natural gas for heating. The outdoor air damper is fixed in these units. The RTU and building characteristics are listed in Table 2. The RTUs were previously managed using simple non-networked programmable thermostats, without the use of a BAS. The existing thermostats were replaced with new Wi-Fi thermostats. The installation of the devices happened in different periods: October 2014 and June 2015. The three zones have variable occupancy, because several students work part-time and other occupants leave for meetings and projects around campus. The whole-building gas and electricity consumption are also metered and the information sent to a central database that could be queried using a RESTful API. This building did not have smart plugs or smart lights installed, but a rich set of data streams were available: Wi-Fi connections (i.e., number of devices connected to the campus Wi-Fi for the access point in the building), the thermal votes from a campus-wide thermal feedback system [89], online weather forecasts, and retail price of electricity. Figure 3.7 shows the three-layer architecture of the BOS implemented in the test bed. Drivers for each source were developed and tested in advance (Tier 1) and a building configuration file loaded (Tier 2).

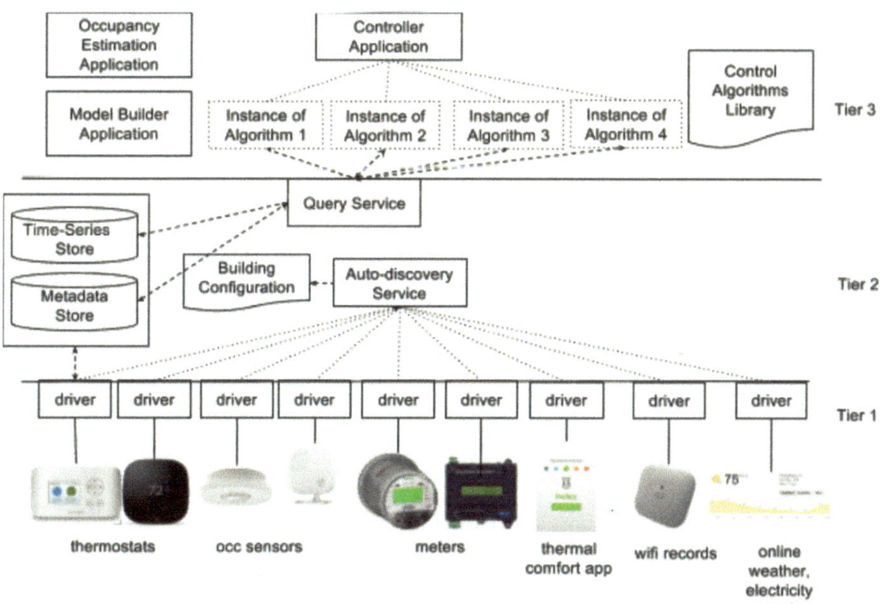


**Fig. 3.7** Three-layer architecture of the control system implemented in the test bed

We tested four of the strategies implemented in the library (Tier 3) in this building during summer of 2016. Since most thermostat APIs do not give direct control of the RTUs, we decided to implement only supervisory control strategies over the thermostats. This means setting the thermostat set points based on a scheme or a higher level control sequence. The strategies tested were

1. Building-level coordinated schedules (# 1 in Table 3.1),
2. Response to Thermal Feedback (# 4),
3. Response to Thermal Feedback with "drifting" algorithm (a variation of #4), and
4. MPC (# 10) .

The simplest strategy used as a baseline consisted in adopting a single set point for all the zones. This was achieved by subscribing the set points of all the devices to a single controller (similar to [49]). The heating and cooling set points mimicked what we consider representative campus set points. We also collected information from the thermal feedback application, described in detail in Sanguinetti et al. [89]. The building occupants were reminded to use the feedback system to vote if they did not feel comfortable.

The second and third strategies used these thermal votes to drive the zone set points. We coded a few options, with differences in how the control votes were accounted for. The simplest logic averaged votes in the previous time step (5 minutes). To avoid oscillation between hot and cold adjustments, a more complex strategy averaged votes in the previous two hours weighting recent votes the most. Since these options did not necessarily save energy, as people mostly vote to improve their comfort, we tested a more energy-efficient strategy. Similar to what was proposed by Purdon et al. [90], the control logic responded to thermal votes, but slowly increased the set point toward outdoor temperature (in the summer) if occupants did not vote. The reduced temperature difference between indoor and outdoor conditions was expected to reduce thermal load on the building and energy use. The first and the last control sequences were deployed in this experiment. Each zone had a separate controller, since the thermal votes were cast based on the occupant zone.

The fourth control sequence tested was an MPC algorithm. MPC is an advanced control strategy that uses a dynamical model a process to predict its future evolution and optimize the control signal to achieve a certain goal, given specified constraints [13]. Figure 3.8 shows the schematic of the controller.

The model represented the relationship between of heating and cooling operation and zone temperature. The zone model was based on simplified models adapted from the literature [85, 86, 91] and is stated below:

$$C\frac{d}{dt}T_{zone} = (T_{adj} - T_{zone})(U\,A_{tot}) + IG + K_H\ runHEAT + K_C\ runCool\ , \quad (3.1)$$

where
$C\frac{d}{dt}T_{zone}$ represents the energy storage term, including the air and part of the wall lumped together;
$T_{zone}$ is the temperature of the zone;

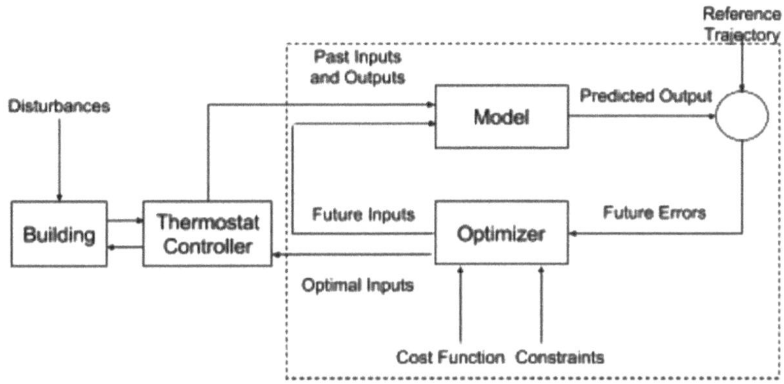


**Fig. 3.8** Representation of the MPC architecture

$T_{adj}$ is the temperature of the adjacent zones or outdoor;
$IG$ are the internal gains in the zone;
$K_H$ and $K_C$ are the coefficients representing the effect of heating and cooling on the internal temperature;
$runHEAT$ and $runCool$ are the runtime of the RTU measured by the thermostat.

At each time step, the controller minimizes the cost for heating and cooling (considering time-variable prices) with a penalty for violation of comfort, as presented in

$$\sum_{k=0}^{N-1}(r_h P_H + r_c P_C) + \rho \sum_{k=1}^{N}(|\underline{\epsilon}_{t+k}| + |\bar{\epsilon}_{t+k}|) \ , \tag{3.2}$$

subjected to the dynamics of the system and

$$T_{set-min} - |\underline{\epsilon}_{t+k}| < T_{k+t} < +|\bar{\epsilon}_{t+k}| + T_{set-max}, \tag{3.3}$$

where $P_H$ and $P_C$ are the power used to heat and cool, $r_h$ and $r_c$ are the cost coefficients for energy use, $\epsilon$ are the temperature violations from the lower and upper bounds, and $\rho$ is the cost attributed to a violation of comfort by one temperature unit.

The details of the formulation can be found in the referenced papers [85, 86, 91] and are out of the scope of this chapter. The optimizer was developed using Jump in Julia Language [3] and called from the control library. The model parameters were estimated using a system identification algorithm implemented in Python, accessing historical data stored in the time series archive. The control strategy was instantiated, the model was learned, and the three zone thermostats were subscribed to the output of these controllers. Each sequence was used for several days during August and September 2016. Results are shown in Fig. 3.9. Each graph presents the cooling set point (green), the cooling runtime (purple), the indoor temperature (red), outdoor

[3]http://julialang.org/.

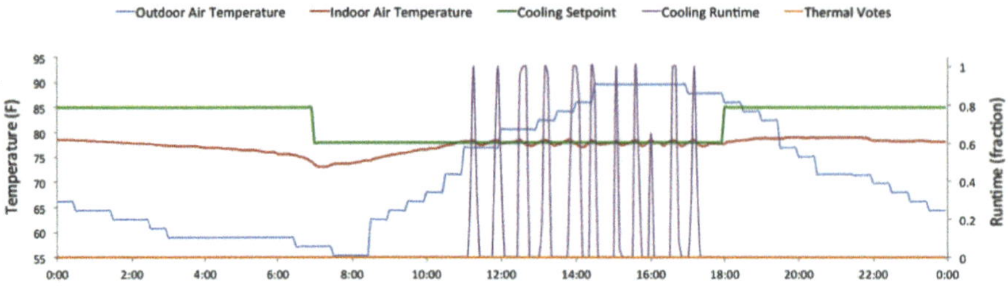


(a) Baseline strategy

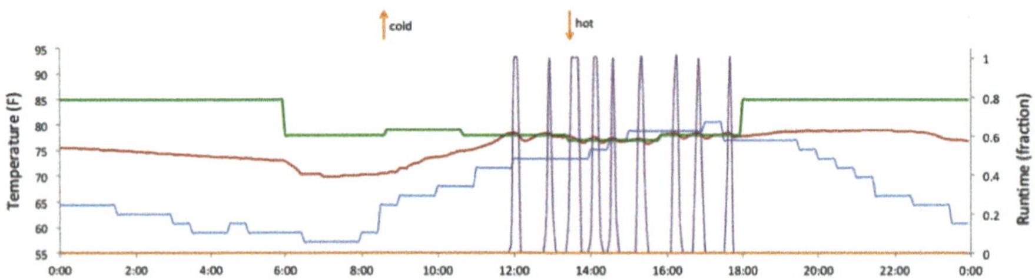


(b) Feedback strategy

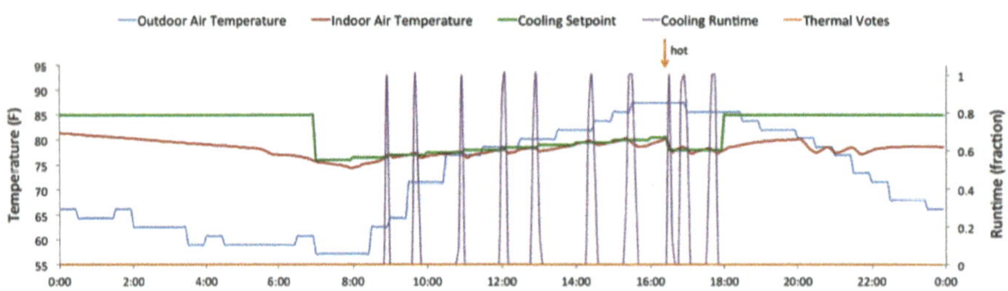


(c) Drifting strategy

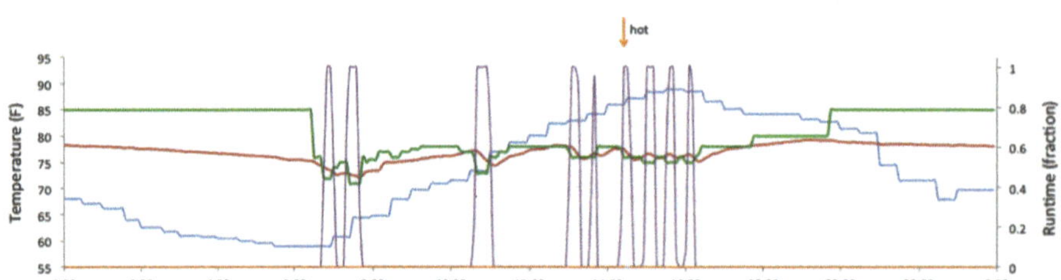


(d) MPC strategy

**Fig. 3.9** Comparison of four thermostat control strategies temperature (blue), and thermal votes (orange). The four days have similar, but not identical weather conditions. The baseline strategy (Fig. 3.9a) has a constant set point from 7 am to 6 pm. Strategy in Fig. 3.9b includes response to feedback. At 8:35 am somebody cast a "cold" vote, when the indoor temperature was 70.5°F, then later, in the afternoon, somebody else cast a "hot" vote, when the indoor temperature reached 77°F. Both votes affected the set point of the thermostat, although only the second one caused the HVAC to turn on. In Fig. 3.9c the "drifting" strategy is clearly visible in the behavior of the cooling set point. When a vote is cast, at 4:30 pm, the set point stops drifting for a few hours, before reverting to the unoccupied set point. Figure 3.9d represents the MPC scheme. The set point is reset continuously by the optimization algorithm that tries to minimize its cost function. Even though not enough data was collected for a direct comparison of energy use, this test shows that the control algorithms operate as expected.

## 3.5 Discussion

The development of the BOS undertaken in this work has provided us with many important lessons. A BOS such as ours is only feasible to implement with a reliable communication network. In our tests, the devices and web services worked sufficiently well; however, some of the networked products suffered from connectivity issues, dropping frequently from the network. The cloud backend of one product, necessary to gather data, was unavailable for two weeks during our experiment, causing significant delays. Other products used gateways to handle protocol conversion, which represented an additional point of failure in the system. All these networked devices had been in the market for just a few years; therefore, we expect their reliability to improve as newer versions are developed. During data collection, unexpected changes in APIs caused data loss that was difficult to track. This suggests that the developers of the BOS should establish relationships with hardware and software manufacturers to be constantly informed about software and firmware updates. In our case, other departments in campus changed database structures and point names, creating problems for our BOS. More coordination is required to make sure these entities communicate any changes in their data streams. These data are essential to the operation of the BOS, and since their availability depends on software, their integration requires more maintenance than older hardware-only components (e.g., wired sensors and actuators).

Our field test focused on connected thermostats, as they control a large fraction of the energy use with a limited number of actuation points. Even though basic features of these thermostats are similar, the functionalities exposed by their APIs are different. For instance, early in the project, we had to exclude a brand of thermostats, because its API exposed only the temperature and the set points, but it did not provide information about the state of the HVAC (e.g., heating/cooling and stages). This information was collected by the device, but the manufacturer did not make it available to the developers. Another device did not allow the specification of both heating and cooling set points at the same time, practically forcing the developers to change the thermostat mode based on prediction of temperature behavior. Some devices allowed overriding set points for unlimited time, some would revert to the scheduled setpoint after a preset time, and some would only switch at epoch changes of internal schedules. We also found other commercial products that did not have open APIs, limiting their use in an open-architecture environment [10, 83]. Some standardization in these elements would be beneficial to multi-device deployments. Further, thermostats were used here as direct-control devices, with our BOS only specifying set points and mode changes. The thermostats run their internal algorithms to translate the set point and temperature information into signals to the RTUs. We noticed that the intelligence in the device sometimes overrode the intent of the supervisory control signal. Features such as minimum compressor runtime, extended fan operation, and heat anticipation[4] delayed or anticipated the actuation of the RTU. The effect was still visible at 5-minute actuation intervals used for supervisory control. The implementation of these features differs among manufacturers and most APIs do not allow setting their parameters. While these features are useful and we would like to preserve them, their effects add uncertainty and lag to the control signal.

At the beginning of the experiment, additional confusion was caused by user interaction with the devices. Most connected thermostats allow users to override set points and select mode (including switching the thermostat off) from the physical interface. These changes conflicted with BOS-driven temperature adjustments and human-driven comfort votes intended to regulate temperature (strategy 2 and 3). After a few attempts to take manual adjustments into account, we decided to lock users out from the physical interface. Users were encouraged to vote using the thermal feedback app, to impact the set points. This is not an ideal solution and future work should look into a more seamless integration between physical interface, native mobile app (frequently provided with the device), and API calls from external applications, such as our BOS. We believe that this integration requires collaboration with device manufacturers.

The field test of the control library yielded interesting results: we were able to discover the devices on the network, load drivers, and start control sequences using configuration files without site-specific recoding. We were also able to simply switch between four control sequences with simple commands and drive the operation of the three zones' RTUs. Other web services not connected to local networks were harder to manage because there was no "discovery" mechanism in place for them. As mentioned above, changes in the data streams addresses and APIs caused data loss that had to be fixed manually. In this case, we did not find an easy solution to reduce the burden for the developer.

Another question we wanted to explore was the degree of automation that can be used to set up a complex control strategy, such as MPC. Recent work has developed "plug-and-play" MPC controls [86], but these applications are meant to address a specific problem (e.g., coordination of cooling stages of multiple RTUs in open spaces) and are difficult to extend to other cases. In a sense, automation in the setup and extensibility (ability to incorporate extensions) are competing objectives. Our

[4]Common features implemented in new connected thermostats.

implementation allows the use of different cost functions that can be simply loaded from a configuration file. This could be useful, in practice, since different utilities may have different price structures, for example, including a demand charge or a time-varying cost of electricity. It also allows to chose weights for the violation of thermal comfort (i.e., Eq. 3.2), adapting the MPC to building manager policies. We could also define the minimum and maximum set points (i.e., $T_{set}$ in Eq. 3.3) in a flexible way either following daily schedules (i.e., occupied and unoccupied periods) or reacting to thermal feedback. Other variables in the zone model (Eq. 3.1) could be adapted to model different sensors available. For instance, the internal gains (IG) can be modeled as a constant, a time-of-the-day constant or, if available, as a function of zone occupancy, measured by sensors. While this architecture can be flexible enough to accommodate a large number of small–medium commercial buildings, several challenges still remain. For instance, to model the thermal interaction between zones, the BOS needs a map of the physical location of rooms and spaces that may be difficult to discover automatically. To add the ability to manage different HVAC systems (e.g., variable capacity or radiant system), more complex models may be required. Each time new combinations of models, sensors, and devices are introduced in the BOS; the developers need to check that these combinations work appropriately. Future work should investigate more deeply the modeling needs for model-based control strategies.

On a final note, we acknowledge that in our implementation we did not address security thoroughly. Even though we implemented authentication to access each component of the system and used secure connections, more attention should be placed on cybersecurity, to minimize the threat of security breaches and hacks. As mentioned above, security is a major concern for commercial application of BOSs.

## 3.6 Conclusion

Energy use in small and medium commercial buildings can be reduced using better controls. Several networked devices and software platforms are now commercially available, but they tend to have low interoperability and mostly target residential applications. By reviewing the literature, we identified missing features of current open-source BOSs for controls of small and medium commercial buildings. Using an existing stack, we extended the software to support developers in writing and testing control sequences for these buildings. The control library allows to reconfigure the same sequence in multiple buildings without additional coding. The software was tested in a real building in the summer period. The implementation shed lights on several problems with these networked devices, including connectivity issues, lack of standardization in the interfaces, and features and need of periodical maintenance due to software updates. Future work should look at better integration with user interfaces and improved cybersecurity. In our test, we were able to automate part of the setup of an MPC algorithm, but future work should investigate how to extend this example to different sensors and system configurations.

## References

1. Administration EI (2012) Commercial buildings energy consumption survey (CBECS). http://www.eia.gov/consumption/commercial/data/2012/
2. Pacific Northwest National Laboratory (2012) Small-and medium-sized commercial building monitoring and controls needs: A scoping study. http://www.pnnl.gov/main/publications/external/technical_reports/PNNL-22169.pdf
3. Mills E, Mathew P (2012) Monitoring based commissioning: Benchmarking of 24 UC/CSU/IOU projects. LBNL Report 1972E, Tech. Rep
4. Center WCE (2014) Multi-Tenant Light Commercial (MTLC) Project Final Report draft. 2014, unpublished
5. Pacific Northwest National Laboratory (2011) Energy savings and economics of advanced control strategies for packaged air-conditioning units with gas heat. http://www.pnnl.gov/main/publications/external/technical_reports/PNNL-20955.pdf
6. Energy savings and economics of advanced control strategies for packaged heat pumps, 2012. http://www.pnnl.gov/main/publications/external/technical_reports/PNNL-21944.pdf
7. Advanced rooftop control (ARC) retrofit: Field-test results, 2013. http://www.pnl.gov/main/publications/external/technical_reports/PNNL-22656.pdf
8. Trane Tracer Concierge (2016) Trane, retrieved Nov 2016. http://commercial.trane.com/content/trane-lp/concierge.html
9. Daintree (2016) Daintree Networks, retrieved Nov 2016. http://www.daintree.net/solutions/building-energy-management-as-a-service/
10. Nest (2016) Nest Labs, retrieved Nov 2016. https://nest.com/
11. Ecobee (2016) ecobee, retrieved Nov 2016. https://www.ecobee.com/
12. Philips huc (2016) Philips Lighting, retrieved Nov 2016. http://www2.meethue.com/
13. Belkin wemo home automation (2016) Belkin, retrieved Nov 2016. http://www.belkin.com/us/Products/home-automation/c/wemo-home-automation/
14. Ford R, Karlin B, Sanguinetti A, Nersesyan A, Pritoni M (2016) Assessing Players, Products, and Perceptions of Home Energy Management. Pacific Gas and Electric, San Francisco, CA, Emerging Technology Report ET15PGE8851, retrieved Nov 2016
15. Works with Nest (2016) Nest Labs, retrieved Nov 2016. https://nest.com/works-with-nest/
16. Works with SmartThings (2016) Samsung, retrieved Nov 2016. https://www.smartthings.com/works-with-smartthings
17. Iris by Lowes: Smart Home + Security (2016) Lowes, retrieved Nov 2016. https://www.irisbylowes.com/products/
18. HomeKit (2016) Apple, retrieved Nov 2016. https://developer.apple.com/homekit/
19. Eclipse: A Flexible Framework for the Smart Home (2016) Eclipse, retrieved Nov 2016. http://www.eclipse.org/smarthome/
20. Temboo:Program the World (2017) Temboo, retrieved Nov 2016. https://temboo.com/
21. The Thing System: Take control of things (2016) the thing system, retrieved Nov 2016. http://thethingsystem.com/
22. openHAB: empowering the smart home (2016) openHAB, retrieved Nov 2016. https://www.openhab.org
23. Home Assistant: Awaken your home (2016) Home Assistant, retrieved Nov 2016. https://home-assistant.io/components/
24. Domoticz: Control at your finger tips (2016) Domoticz, retrieved Nov 2016. https://domoticz.com/
25. OpenMotics: Open Source Home Automation and Power Measurement Platform (2016) OpenMotics, retrieved Nov 2016. https://www.openmotics.com/
26. Calaos: Open Source Home Automation (2016) Calaos, retrieved Nov 2016. https://calaos.fr/en/
27. Tridium: Harness the power of the Internet of Thing with Niagara (2016) Tridium, retrieved Nov 2016. https://www.tridium.com/

28. Kite and Lighting: UNITY Software Suite (2016) Kite and Lighting, retrieved Nov 2016. http://kiteandlightning.com/
29. VOLTTRON Distributed Control System Platform (2016) Pacific Northwest National Laboratory, retrieved Nov 2016. https://github.com/VOLTTRON/volttron
30. AllSeen Alliance: AllJoyn Framework (2016) AllSeen Alliance, retrieved Nov 2016. https://allseenalliance.org/framework/
31. Open Connectivity Foundation (2016) Open Connectivity Foundation (OCF), retrieved Nov 2016. https://openconnectivity.org/
32. IoT Eclipse:IoT development made simple (2016) Eclipse, retrieved Nov 2016. http://iot.eclipse.org/
33. Industrial Internet Consortium (2016) Industrial Internet Consortium (IIC), retrieved Nov 2016. http://www.iiconsortium.org/
34. RTI Connext DDS platform (2016) Real Time Innovation (RTI), retrieved Nov 2016. https://www.rti.com/products/dds/
35. Prismtech Vortex platform (2016) Prismtech, 2016, retrieved Nov 2016. http://www.prismtech.com/vortex
36. Gabriel F, Culler DE, Peffer T (2016) Enabling portable energy applications with a building operating system. In: Proceedings of the ACEEE Summer Study on Energy Efficiency in Buildings, Asilomar, CA. ACEEE
37. Weng T, Nwokafor A, Agarwal Y (2013) Buildingdepot 2.0: An integrated management system for building analysis and control. In: Proceedings of the 5th ACM Workshop on Embedded Systems For Energy-Efficient Buildings. ACM, pp 1–8
38. Arjunan P, Batra N, Choi H, Singh A, Singh P, Srivastava MB (2012) SensorAct: a privacy and security aware federated middleware for building management. In: Proceedings of the Fourth ACM Workshop on Embedded Sensing Systems for Energy-Efficiency in Buildings. ACM, pp 80–87
39. Rowe A, Berges M, Bhatia G, Goldman E, Rajkumar R, Soibelman L (2009) The sensor andrew infrastructure for large-scale campus-wide sensing and actuation. In: International Conference on Information Processing in Sensor Networks (IPSN). IEEE, pp 415–416
40. Rowe A, Berges ME, Bhatia G, Goldman E, Rajkumar R, Garrett JH, Moura JM, Soibelman L (2011) Sensor Andrew: Large-scale campus-wide sensing and actuation. IBM J Res Dev 55(1.2):6.1–6.14
41. Dawson-Haggerty S, Jiang X, Tolle G, Ortiz J, Culler D (2010) sMAP: a simple measurement and actuation profile for physical information. In: Proceedings of the 8th ACM Conference on Embedded Networked Sensor Systems. ACM, pp 197–210
42. Santanche A, Nath S, Liu J, Priyantha B, Zhao F (2006) Senseweb: Browsing the physical world in real time. Demo Abstract, ACM/IEEE IPSN06. Nashville, TN, pp 1–2
43. Chu X, Kobialka T, Durnota B, Buyya R (2006) Open sensor web architecture: Core services. In: Fourth International Conference on Intelligent Sensing and Information Processing (ICISIP). IEEE, pp 98–103
44. Aberer K, Hauswirth M, Salehi A (2006) Global sensor networks. Ecole Polytechnique Fédérale de Lausanne (EPFL), Lausanne, Switzerland, LSIR-Report 2006-001
45. Krioukov A, Fierro G, Kitaev N, Culler D (2012) Building application stack (BAS). In: Proceedings of the Fourth ACM Workshop on Embedded Sensing Systems for Energy-Efficiency in Buildings. ACM, pp 72–79
46. Dawson-Haggerty S, Krioukov A, Taneja J, Karandikar S, Fierro G, Kitaev N, Culler DE (2013) BOSS: Building operating system services. In: NSDI, vol 13, pp 443–458
47. Khamphanchai W, Saha A, Rathinavel K, Kuzlu M, Pipattanasomporn M, Rahman S, Akyol B, Haack J (2014) Conceptual architecture of building energy management open source software (BEMOSS). In: Innovative Smart Grid Technologies Conference Europe (ISGT-Europe), (2014) IEEE PES. IEEE, pp 1–6
48. Blumstein C, Culler D, Fierro G, Peffer T, Pritoni M (2015) Open software-architecture for building monitoring and control. Paper and Presentation at Sustainable Places, pp 16–18

49. Fierro G, Culler DE (2015) XBOS: An Extensible Building Operating System. In: Proceedings of the 2nd ACM International Conference on Embedded Systems for Energy-Efficient Built Environments. ACM, pp 119–120
50. RFC 4862: IPv6 Stateless Address Autoconfiguration (2016) Internet Engineering Task Force (IETF), retrieved Nov 2016. https://tools.ietf.org/html/rfc4862
51. Guttman E (2001) Autoconfiguration for IP networking: Enabling local communication. IEEE Internet comput 5(3):81–86
52. Deri L, Mainardi S, Fusco F (2012) tsdb: A compressed database for time series. In: International Workshop on Traffic Monitoring and Analysis. Springer, pp 143–156
53. OpenTSDB: A Distributed, Scalable Monitoring System (2016) OpenTSDB, retrieved Nov 2016. http://opentsdb.net/
54. Buevich M, Wright A, Sargent R, Rowe A (2013) Respawn: A distributed multi-resolution time-series datastore. In: Real-Time Systems Symposium (RTSS), (2013) IEEE 34th. IEEE, pp 288–297
55. Andersen MP, Culler DE (2016) BTrDB: Optimizing Storage System Design for Timeseries Processing. In: FAST, pp 39–52
56. KairosDB: Fast scalable time series database (2016) KairosDB, 2015, retrieved Nov 2016. http://kairosdb.github.io/
57. Prometheus: Monitoring system and time series database (2016) Prometheus, retrieved Nov 2016. https://prometheus.io/
58. InfluxDB: Open Source Time Series Database (2016) InfluxData, retrieved Nov 2016. https://www.influxdata.com/
59. Periscope Data: from SQL query to analysis in seconds (2016) Periscope Data, retrieved Nov 2016. https://www.periscopedata.com/
60. IBM Informix TimeSeries (2016) IBM, retrieved Nov 2016. http://www.ibm.com/analytics/us/en/technology/informix/
61. Bhattacharya AA, Hong D, Culler D, Ortiz J, Whitehouse K, Wu E (2015) Automated metadata construction to support portable building applications. In: Proceedings of the 2nd ACM International Conference on Embedded Systems for Energy-Efficient Built Environments. ACM, pp 3–12
62. Project Haystack (2016) Project Haystack, retrieved Nov 2016. http://project-haystack.org/
63. Palmer C, Lazik P, Buevich M, Gao J, Berges M, Rowe A, Pereira RL, Martin C (2014) Mortar.io: A concrete building automation system. In: Proceedings of the 1st ACM Conference on Embedded Systems for Energy-Efficient Buildings. ACM, pp 204–205
64. The Atom Syndication Format Proposed Standard (2016) Internet Engineering Task Force (IETF), retrieved Nov 2016. https://tools.ietf.org/html/rfc4287
65. The Atom Publishing Protocol Proposed Standard (2016) Internet Engineering Task Force (IETF), retrieved Nov 2016. https://tools.ietf.org/html/rfc5023
66. RSS 0.90 Specification (2016) RSS Advisory Board, retrieved Nov 2016. http://www.rssboard.org/rss-0-9-0
67. PubSubHubbub Core 0.4 – Working Draft (2016) PubSubHubbub, retrieved Nov 2016. http://pubsubhubbub.github.io/PubSubHubbub/pubsubhubbub-core-0.4.html
68. Glantz CS, Somasundaram S, Mylrea ME, Underhill RM, Nicholls AK (2016) Evaluating the Maturity of Cybersecurity Programs for Building Control Systems. Tech. Rep, Pacific Northwest National Laboratory (PNNL), Richland, WA (US)
69. Dyn Analysis Summary of Friday October 21 Attack (2016) Dyn, retrieved Nov 2016. http://dyn.com/blog/dyn-analysis-summary-of-friday-october-21-attack/
70. Akyol B, Haack J, Carpenter B, Ciraci S, Vlachopoulou M, Tews C (2012) Volttron: An agent execution platform for the electric power system. In: Third international workshop on agent technologies for energy systems valencia, Spain
71. Zhang H, Arens E, Zhai Y (2015) A review of the corrective power of personal comfort systems in non-neutral ambient environments. Build Environ 91:15–41
72. Andersen MP, Fierro G, Kumar S, Chen M, Truong L, Kim J, Arens EA, Zhang H, Raftery P, Culler DE (2015) Well-connected microzones for increased building efficiency and occupant

comfort. In: Proceedings of the 2nd ACM International Conference on Embedded Systems for Energy-Efficient Built Environments. ACM, pp 121–122
73. Hoyt T, Arens E, Zhang H (2015) Extending air temperature setpoints: Simulated energy savings and design considerations for new and retrofit buildings. Build Environ 88:89–96
74. Bhattacharya A, Ploennigs J, Culler D (2015) Analyzing metadata schemas for buildings: The good, the bad, and the ugly. In: Proceedings of the 2nd ACM International Conference on Embedded Systems for Energy-Efficient Built Environments. ACM, pp 33–34
75. Shaikh PH, Nor NBM, Nallagownden P, Elamvazuthi I, Ibrahim T (2014) A review on optimized control systems for building energy and comfort management of smart sustainable buildings. Renew Sustain Energy Rev 34:409–429
76. Dounis AI, Caraiscos C (2009) Advanced control systems engineering for energy and comfort management in a building environment - a review. Renew Sustain Energy Rev 13(6):1246–1261
77. BuildingIQ: Building Energy and Operating Intelligence (2016) BuildingIQ, retrieved Nov 2016. https://buildingiq.com/
78. Optimum Energy (2016) Optimum Energy, retrieved Nov 2016. http://optimumenergyco.com/
79. MelRok: Energy IoT(2016) MelRok, retrieved Nov 2016. https://www.melrok.com/#/pages/main.html
80. Armstrong PR, Leeb SB, Norford LK (2006) Control with building mass-Part I: Thermal response model. Trans Am Soc Heat Refrig Air Cond Eng (ASHRAE) 112(1):449
81. Braun JE, Chaturvedi N (2002) An inverse gray-box model for transient building load prediction. HVAC R Res 8(1):73–99
82. Lee K-H, Braun JE (2016) Development and application of an inverse building model for demand response in small commercial buildings. IBPSA-USA J 1(1)
83. Cai J, Braun JE (2012) Efficient and robust training methodology for inverse building modeling. IBPSA-USA J 5(1):456–461
84. Bengea SC, Kelman AD, Borrelli F, Taylor R, Narayanan S (2014) Implementation of model predictive control for an HVAC system in a mid-size commercial building. HVAC R Res 20(1):121–135
85. Kelman A, Ma Y, Borrelli F (2013) Analysis of local optima in predictive control for energy efficient buildings. J Build Perform Simul 6(3):236–255
86. Kim D, Braun J, Cai J, Fugate D (2015) Development and experimental demonstration of a plug-and-play multiple RTU coordination control algorithm for small/medium commercial buildings. Energy Build 107:279–293
87. Peffer T (2015) Software-defined solutions for managing energy use in small to medium sized commercial buildings. Regents of the University of California, Berkeley, CA (United States), Tech. Rep
88. If This Then That (IFTTT) (2016) IFTTT, retrieved Nov 2016. https://ifttt.com/
89. Sanguinetti A, Pritoni M, Salmon K, Morejohn J (2016) Using occupant feedback to drive energy efficiency across an entire university campus. In: ACEEE Summer Study on Energy Efficiency in Buildings, Asilomar. CA, ACEEE
90. Purdon S, Kusy B, Jurdak R, Challen G (2013) Model-free HVAC control using occupant feedback. In: 38th Conference on Local Computer Networks Workshops (LCN Workshops). IEEE, pp 84–92
91. Beltran A, Cerpa AE (2014) Optimal HVAC building control with occupancy prediction. In: Proceedings of the 1st ACM Conference on Embedded Systems for Energy-Efficient Buildings. ACM, pp 168–171

# Part II
# The Heating, Ventilation, Air Conditioning (HVAC) System

# Chapter 4
# HVAC System Modeling and Control: Vapor Compression System Modeling and Control

**Bryan P. Rasmussen, Christopher Price, Justin Koeln, Bryan Keating and Andrew Alleyne**

## Nomenclature

### *Variables*

| | | |
|---|---|---|
| $c_p$ | Specific Heat Capacity | $\frac{J}{kg \cdot K}$ |
| $h$ | Specific Enthalpy | $\frac{J}{kg}$ |
| $k_o$ | Steady-State Gain | |
| $l$ | System Operating Load | |
| $\dot{m}$ | Mass Flow Rate | $\frac{kg}{s}$ |
| $s$ | Specific Entropy | $\frac{J}{kg \cdot K}$ |
| $u$ | Specific Internal Energy | $\frac{J}{kg}$ |
| $A$ | Area | $m^2$ |
| $C_d$ | Discharge Coefficient | |
| $E$ | Energy | J |
| $L$ | Length | m |
| $K_L$ | CascadedInner Loop Gain | |
| $P$ | Pressure | kPa |
| $Q$ | Heat | J |
| $T$ | Temperature | K |
| $V$ | Volume | $m^3$ |
| $\dot{W}$ | Power | $\frac{J}{s}$ |
| $\alpha$ | Heat Transfer Coefficient | $\frac{W}{m^2 \cdot K}$ |
| $\bar{\gamma}$ | Mean Void Fraction | |

B.P. Rasmussen (✉) · C. Price
Texas A&M University, College Station, TX, USA
e-mail: brasmussen@tamu.edu

J. Koeln · B. Keating · A. Alleyne
University of Illinois Urbana-Champaign, Champaign, IL, USA



J.T. Wen and S. Mishra (eds.), *Intelligent Building Control Systems*,
Advances in Industrial Control, https://doi.org/10.1007/978-3-319-68462-8_4

| | | |
|---|---|---|
| $\eta$ | Efficiency | |
| $\theta$ | Actuator Position | % |
| $\lambda$ | Normalized Length | |
| $\rho$ | Density | $\frac{kg}{m^3}$ |
| $\tau$ | Time Constant | s |
| $\psi$ | System Nonlinearity | |
| $\omega$ | Angular Velocity | $\frac{rad}{s}$ |

### *Subscripts*

| | |
|---|---|
| $a$ | Air |
| $c$ | Condenser |
| $f$ | Liquid |
| $g$ | Vapor |
| $i$ | Inside/Inner |
| $k$ | Compressor |
| $o$ | Outside/Outer |
| $r$ | Refrigerant |
| $w$ | Wall |
| $amb$ | Ambient |
| $cs$ | Cross-Sectional |
| $dis$ | Disturbance |
| $int$ | Interface |
| $isen$ | Isentropic |
| $rm$ | Room |
| $vol$ | Volume |

## 4.1 Introduction

Studies estimate that most North Americans and Europeans spend nearly 90% of their lives indoors [1, 2] in environmentally conditioned spaces. Heating, Ventilation, and Air Conditioning (HVAC) technology is considered to be one of the greatest engineering achievements of the twentieth century [3]. It is ubiquitous and almost a necessity for occupant comfort, health, safety, and productivity. The primary energy system in building air conditioning, heat pump, chilled water, and refrigeration systems is the Vapor Compression Cycle system (VCC).

A basic VCC consists of two heat exchangers, an expansion valve and a compressor (Fig. 4.1a). The ideal form of this thermodynamic cycle utilizes four processes to transfer heat between the external environment and the conditioned space: (1) isentropic compression, (2) isobaric heat rejection and condensation, (3) isenthalpic expansion, and (4) isobaric heat absorption and evaporation (Fig. 4.1b). During steady-state operation, refrigerant flows through the system to efficiently transfer

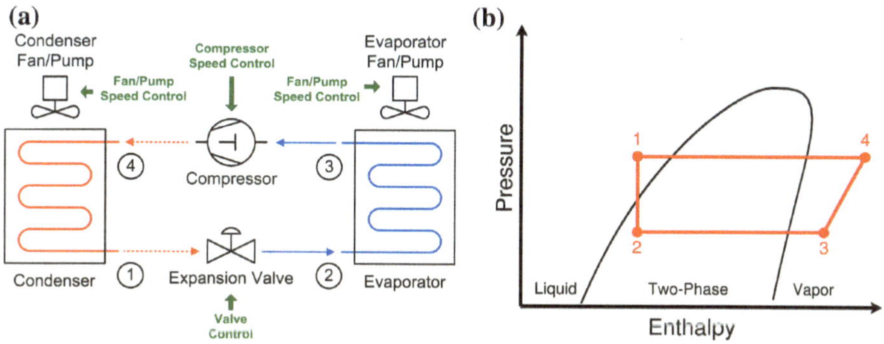


**Fig. 4.1** **a** Diagram of a simple VCC. **b** Pressure-enthalpy diagram for an ideal VCC

energy and maintain occupied spaces within desired temperature and humidity levels.

Effective control of these systems is essential in order to maximize efficiency and ensure safe operation, particularly during system start-up and when responding to changes in heat load or external conditions. The dynamics of these systems evolve at different time scales. Changes in compressor speed or valve opening can result in relatively rapid changes in refrigerant flow rates and induce quick changes in operating pressures. These dynamics are primarily related to the mass and energy of the refrigerant in the system. On the other hand, fluctuations in system temperatures are generally slower, such as those induced by changes in external temperature or flow rate. These dynamics are most closely linked to energy stored in the heat exchanger metal, i.e., the thermal mass. The coupling of such dynamic modes lead to models that can be represented by a "stiff" set of differential equations, or simplified to a nonlinear set of Differential Algebraic Equations (DAEs).

### *4.1.1 System Configurations*

In contrast with idealized VCC, physical realizations of VCC systems often include auxiliary components that enable reliable operation. For example, many systems utilize an accumulator placed after the evaporator or a receiver placed after the condenser. These serve not only to store excess refrigerant, but also ensure that the desired single phase fluid enters the compressor or expansion valve, respectively, (Fig. 4.2a). More complex systems may utilize multiple compression stages (Fig. 4.2b), or use alternative refrigerants, such as $CO_2$, that has a lower environmental impact but whose cycle includes supercritical fluid phases (Fig. 4.2c).

Other VCC variations are numerous and have common or uncommon applications (e.g., cryogenics, thermal storage, etc.) [4]. This variety in system configuration has

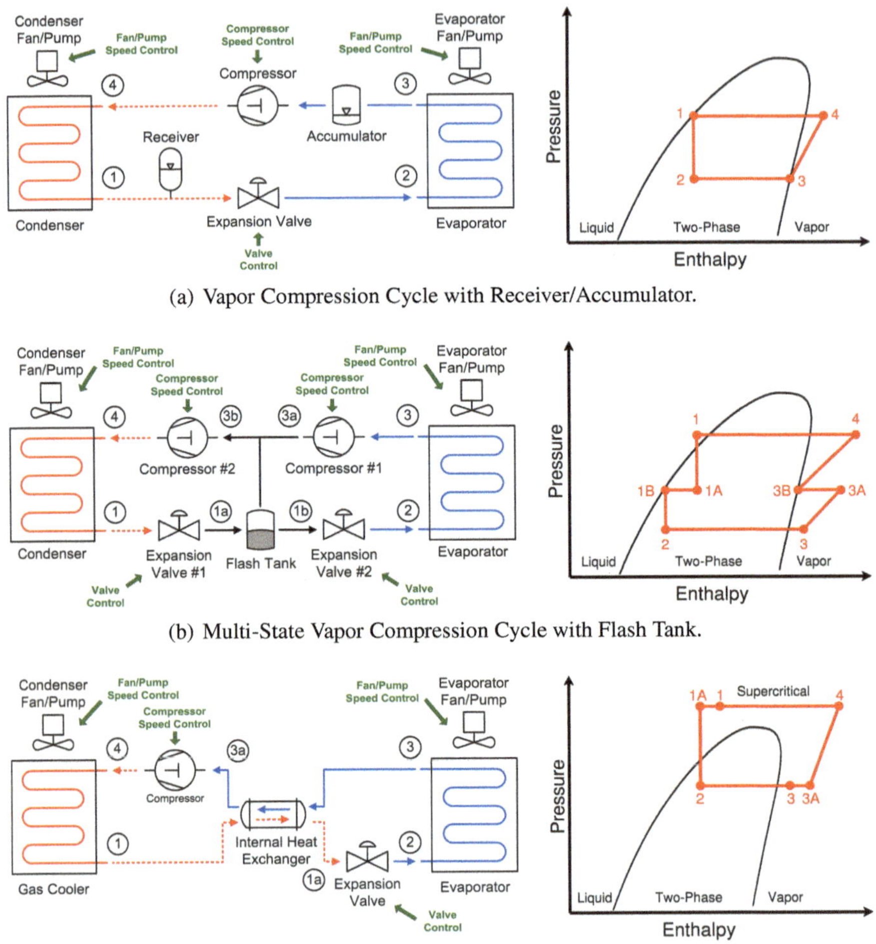


(a) Vapor Compression Cycle with Receiver/Accumulator.

(b) Multi-State Vapor Compression Cycle with Flash Tank.

(c) Transcritical Vapor Compression Cycle with Internal Heat Exchanger.

**Fig. 4.2** Schematics and pressure-enthalpy diagrams for several VCC system configurations

led to *component-focused* modeling paradigms with boundary conditions that define the fluid state of the refrigerant communicated between component models (e.g., refrigerant pressure, enthalpy, and flow rate).

## 4.1.2 Literature Surveys

For over fifty years, numerous models to capture the salient dynamics of evaporating or condensing flows have been proposed in the literature. These models have increased in sophistication and complexity, supported by advances in computation and measurement technology. Several excellent published survey papers have helped to characterize, categorize, and critique the various modeling approaches. The surveys [5, 6] present a detailed review of published modeling approaches and emphasize that the bulk of research efforts are focused on capturing two-phase flow dynamics in heat exchangers, while attempting to balance computational simplicity with predictive accuracy and physical fidelity. Subsequent surveys have focused on modeling innovations and outline simulation challenges, specifically with regard to heat exchangers [7–9].

## 4.2 Dynamic Modeling of Heat Exchangers

Heat exchangers are incredibly diverse in their design, with sizing, geometries, and flow patterns intended to optimize heat transfer for specific operating conditions (e.g., Fig. 4.3). This is particularly true for vapor compression systems, where the expanding/contracting volume of the evaporating/condensing refrigerant flow and the potentially large changes in heat flux due to phase changes adds additional complexity to the design problem. This also leads to challenges when attempting to develop a generalized modeling approach for these components.

### 4.2.1 *Multi-phase Flow*

Additional modeling complexities arise from the evaporating and condensing flows that are the primary means of heat transfer in VCC. Many different two-phase flow regimes are possible, depending on refrigerant conditions and heat exchanger geometry (Fig. 4.4).

Some steady-state modeling software packages attempt to explicitly account for complex heat exchanger geometries and two-phase flow regimes to accurately predict heat exchanger performance using Computational Fluid Dynamics (CFD) methods.

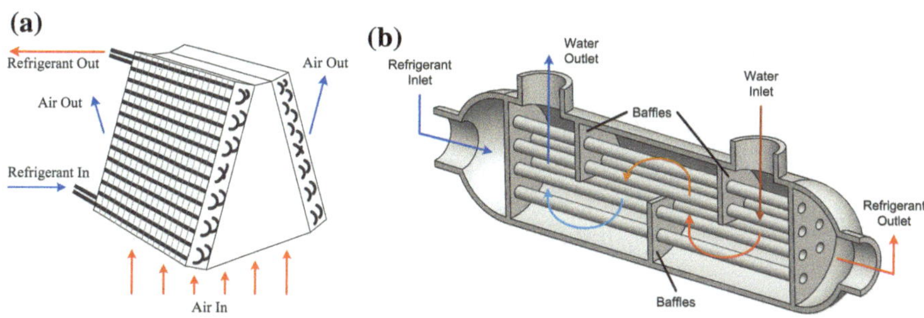


**Fig. 4.3** **a** Cross-flow tube-and-fin heat exchanger used in residential air-conditioning. **b** Counter-flow shell-and-tube heat exchanger used in industrial refrigeration chillers

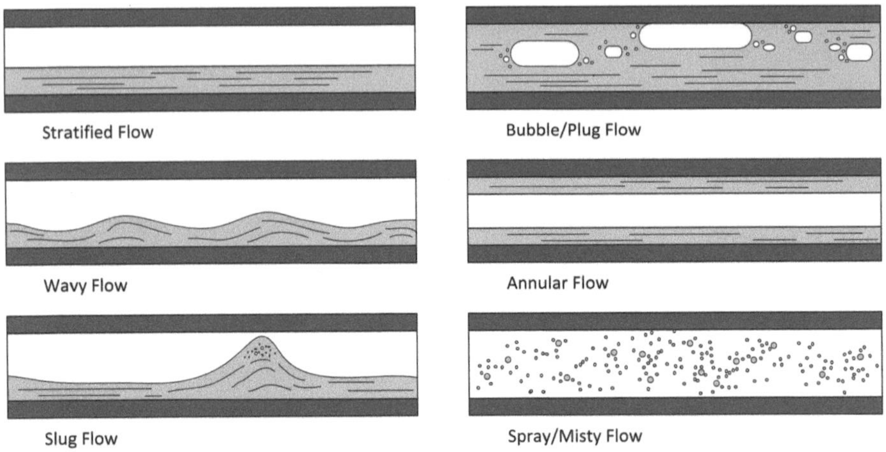


**Fig. 4.4** Common two-phase flow regimes

These approaches can be extended to predict the dynamic response of heat exchangers, but are generally computationally prohibitive for the vast majority of control and diagnostic tasks. Of necessity, complex heat exchanger geometries are simplified to a model of uniform one-directional flow with corresponding effective parameters (i.e., surface areas, volume, length, etc.) extracted from the actual heat exchanger dimensions.

Early dynamic modeling efforts used a single control volume for the entire heat exchanger with lumped parameters. These models can potentially capture gross transient responses (e.g., cooling capacity) for a restricted set of dynamic inputs. However, more detailed modeling paradigms are required to capture the higher order dynamic behavior of key refrigerant temperatures and pressures. The two dominant approaches are a discretization-based Finite Control Volume (FCV) model and the phase-based Moving Boundary (MB) model.

## 4.2.2 *Fixed Control Volume Models*

Finite control volume models discretize the heat exchanger, using average parameter values within each volume and applying equations for the conservation of energy and mass. Alternatively, governing partial differential equations can be directly discretized using finite difference methods. Highly discretized heat exchangers can thus capture gradients in thermo-physical properties and permit detailed representations of complex heat exchanger geometries. Reportedly, this leads to greater model fidelity when compared to lumped parameter approaches that rely on effective parameters and only describe the dominant physical mechanisms for heat transfer and fluid flow. However, the burden of these models can be quite high, as authors report

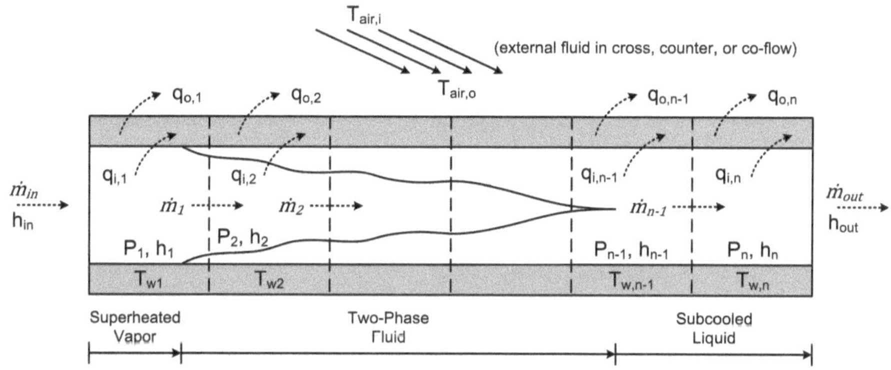


**Fig. 4.5** Finite Control Volume condenser model diagram

using hundreds of control volumes to ensure dynamic predictions are independent of the level of discretization (see [8] for a summary of reported discretization levels). Moreover, if a pressure gradient is included within the heat exchanger model, the inherently stiff system dynamics become even more numerically sensitive. In this case a staggered grid approach is commonly employed [10], where pressure differentials dictate mass flow rates, and mass flow rate differentials drive the pressure and energy dynamics (Fig. 4.5).

In this modeling paradigm, the governing equations for each control volume region of a condensor include the following:

- Conservation of refrigerant mass (4.1), where the mass flow rate between volumes is determined using (4.2), and the coefficient $C_d$ is a function of the tube diameter, length, friction factor, etc.
- Conservation of refrigerant energy (4.3), where the storage rate of specific energy, $\dot{u}_{c,j}$, is expanded in terms of pressure and enthalpy.
- Conservation of heat exchanger wall energy (4.4), where $q_{i,j} = \alpha_{i,j} A_{i,j}(T_{w,j} - T_{r,j})$ and $q_{o,j} = \alpha_{o,j} A_{o,j}\left(T_{air,j} - T_{w,j}\right)$.

$$\dot{m}_{c,j} = \dot{m}_{j-1} - \dot{m}_j \tag{4.1}$$

$$\dot{m}_j = C_d A\sqrt{\rho(P_j - P_{j+1})} \tag{4.2}$$

$$\dot{U}_j = \dot{m}_{c,j} u_{c,j} + m_{c,j}\left(\frac{du_{c,j}}{dP_c}\dot{P}_j + \frac{du_{c,j}}{dh_j}\dot{h}_j\right) = \dot{m}_{j-1}h_{j-1} + \dot{m}_j h_j + q_{i,j} \tag{4.3}$$

$$\dot{E}_{w,j} = (mc_p)_w \dot{T}_{w,j} = q_{o,j} - q_{i,j} \tag{4.4}$$

These can be organized into a nonlinear differential equation, of the form $Z(x)\dot{x} = f(x, u)$:

$$\underbrace{\begin{bmatrix} \dot{U}_1 \\ \vdots \\ \dot{U}_n \\ \dot{m}_{c,1} \\ \vdots \\ \dot{m}_{c,n} \\ \dot{E}_{w,1} \\ \vdots \\ \dot{E}_{w,n} \end{bmatrix}}_{\substack{\text{mass and energy}\\ \text{storage rates}}} = \underbrace{\begin{bmatrix} Z_{11} & Z_{12} & 0 \\ Z_{21} & Z_{22} & 0 \\ 0 & 0 & Z_{33} \end{bmatrix}}_{\substack{\text{pseudo-state}\\ \text{transformation}\\ \text{matrix}}} \underbrace{\begin{bmatrix} \dot{P}_1 \\ \vdots \\ \dot{P}_n \\ \dot{h}_1 \\ \vdots \\ \dot{h}_n \\ \dot{T}_{w,1} \\ \vdots \\ \dot{T}_{w,n} \end{bmatrix}}_{\substack{\text{dynamic state}\\ \text{variables}}} = \underbrace{\begin{bmatrix} \dot{m}_i h_i - \dot{m}_1 h_1 + q_{i,1} \\ \vdots \\ \dot{m}_{n-1} h_{n-1} - \dot{m}_o h_o + q_{i,n} \\ \dot{m}_i - \dot{m}_1 \\ \vdots \\ \dot{m}_{n-1} - \dot{m}_o \\ q_{o,1} - q_{i,1} \\ \vdots \\ q_{o,n} - q_{i,n} \end{bmatrix}}_{\substack{\text{steady-state mass and}\\ \text{energy balances}}} \tag{4.5}$$

If isobaric evaporation/condensation is assumed, $P_j = P_c, \forall j$, thus neglecting the conservation of momentum, the conservation equations may be simplified to eliminate the intermediate mass flow variables as: $\dot{m}_j = \dot{m}_{in} - \sum_{k=1}^{j} \dot{m}_{c,k}$. For example, the conservation of refrigerant energy (4.3) now becomes:

$$\begin{aligned} \dot{U}_j &= \dot{m}_{c,j} u_{c,j} + m_{c,j} \left( \frac{du_{c,j}}{dP_c} \dot{P}_c + \frac{du_{c,j}}{dh_j} \dot{h}_j \right) \\ &= \left( \dot{m}_{in} - \sum_{k=1}^{j-1} \dot{m}_{c,k} \right) (h_{j-1} - h_j) + \dot{m}_{c,j} h_j + q_{i,j} \end{aligned} \tag{4.6}$$

where the rate of mass storage in each region is given as: $\dot{m}_{c,j} = \dot{\rho}_{c,j} V_j = \left( \frac{d\rho_{c,j}}{dP_c} \dot{P}_c + \frac{d\rho_{c,j}}{dh_j} \dot{h}_j \right) V_j$. After these substitutions, the conservation equations in (4.5) are transformed into (4.7):

$$\underbrace{\begin{bmatrix} \tilde{Z}_{11} & \tilde{Z}_{12} & 0 \\ \tilde{Z}_{21} & \tilde{Z}_{22} & 0 \\ 0 & 0 & \tilde{Z}_{33} \end{bmatrix}}_{\substack{\text{pseudo-state}\\ \text{transformation}\\ \text{matrix}}} \underbrace{\begin{bmatrix} \dot{P}_c \\ \dot{h}_1 \\ \vdots \\ \dot{h}_n \\ \dot{T}_{w,1} \\ \vdots \\ \dot{T}_{w,n} \end{bmatrix}}_{\substack{\text{dynamic state}\\ \text{variables}}} = \underbrace{\begin{bmatrix} \dot{m}_{in}(h_{in} - h_1) + q_{i,1} \\ \vdots \\ \dot{m}_{in}(h_{n-1} - h_{out}) + q_{i,n} \\ \dot{m}_{in} - \dot{m}_{out} \\ q_{o,1} - q_{i,1} \\ \vdots \\ q_{o,n} - q_{i,n} \end{bmatrix}}_{\substack{\text{modified steady state mass}\\ \text{and energy balances}}} \tag{4.7}$$

This model has $2n + 1$ dynamic states, where $n$ is the number of control volumes (conservation of refrigerant/heat exchanger energy for each region and one equation for the conservation of refrigerant mass). If conservation of momentum is included, the model includes $3n$ dynamic equations, with the additional algebraic constraints that define the mass flow rate between regions. Publications report using a minimum of $n > 20$ control volumes to achieve grid independence, although the dominant dynamics can be accurately captured using much lower order models [11]. Thus, this spatially discretized paradigm is useful for accurately modeling subtle parameter variations, but results in high order dynamic models more useful for controller evaluation, rather than for dynamic analysis or model-based controller design. Additionally, this strategy is inherently better suited to detailed modeling of multi-pass heat exchangers, as these designs often involve complex flow arrangements (both internal and external). The discretization can be selected to correspond to individual tubes or channels, where each discretized region has a different value for surface area, internal volume, etc. Furthermore, the model can be formulated to capture the direction of flow from one control volume to the next, and which may be different for internal and external flows.

The published literature using the FCV approach to model VCC systems is extensive. Representative efforts include the early work of [12, 13] which also included extensive experimental validation [14]. More recent efforts reflect the usefulness of these models for a variety of end uses. For example: modeling of liquid chillers [15], automotive air conditioning [16], transcritical VCC systems [17–19], predicting system start-up responses [20], and controller tuning [21]. Discretized dynamic models are also commonly used for heat exchangers in other HVAC related systems, such as absorption systems [22–24], adsorbption systems [25], chilled water coils [26–28], etc.

### *4.2.3 Moving Boundary Models*

In contrast with FCV models that achieve accuracy using a discretization approach that results in models that are both parametrically complex and of high dynamic order. The moving boundary (MB) method employs separate control volumes and lumped parameters for each fluid phase (i.e., subcooled liquid, two-phase, or superheated vapor), but it assumes that these control volumes have time-varying boundaries (Fig. 4.6). Thus moving boundary models seek to capture the specific dynamic behavior due to time-varying evaporating (condensing) flows while minimizing the number of differential equations required. This results in a model more amenable to insightful dynamic analysis, model-based control design, or fast dynamic simulation.

To derive governing differential equations suitable for simulation or analysis, the most common method is to begin with the governing partial differential equations (PDEs) for one-dimensional fluid flow in a tube [29]. Isobaric conditions are assumed and axial conduction is neglected. The PDEs are integrated along the length of the heat exchanger to remove the spatial dependence, resulting in ordinary differential equations (ODEs) [30, 31].

As an example, for the condensor shown in Figs. 4.6, (4.8)–(4.10) are the conservation of refrigerant mass for each region, (4.11)–(4.13) are the conservation of refrigerant energy, and (4.14)–(4.16) are the conservation of heat exchanger

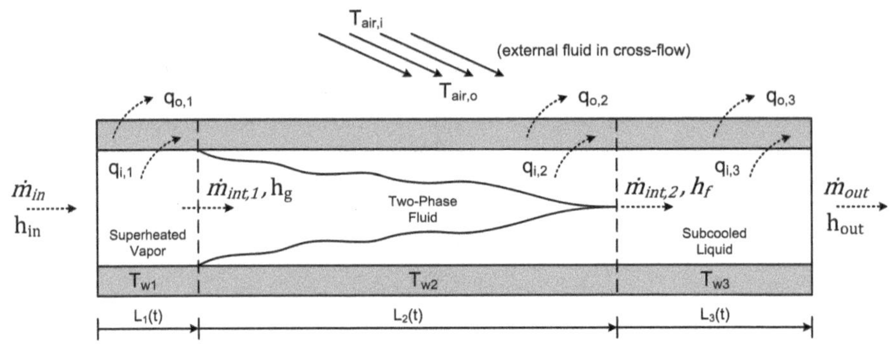


**Fig. 4.6** Moving boundary condenser model diagram

wall energy. In these equations $q$ is the heat transfer to/from the wall, such as $q_{i1} = \alpha_{i1} A_i \lambda_1 (T_{w1} - T_{r1})$ and $q_{o2} = \alpha_o A_o \lambda_2 (T_{air} - T_{w2})$, and $\lambda$ is the normalized length of each fluid region, i.e., $\lambda_2 = L_2/L_T$ where $L_T = L_1 + L_2 + L_3$.

$$A_{cs} L_1 \left( \frac{d\rho_1}{dP_c} \dot{P}_c + \frac{d\rho_1}{dh_1} \dot{h}_1 \right) + A_{cs} \dot{L}_1 \left( \rho_1 - \rho_g \right) = \dot{m}_i - \dot{m}_{int,1} \tag{4.8}$$

$$A_{cs} L_2 \left( \frac{d\rho_2}{dP_c} \right) \dot{P}_c + A_{cs} \left( \rho_g - \rho_f \right) \left( \dot{L}_1 + \bar{\gamma} \dot{L}_2 \right) = \dot{m}_{int,1} - \dot{m}_{int,2} \tag{4.9}$$

$$A_{cs} L_3 \left( \frac{d\rho_3}{dP_c} \dot{P}_c + \frac{d\rho_3}{dh_3} \dot{h}_3 \right) + A_{cs} \dot{L}_3 \left( \rho_3 - \rho_f \right) = \dot{m}_{int,2} - \dot{m}_o \tag{4.10}$$

$$A_{cs} L_1 \left[ \left( \frac{d\rho_1 h_1}{dP_c} - 1 \right) \dot{P}_c + \frac{d\rho_1 h_1}{dh_1} \dot{h}_1 \right] + A_{cs} \dot{L}_1 \left( \rho_1 h_1 - \rho_g h_g \right) = \dot{m}_i h_i - \dot{m}_{int,1} h_g + q_{i1} \tag{4.11}$$

$$A_{cs} L_2 \left[ \frac{d\rho_2 h_2}{dP_c} - 1 \right] \dot{P}_c + A_{cs} \left( \rho_g h_g - \rho_f h_f \right) \left( \dot{L}_1 + \bar{\gamma} \dot{L}_2 \right) = \dot{m}_{int,1} h_g - \dot{m}_{int,2} h_f + q_{i2} \tag{4.12}$$

$$A_{cs} L_3 \left[ \left( \frac{d\rho_3 h_3}{dP_c} - 1 \right) \dot{P}_c + \frac{d\rho_3 h_3}{dh_3} \dot{h}_3 \right] + A_{cs} \dot{L}_3 \left( \rho_3 h_3 - \rho_3 h_3 \right) = \dot{m}_{int,2} h_f - \dot{m}_o h_f + q_{i3} \tag{4.13}$$

$$(mc_p)_w \left[ \left( \frac{T_{w1} - T_{w,int,1}}{L_T} \right) \dot{L}_1 + \lambda_1 \dot{T}_{w1} \right] = q_{o1} - q_{i1} \tag{4.14}$$

$$(mc_p)_w \left[ \dot{T}_{w2} \right] = q_{o2} - q_{i2} \tag{4.15}$$

$$(mc_p)_w \left[ \left( \frac{T_{w3} - T_{w,int,2}}{L_T} \right) \dot{L}_3 + \lambda_3 \dot{T}_{w3} \right] = q_{o3} - q_{i3} \tag{4.16}$$

Similar to the FCV models, these equations can be algebraically combined to eliminate the intermediate mass flow rate(s), and again organized into the form $Z(x)\dot{x} = f(x, u)$, as in (4.17). The matrix $Z(x)$ remains full rank and invertible as long as $\lambda_{1,2,3} > 0$, or in other words, as long as both fluid regions exist.

$$\begin{bmatrix} Z_{11} & 0 & Z_{13} & 0 & 0 & 0 & 0 \\ Z_{21} & Z_{22} & Z_{23} & Z_{24} & 0 & 0 & 0 \\ Z_{31} & Z_{32} & Z_{33} & Z_{34} & 0 & 0 & 0 \\ Z_{41} & Z_{42} & Z_{43} & Z_{44} & 0 & 0 & 0 \\ Z_{51} & 0 & 0 & 0 & Z_{55} & 0 & 0 \\ 0 & 0 & 0 & 0 & 0 & Z_{66} & 0 \\ Z_{71} & Z_{72} & 0 & 0 & 0 & 0 & Z_{77} \end{bmatrix} \begin{bmatrix} \dot{L}_1 \\ \dot{L}_2 \\ \dot{P}_c \\ \dot{h}_o \\ \dot{T}_{w1} \\ \dot{T}_{w2} \\ \dot{T}_{w3} \end{bmatrix} = \begin{bmatrix} \dot{m}_i(h_i - h_g) + q_{i1} \\ \dot{m}_i h_g - \dot{m}_o h_f + q_{i2} \\ \dot{m}_o(h_f - h_o) + q_{i3} \\ \dot{m}_i - \dot{m}_o \\ q_{o1} - q_{i1} \\ q_{o2} - q_{i2} \\ q_{o3} - q_{i3} \end{bmatrix} \tag{4.17}$$

The final form of governing equations differ slightly in the literature depending on assumptions regarding the lumped fluid properties, void fraction $\gamma$, and interface wall temperatures $T_{w,int}$. While average fluid properties in the single phase region(s) are generally assumed, properties in the two-phase region typically employ a mean void fraction $\bar{\gamma}$, i.e., $\rho_2 h_2 = \rho_f h_f (1 - \bar{\gamma}) + \rho_g h_g \bar{\gamma}$. Two-phase void fraction is defined as the ratio of vapor volume to total volume. Various experimental correlations are available for predicting void fraction; reviews of well-known correlations can be found in [32, 33].

The use of a mean void fraction was first proposed in [34, 35], and this technique is used almost universally by every proponent of MB models, with subtle differences. In general, authors employ a time-invariant mean void fraction, which is a reasonable assumption when considering small dynamic responses around a single operating condition. The predicted amount of refrigerant charge in a two-phase heat exchanger is strongly correlated to the assumed void fraction. Thus simulating system responses across a large range of operating conditions, using a void fraction correlation, or expanding the model to explicitly include a time-varying void fraction is more appropriate.

A complete description of the modeling procedure can be found in [31, 36], among others. Due to its relatively compact form, and ability to predict the transient response of evaporator superheat (condenser subcool), as well as the system pressures and overall heat load, this approach has found particular use in multivariable control design efforts.

When the FCV and MB models are initialized with the same physical parameters, the resulting simulations exhibit only small differences [37–39]. Primary differences are in the relative speed of the response and in the steady-state predicted values. To compensate for the modeling errors introduced by the lumped parameters employed by the MB model, some adjustment of the effective parameters may be appropriate. However, these small differences are within the expected prediction error when compared to experimental data, and therefore, may not be critical for most modeling purposes.

### 4.2.4 Extreme Transient Scenarios

If properly configured, FCV models are capable of handling extreme transients, such as reversing flows or system start-up and shutdown. This also includes transient conditions where different fluid phases may appear or disappear (e.g., the loss of evaporator superheat). MB models, however, require modifications to do the same. In these situations Switched Moving Boundary (SMB) models [40–47] avoid the numerical singularities experienced by MB models by replacing nonactive differential equations with appropriate substitutions, or reducing the number of differential equations in real time, resulting in a set of Differential Algebraic Equations [48]. Alternatively, small deviations from saturated outlet conditions can be handled by extending the concept of the mean void fraction [49].

### 4.2.5 Pressure Drop

As mentioned earlier, there are a variety of methods for capturing the drop in pressure as refrigerant flows through the heat exchanger. For FCV models, the pressure gradient may be explicitly included in the spatial discretization scheme. This results in additional dynamic modes, typically at much faster time scales. A second approach is to compute static pressure drops between each region, resulting in a set of DAEs. Finally, the aggregate pressure drop can be computed and included only at the end of the heat exchanger model, or simply neglected completely. This final method is generally the only option appropriate for MB models. Reference [50] presents a direct comparison of these methods, and clearly illustrates that including *dynamic* momentum effects is not critical for large transient simulations. Discernible differences are only present when instantaneous compressor or valve actuation is assumed. However, the *static* contribution of momentum effects are often critical for accurate simulation of large, multi-pass heat exchangers. A listing of methods employed in published models can be found in [8].

### 4.2.6 Model Simplifications

Analysis of either FCV or MB heat exchanger models reveal opportunities for further dynamic simplification [31, 51, 52]. A key consideration is which system signals are needed for prediction or feedback control. Published literature clearly indicates that the overall cooling/heating load and system pressures can be predicted accurately with low order, lumped parameter models that do not attempt to capture the varying two-phase flow lengths or evaporator superheat conditions. If predicting these subtler transient effects are also needed, the required dynamic model order is still less than produced by the FCV or MB modeling methods. The physical justification for model order reduction varies slightly. Some authors neglect the effects of single phase heat transfer [51–53], others assume quasi-steady-state flow conditions [54], and others propose singular perturbation techniques for reducing the conservation of refrigerant energy techniques [31, 55]. Each of these approaches has been experimentally validated, indicating that each may be appropriate depending on the predictive capabilities desired.

## 4.3 Dynamic Modeling of Actuators and Ancillary Components

### *4.3.1 Compressors*

Compressors are the prime mover of the vapor compression cycle, and a key actuator of the system. Compressors drive the flow of refrigerant, and establish the pressure differential that allows energy to be absorbed by the evaporator, and rejected by the condenser. Compressors may be fixed-, multi-, or variable-speed. While detailed dynamic models of various types of compressors exist in the literature, their inherent mechanical dynamics evolve on time scales much faster than those of the heat exchanger. Thus, when compressors are integrated into a complete VCS model then the simple static models are generally used to define the refrigerant mass flow rate and the outlet fluid enthalpy. As needed, a simple first-order dynamic model of the thermal capacitance of the compressor shell may be included.

The mass flow rate (4.18) is characterized by the speed, $\omega_k$, the volume, $V_k$, the inlet density, $\rho_k = \rho(P_{k,in}, h_{k,in})$, and a volumetric efficiency, typically modeled using a semi-empirical relationship based on the pressure ratio and speed, $\eta_{vol} = f(P_{k,out}/P_{k,in}, \omega_k)$. The state of the outlet fluid is determined assuming an adiabatic efficiency (4.19), where the isentropic enthalpy is determined as $h_{out,isen} = h(P_{k,out}, s_{k,in})$ and $s_{k,in} = s(P_{k,in}, h_{k,in})$.

$$\dot{m}_k = \omega_k V_k \rho_k \eta_{vol} \tag{4.18}$$

$$\frac{h_{k,out,isen} - h_{k,in}}{h_{k,out} - h_{k,in}} = \eta_k \tag{4.19}$$

## 4.3.2 Expansion Valves

While compressors provide the energy to build a pressure differential between the heat exchangers, the expansion valve is the actuator that regulates the amount of refrigerant entering the evaporator. Orifice valves and capillary tubes are obstruction devices with a fixed area that allow the refrigerant to expand into a two-phase fluid, but are designed for a single operating condition. While this limits their potential applications, specific models for these components have been developed [56–58]. Thermostatic Expansion Valves (TEVs) use a sensing bulb filled with a two-phase refrigerant placed at the evaporator outlet to measure superheat temperature. The bulb pressure acts on a diaphragm inside the valve to vary the flow area and thus provide a mechanical feedback mechanism to regulate the refrigerant flow. TEVs are an established technology, but can suffer from a well-documented problem of "hunting" characterized by oscillations in the length of two-phase flow in the evaporator, and subsequent oscillations in the amount of superheated vapor at the evaporator exit. This undesirable behavior is caused by the interaction of the evaporator and TEV at off-design conditions, or because of poor valve selection.

Various dynamic models for TEVs have been proposed, attempting to explain or emulate the valve hunting problem. Unfortunately, relatively few include experimental validation. Early models assumed linear dynamics [59–62] and often used data-based modeling techniques [63, 64]. Increasing model complexity was reported, but rarely included convincing validation [65–67]. Although later research improved validation [68], this technology is steadily being replaced, and additional model development is unlikely.

Electronic expansion valves (EEVs) modify the flow area using an externally controlled mechanism. The most common is a stepper motor, but solenoids and Micro-Electro-Mechanical systems (MEMs)-based EEVs are also available commercially. Although most commercial EEVs utilize a PID-style controller, the ability to directly actuate the expansion valve electronically is assumed by researchers proposing more advanced control strategies, including gain scheduled [69, 70] or coordinated control of all VCC actuators [30, 71, 72].

In nearly all publications, valve models assume ideal (isenthalpic) throttling, $h_{v,in} = h_{v,out}$. For fixed orifice expansion valves, refrigerant flow is typically modeled using a form of Bernoulli's equation (4.20), where $\rho_v = \rho(P_{v,in}, h_{v,in})$. The discharge coefficient is determined empirically or from manufacturer information, $C_d = f(A_v, \Delta P)$. For fixed orifice valves the area is constant, $A_v = a_0$, whereas for electronic expansion valves the orifice area is directly controlled, $A_v = f(u_v)$. In thermostatic expansion valves, the orifice area depends on internal geometry and is adjusted by fluctuating bulb and evaporator pressures, $A_v = f(P_{bulb}, P_e)$. Pressure in the sensing bulb is typically modeled by performing an energy balance (4.21), leading to a simple first order dynamic model (4.22) with a time constant, $\tau$. By including separate energy balance equations for the bulb, pipe wall, refrigerant, etc., it is possible to develop higher order models of TEV behavior.

$$\dot{m}_v = A_v C_d \sqrt{\rho_v (P_{v,in} - P_{v,out})} \tag{4.20}$$

$$m_{bulb} C_{p,bulb} \dot{T}_{bulb} = \alpha_o A_o (T_{amb} - T_{bulb}) - \alpha_i A_i (T_{bulb} - T_{ero}) \tag{4.21}$$

$$\frac{T_b(s)}{T_{ero}(s)} = \frac{k_o}{\tau s + 1} \tag{4.22}$$

### *4.3.3 Fans and Pumps*

For most air-conditioning systems, fans are used to move air across the evaporator and condenser. For industrial or large commercial cooling applications, chillers may be employed. These systems use water as the secondary fluid, and employ pumps to move water across the evaporator, and subsequently to locations where cooling is needed. Chillers can either use air or water to reject heat from the condenser. These fans and pumps may operate at fixed speed(s) or be designed for fully variable speed. Static models are typically employed for these components, with affinity laws used to relate the fan/pump speed to flow rate and power consumption.

### *4.3.4 Pressure Vessels*

While some VCS are critically charged and operate using a specified amount of refrigerant in the system, many systems employ receivers and accumulators to store excess refrigerant. Receivers are placed before the expansion valve and ensure saturated or subcooled liquid enters the valve. Similarly, accumulators are placed before the compressor inlet and ensure that the refrigerant entering the compressor is a saturated or superheated vapor. These components are used, respectively, to prevent slug flow from entering the expansion valve and to prevent liquid refrigerant from damaging the compressor.

Models for these components apply conservation of mass and energy to the refrigerant, and conservation of energy to the vessel shell, resulting in dynamic models with only a few differential equations (e.g., [38]). Several dynamic model variations were developed and compared [73]. The primary modeling challenge arises from the tight coupling between the heat exchanger and receiver/accumulator dynamics, as the pressure difference is typically small. One possible solution is a combined model, assuming isobaric conditions and neglecting the intermediate pressure drop [49]. Flash tanks are used in multi-stage VCS (Fig. 4.2b), and while functionally similar to receivers and accumulators, modeling the splitting of refrigerant flows offers unique numerical and simulation challenges [74].

### 4.3.5 *Pipe Losses*

For compact or packaged systems, the piping between components is intentionally minimized and the impact on system dynamics can be negligible. For other applications, such as industrial and transport refrigeration, the pipe lengths may be significant. This can effectively slow the pressure dynamics, as the internal pipe volume must be included. Additionally, secondary effects of parasitic heat loss/gain through natural convection, pressure drop due to fluid friction, and transport delay may need to be incorporated. Although these effects are not addressed specifically in the published literature, as they are considered minor effects, they will be necessities for high fidelity simulation models of large, complex vapor compression systems.

## 4.4 Modeling Challenges

VCC simulations present challenges stemming from the multi-phase fluid dynamics. These can broadly be categorized as challenges with multi-time scale simulation and other computational challenges, and the difficulties with parameter selection and experimental validation.

### 4.4.1 *Time Scale Selection*

As noted previously, the dynamics of VCS span multiple time scales. For most applications, the dynamics of greatest interest are the thermal dynamics that evolve on the scale of seconds or minutes. The slower dynamics of the conditioned space or external environment are treated as constant or slowly varying disturbances to the system. Likewise, the propagation of pressure waves through refrigerant flow as captured by the conservation of momentum equations are considered as instantaneous changes, relative to the dominant thermal dynamics. For this reason, the pressure in heat exchanger models is assumed to be isobaric, or if the associate pressure drop is calculated, it is simply a static offset that is included after the dynamic calculations are complete. Zhang [50] presents a direct comparison of the most common assumptions regarding how the momentum equation is handled for VCC simulations, demonstrating that for most applications they do not discernibly affect transient predictions.

However, the coupled dynamics associated with the conservation of mass and energy of the refrigerant and heat exchanger result in multi-time scale dynamics [31]. This has prompted researchers to utilize reduced-order models that eliminate dynamics outside of the time scale of interest. This simplifies the model for both simulation and control design. Authors have proposed purely numerical methods for model reduction [30, 75], as well as various physical justifications for eliminating fast dynamic modes [36, 53]. This has extended to attempts to reformulate the governing equations to virtually decouple fast and slow modes [31], and utilize nonlinear model reduction techniques to preserve accuracy while simplifying the models for simulation [55].

### 4.4.2 Computation

The ability to simulate the dynamics of VCS in real time can be important when employing model-based control designs or fault detection and diagnosis algorithms. The computational requirements for VCS simulations are largely dependent on the heat exchanger modeling paradigm. The model for these components is given by the nonlinear descriptor form $Z(x)\dot{x} = f(x, u)$, and requires a matrix inversion, where the size of the matrix depends on the number of dynamic states. Moving Boundary models typically have 5–10 states, while Finite Control Volume models have $2n + 1$ dynamic states, where $n$ is the number of control volumes. Published literature generally recommends a minimum of 20 volumes to avoid significant discretization errors. Linearized model approximations are much faster as they avoid performing this matrix inversion at each simulation time step.

Calculation of fluid properties and heat transfer correlations can add to the computational burden. External software calls to property databases, or recursive solutions to Thermodynamic Equations of State significantly reduce the speed of simulations. Most simulations interpolate property tables to ensure fast calculations, although care must be taken to avoid interpolation errors at the discontinuity of a fluid phase change boundary [38]. Similar challenges may arise when using complex correlations for determining heat transfer coefficients, frictional pressure drop, or two-phase void fraction.

Few publications report the details regarding computational capabilities of their proposed VCC system models, or the type of differential equation solver used. While many use numerical solvers specifically suited for "stiff" dynamic systems with multi-time scale behavior or Differential Algebraic Equation solvers, others use more traditional fixed step solvers. [8] provides an overview of these details.

### 4.4.3 Parametric Sensitivities

Dynamic models of VCS exhibit distinctly different parametric sensitivities when compared to their steady-state counterparts. For example, accurate heat transfer correlations are important to predicting the steady-state heating/cooling capacity over the range of anticipated operating conditions. In contrast, the dynamic response of the heat exchanger is relatively insensitive to this parameter. Heat exchanger dynamics are driven primarily by the transient difference in inlet and outlet refrigerant mass flow rates. As a result, accurate models of the expansion valve and compressor are critical. Thus, although the majority of the model complexities are due to the two-phase flow in the heat exchangers, it is the accuracy of the relatively simple actuator models that have the greatest impact on the transient prediction quality.

Additional attention should be given to selection of the void fraction correlation used in the two-phase heat exchanger models. Various correlations for void fraction are available, and this choice impacts the dynamic response of the system, but not steady-state values for system pressures or temperatures. The amount of refrigerant charge is highly sensitive to the void fraction [52] and thus tuning of the void fraction correlation is important if quantitative charge prediction is needed [76].

### *4.4.4 Model Tuning and Validation*

Experimental validation of VCC models can pose several specific challenges. Beyond the aforementioned sensitivity issues associated with the actuators, heat exchangers have a large number of parameters, including measurable physical parameters (areas, volumes, etc.) as well as unmeasurable or uncertain parameters (void fraction, heat transfer correlations, etc.). Moreover, many model parameters are lumped (effective surface areas, effective volumes, etc.) and must be adjusted depending on the system configuration. An example is how the effective internal volume of the heat exchanger should include the volume of the heat exchanger, as well as the volume of all connected pipes, as this entire volume participates in the increase/decrease in system pressure.

Despite prolific research and many proposed modeling innovations, the bulk of reported dynamic models for VCS do not include adequate experimental validation. Particularly when using VCC models for multivariable control design, care must be taken to validate the response of all system outputs for expected transient input signals. Steady-state experimental validation may prove helpful for tuning a subset of key variables, such as heat transfer correlations or compressor efficiencies, but many variables exhibit little influence on steady-state performance, but strongly affect the dynamic response of critical signals (e.g., internal system volumes, void fraction).

## 4.5 Control of HVAC Systems

### *4.5.1 Nonlinearity in HVAC Components*

There are many challenges associated with HVAC control stemming from the components themselves to the fundamental characteristics of a heat transfer process. HVAC actuators come in two main categories, obstruction and displacement. Obstruction actuators include probably the most well-known HVAC components such as valves and dampers. These components open and close to impinge the flow of a working

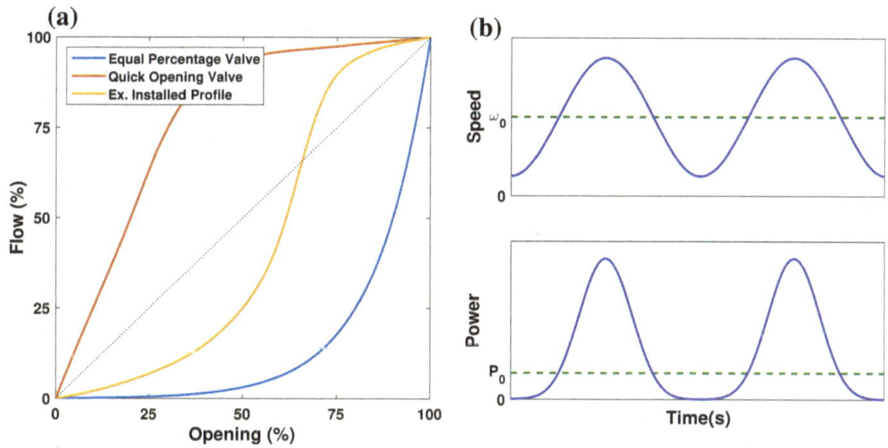


**Fig. 4.7** **a** Many valves are designed to have nonlinear opening-to-flow relationships. Profiles can be used to reduce nonlinear behavior of the installed system. **b** Affinity laws cause energy consumption to be significantly higher above a median speed than below. This is why oscillating controllers waste energy

fluid (e.g., air, water, refrigerant) thereby controlling the rate of energy transfer. For example, an expansion valve will reduce outlet superheat by opening and allowing more refrigerant to enter the evaporator. Obstruction actuators are inherently wasteful; energy must be expended to pressurize the working fluid and deliver it to its destination. That effort is wasted by obstructing actuators if the delivered pressure produces excess flow. Displacement actuators actively consume electrical energy to displace a working fluid. These actuators in HVAC systems include components such as fans, pumps, and compressors. While most energy consumption of an HVAC system will take place in these components, their use is necessary but can be minimized.

Both types of actuators can have nonlinear performance characteristics in terms of flow and energy consumption. Valves and dampers typically have nonlinear relationships between position and flow (Fig. 4.7a). For example, a VAV damper will induce more change in flow rate when mostly closed than when mostly open. While valves can intentionally be designed to have specific flow characteristics (e.g., equal percentage valves), many are sized incorrectly or have the wrong profile to provide any benefit. Fans, pumps, and compressors are governed by well-known affinity laws (4.23) that feature quadratic and cubic relationships between power ($\dot{W}$), pressure head ($P$), flow rate ($\dot{V}$), and speed ($\omega$). These relationships mean that a 10% reduction in average speed can reduce power input by more than 25%. Small reductions in use for these components can therefore reap large energy savings (Fig. 4.7b).

$$\frac{\dot{W}_1}{\dot{W}_2} = \left(\frac{\omega_1}{\omega_2}\right)^3 \quad \& \quad \frac{P_1}{P_2} = \left(\frac{\omega_1}{\omega_2}\right)^2 \quad \& \quad \frac{\dot{V}_1}{\dot{V}_2} = \frac{\omega_1}{\omega_2} \tag{4.23}$$

HVAC actuators also display many classic nonlinear behaviors. Hysteresis is the dependence of an actuator's output on time and previous input. For HVAC systems, faulty or loose linkages may cause immediate action in one direction with delayed action in another. This effect also manifests as actuator drift where actuator position settings change over time. Actuators can experience stiction effects whereby large input is required to initiate motion. This effect can cause excessive control input as controllers must windup to overcome initial resistance. HVAC systems are also prone to faults and failures in sensing and control. Communication of temperature, occupancy, and ventilation data for multiple rooms and systems requires a large number of sensors any of which can fail or provide erroneous readings. Detailed descriptions of the behavior of specific HVAC equipment can be found in [77].

Other nonlinear effects in HVAC components stem from fundamental properties of heat and mass transfer. The main modes of heat transfer used for temperature control by these systems (free convection, forced convection, and radiation) all display complex and time-varying properties. Because the relationship between temperature differentials and actuator flow rates is multiplicative (e.g., as in Eqs. (4.14) and (4.15)), HVAC systems will have dynamics heavily dependent on operating conditions. This is the case with Air Handling Units (AHUs) which can see large changes in dynamic response with load (Fig. 4.8). Another example is Variable Air Volume (VAV) dynamics that are dependent on the relative difference between the supply air, current room air, and outside air temperatures, as well as, the current system static pressure and damper position as in (4.24) from [78].

$$C_{rm} \cdot \frac{dT_{rm}}{dt} = \rho_a \cdot c_{p,a} \cdot \underbrace{\dot{V}(P_s, \theta)}_{\text{Damper}} \cdot \underbrace{(T_s - T_{rm})}_{\Delta T_{rm}} + \alpha_{rm} \cdot \underbrace{(T_{oa} - T_{rm})}_{\Delta T_{oa}} + Q_{dis} \quad (4.24)$$

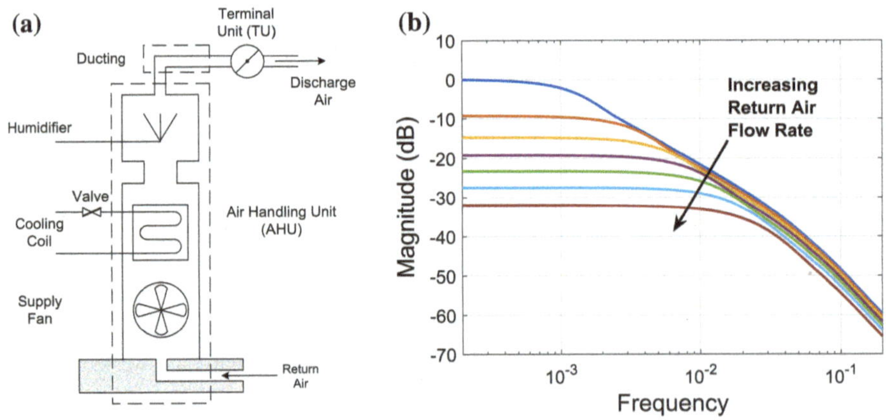


**Fig. 4.8** **a** An Air Handling Unit (AHU) consists of a fan, cooling coil, and ducting. **b** AHU dynamics are heavily dependent on operating conditions. The bode plot shows that steady-state gains for the model from [79] vary by more than 40x for a range of return air flow rates

## 4.5.2 Linear Control Strategies for HVAC Systems

The nonlinear effects discussed in the previous section make control of HVAC systems and components inherently difficult. However, a large body of work has been generated to adapt traditional Single-Input-Single-Output (SISO) control techniques for HVAC applications. There have also been recent developments that take advantage of nonlinear compensation methods to eliminate some or all nonlinear effects. This section provides a short history of component control, presents common control techniques, and discusses best practices for control in buildings today.

**Mechanical Control**

As in most fields, control of HVAC systems began with purely mechanical regulation. Until the advent of computer technology, thermostats utilized bimetallic strips and relay control, also known as "bang-bang" control, to regulate temperature. These controllers used the difference in thermal expansion between two metals to operate a switch powering heating and cooling equipment. Although effective, this type of control can only regulate temperature between a substantial dead-band. Such on/off control also extended to the control of compressors and fans. Typical refrigeration cabinets found in supermarkets and convenience stores use this strategy and are toggled by local thermostats. However, on/off of control has documented issues with synchronization for parallel systems which can lead to higher electrical demand and poor temperature regulation [80, 81].

Other mechanical control devices offer the ability to provide analog control input. For example, VCSs can regulate evaporator pressure using a mechanical valve known as a Pressure Regulating Valve (PRV). The valve uses a flexible diaphragm to adjust valve stem position based on fluctuations in system pressure. PRVs provide superior pressure disturbance rejection but are only suited for applications with stable load requirements. The Thermostatic Expansion Valve (TEV or TXV) is an extension of the PRV that uses a sensing bulb filled with saturated refrigerant at the evaporator outlet to adjust valve position. As the evaporator discharge temperature changes, pressure inside the bulb fluctuates causing pressure differentials across the flexible diaphragm and thereby metering refrigerant. This configuration makes the TXV a superheat regulating device allowing it to adjust to changes in system demand. Due to the physical separation of the bulb and valve, these valves will often oscillate around their intended setpoint [82]. Schematics of both types of valves are shown in Fig. 4.9 which demonstrate key principles of mechanical control.

Over the past few decades, mechanical control devices have steadily been replaced by digital equivalents. These devices offer the ability to provide adjustable, remote, and modular control for many HVAC systems. For example, Electronic Expansion Valves (EEVs) for VCC systems use stepper motors to precisely change valve position based on digitally acquired signals. Likewise, Variable Frequency Drive (VFD) technology allows compressors, motors, and fans to operate at intermediate speeds based on system demand. Not surprisingly, digitization and miniaturization of technology has led to the expansion of digital control strategies for HVAC systems.

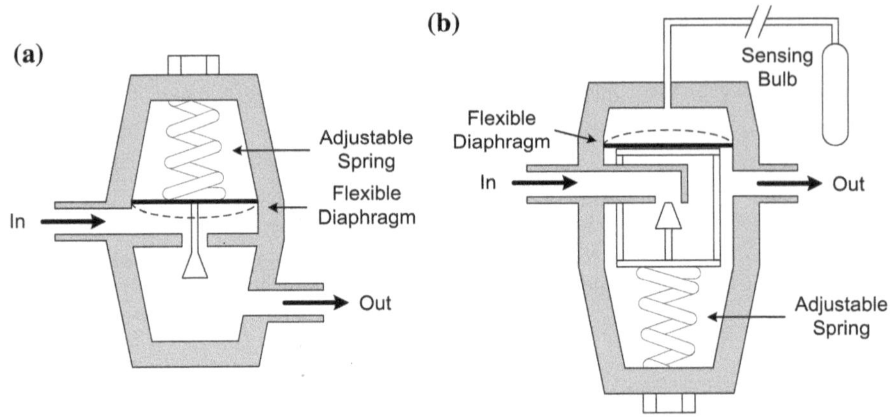


**Fig. 4.9** Diagrams of **a** Pressure Regulative Valve (PRV) **b** Thermostatic Expansion Valve (TXV). Both valves have flexible diagrams that regulate the position of the valve stem and an adjustable spring that determines control aggression. The TXV also has a sensing bulb attached to the evaporator outlet to meter refrigerant based on superheat

### Standard Control Techniques

The application of computer hardware to HVAC control has allowed for more complex algorithms to be implemented. Equipment scheduling and other rule-based control algorithms have been shown to significantly improve system performance. These controllers can be as simple as temperature setbacks during evening hours [83] or a detailed set of rules based on occupancy, outside weather conditions, and comfort [84]. Note that depending on the complexity of the rule set, rule-based controls can be classified as fuzzy algorithms.

The control algorithm with the most penetration in the HVAC field is the eponymous Proportional-Integral-Derivative (PID) controller. The PID controller is low order, intuitive, and robust making it ideal for HVAC control due widely varying systems and components. From (4.25), a PID controller has a proportional component that directly responds to system errors, an integral component that eliminates steady-state errors, and a derivative component that prevents violent changes in control input. Note that the derivative term is rarely used in most HVAC controllers due to its sensitivity to noise and perceived implementation difficulties. Despite its simplicity, the PID controller has demonstrated effective control in a wide range of applications and even acceptable performance despite improper tuning.

$$u(t) = k_P \cdot e(t) + k_I \cdot \int_0^{\infty} e(t)dt + k_D \cdot \frac{d}{dt}e(t) \tag{4.25}$$

Due to its prominence in HVAC control, there has been much literature devoted to development of tuning techniques for PID parameters. Many HVAC technicians today still use the tuning method proposed by Ziegler and Nichols [85]. However, this tuning procedure was designed for process industries and can provide inconsistent performance when applied to HVAC systems. Other tuning rules have been developed that offer certain benefits over the ZN method including Approximate $M$-constrained Integral Gain Optimization (AMIGO) [86] and Internal Model Control (IMC) [87].

More advanced tuning methods have also been proposed. Lim et al. [88] proposed using Hurwitz criteria and the Hermite–Biehler Theorem to identify stability regions where gain and phase margin requirements are satisfied. This procedure fixes one control gain and sweeps predetermined planes of gains for possible combinations. While more computationally complex, Lim's method allowed for robust tuning of a VCC system by intersecting stability regions across multiple operating conditions. Robust PID gains can also be determined using more traditional mixed sensitivity techniques as was done by [89] to improve performance of a VAV system.

While effective in many cases, PID controllers and other simple control techniques are susceptible to actuator hunting. This phenomenon manifests as a sustained oscillation in control input solely due to the controller and not system disturbances or loads. Hunting has been well documented in the HVAC field [60, 90] and can lead to wasted energy and excessive actuator wear. The behavior is due to the wide range of conditions and nonlinearities that are inherent to HVAC systems. As operating conditions change, simple controllers tuned under different loads can push the system towards instability resulting in unwanted fluctuations.

The degraded behavior of PID controllers over wide HVAC operating conditions has prompted the development of more sophisticated techniques. Auto-tuning PID controllers use periodic testing procedures to determine process models and update control gains. Great care is taken to ensure system stability during and after identification as well as when switching gains. Bi et al. [91] applied the auto-tuning technique to several HVAC control loops and showed improved performance. Adaptive tuning techniques can identify system models in real time during normal closed loop operation. Process gains can be modified according to identified models as was done by [92].

Gain scheduling is a nonlinear control technique that takes advantage of a large family of local linear controllers. The process essentially divides an HVAC system's operating conditions and associated dynamics into several problems each of which can be solved by traditional control methods including PID. Gain scheduled controllers have been successfully applied to EEV control for VCC systems as in [69, 70]. These methods also include Linear Parameter Varying (LPV) control which has been applied successfully on hydronic radiator systems [93] and Multi-Input-Multi-Output (MIMO) variants that have been applied to VCC systems [94].

The focus so far has been on SISO control techniques. However, HVAC systems are often MIMO systems with strong coupling between inputs and outputs. One solution to this problem is applying decoupling controllers to untangle system system I/O pairs and recover SISO behavior. Dynamic decoupling seeks to separate pairs for all frequencies and has shown promise in HVAC systems including independent control of temperature and relative humidity [95]. However, dynamic decoupling analysis is an intensive process and not applicable to all systems [71]. A simpler version seeks to decouple I/O pairs at steady-state using a static decoupling matrix. This approach uses only steady-state gains to uncouple pairs but can interact poorly with high-frequency modes of a system [96]. Another approach uses cascaded PID controllers to decouple steady-state behavior. Cascaded control has successfully decoupled a multi-evaporator VCC system without needing to model complex dynamics [97].

The MIMO strategy that is the focus of most intense study today is Model Predictive Control (MPC). The goal of MPC is to coordinate disparate, coupled building systems to reduce overall energy use and ensure occupant comfort. These methods use system models to simulate along a receding time horizon and predict optimal control actions for a given cost function [98]. MPC has always shown promise in improving overall building efficiency in simulation [99] but has increasingly been applied in practice [100, 101]. Variants of MPC have even been used to optimally adapt PID control gains [102]. However, centralized MPC has intense communication and computational demands which can make implementation prohibitive. Strategies to limit network demand include steady-state optimization (NC-OPT) [103], Decentralized MPC (DMPC) [104], and Limited-Communication-Decentralized MPC (LC-DMPC) [105]. Each requires less communication but can still converge to centralized optimums given certain conditions. For a complete discussion of MPC for HVAC equipment, see Chaps. 5 and 7.

### *4.5.3 Nonlinear Compensation Techniques*

In some cases, nonlinear compensation in HVAC systems can be achieved by utilizing specially designed actuators [106, 107]. All actuators have two characteristic curves: inherent and installed. Inherent characteristics are determined under laboratory conditions with constant pressure drop, while installed characteristics can only be determined during operation where pressure varies constantly. In many cases, actuators with linear inherent characteristics can be nonlinear once installed resulting in instabilities and generally poor performance. Equal percentage valves or equivalent actuators with nonlinear inherent flow profiles can be used to ensure linear installed characteristics given proper sizing and system characterization.

Nonlinear characteristics can also be dealt with using a range of feedback linearization techniques. Manufacturer specifications or installed characteristic data can be used to determine exact nonlinear characteristics. Provided that these relations are invertible, nonlinearities can be dealt with explicitly in the feedback or feed-forward loop. Singal and Salsbury [108] showed that exact knowledge, however, is not required and instead used an invertible S-shaped exponential curve to characterize nonlinearity. Other techniques use low order models to identify and correct for nonlinear behavior including [109]. A survey of modern feedback linearization techniques can be found in [110].

#### Cascaded Control

The methods discussed in the previous sections all have the ability to improve HVAC system performance. A major difference between methods is the investment required

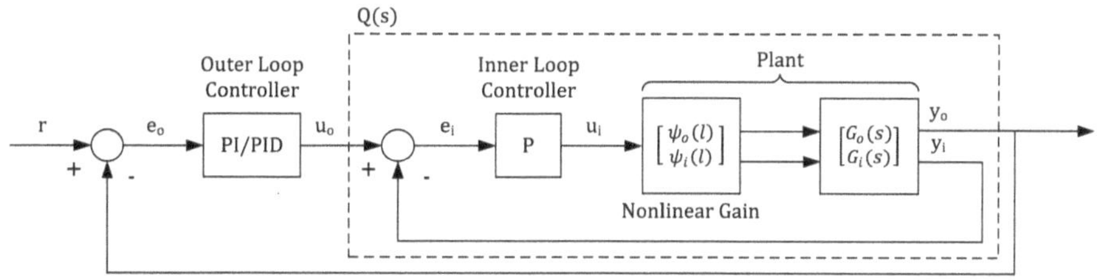


**Fig. 4.10** The cascaded loop for a process $G(s)$ with nonlinearity $\psi(l)$ consists of an inner loop proportional controller and an outer loop PI/PID controller. The nonlinearity is contained inside the inner loop where it is approximately linearized for the outer loop controller. Figure adapted from [111] and reprinted by permission of the publisher (Taylor & Francis Ltd, http://www.tandfonline.com)

to make them function properly. Advanced techniques such as gain scheduling and MIMO control all require detailed modeling and heavy computational analysis. These barriers can make large scale implementation prohibitive. However, recent work into cascaded control architectures has proven it to be a simple and effective alternative for nonlinear HVAC control. The technique utilizes nested PID loops to compensate for nonlinear behavior associated with wide operating ranges [79]. As PID loops are standard in most building control software and technicians already have an intuitive understanding of their operation, adoption of cascaded techniques can be accomplished relatively quickly and easily.

As illustrated in Fig. 4.10, the basic cascaded architecture consists of a proportional inner loop controller and an outer loop PI/PID controller that use measurement feedback from the signals $y_i$ and $y_o$, respectively. The nonlinear characteristics of the controlled system ($\psi_i$ and $\psi_o$) are placed within the inner loop process. System nonlinearities appear in the numerator and denominator of the inner loop transfer function (4.26) which allows them to counteract provided that both share the same trend with respect to system load ($l$). In addition, as the magnitude of the inner loop gain $K_L$ increases, differences between dynamic responses at different loads are reduced (Fig. 4.11).

$$Q(s, K_L, l) = \frac{K_L \psi_o(l) G_o(s)}{1 + K_L \psi_i(l) G_i(s)} \tag{4.26}$$

The reduced variation in system dynamics and overall faster response times associated with cascaded control are realized without the need for detailed models of system nonlinearities. As both $\psi_i$ and $\psi_o$ are inherently placed in the numerator and denominator of the inner loop transfer function, the nonlinearities counteract each other no matter their structure. This makes the cascaded architecture highly modular which is crucial for HVAC systems. For example, a single building can have one or more air handling units, dozens of VAV terminal boxes, and several pumps. Developing detailed models for each of these systems can quickly become infeasible especially when across an entire campus [112]. Inserting additional proportional

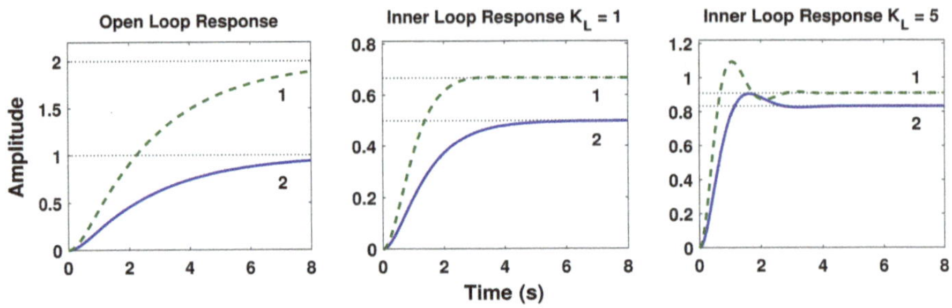


**Fig. 4.11** The cascaded inner loop control reduces the difference between steady-states at different operating conditions as well as speeds the overall response

control loops to existing software represents minimal effort while offering reduced hunting, better performance, and improved comfort.

The addition of extra loops to the control architecture adds complexity to the tuning process. Tools for cascaded loop tuning were developed in [111] to tune controller gains for linearization effect and performance. Inner and outer loop gap metrics (NGMs) quantified linearization by the inner loop control and allowed for target gains to be established. A Linear Quadratic (LQ) framework was presented in [111] to optimally tune inner and outer loop controllers simultaneously and robustly while the simple methods described in [113] allow for widespread adoption in the field. These tuning help to make cascaded control accessible throughout the HVAC field.

## 4.6 Conclusion

This chapter presented an overview of current modeling techniques for VCCs commonly used in HVAC systems. The varied geometry, configurations, and cycles utilized by VCC systems has led to the development of component-focused modeling of heat exchangers and ancillary elements. A discussion of the two most prominent modeling techniques, FCV and MB modeling, detailed their relative complexity and ability to capture dynamic characteristics including extreme transients (e.g., start-up/shutdown, reversing flow, etc.). Details on common approaches to model system actuators and other equipment were also presented.

Each technique discussed has to compensate for difficulties inherent to modeling of VCC systems. These cycles include vastly different time scales that result in stiff sets of differential equations that require extremely small time steps in simulation. Depending on the discretization or level of desired fidelity, the computational burden may become impractical for real-time simulation or model-based control. Tuning and validation of complex models is also difficult due to the large number of model parameters and their associated sensitivities. However, guidelines are provided in the associated references that aid in the development of successful models.

The complexity of VCC systems results in nonlinear and load-dependent behavior. These nonlinearities are present in the fundamental heat transfer process and in system actuators such as valves and pumps. A large body of work has been generated to enable application of traditional SISO control techniques to work effectively with VCC and other HVAC systems. Modern computational technology has also allowed for more complex, nonlinear control techniques to be applied to HVAC control. However, recent work has shown that nonlinear compensation for VCC and building system control can be accomplished through cascaded control using standard PID loops. Simplicity and flexibility make cascaded control ideal for implementation in existing HVAC building control software.

## References

1. Klepeis N, Nelson W, Ott W, Robinson J, Tsang A, Switzer P, Behar J, Hern S, Engelmann W (2001) The National Human Activity Pattern Survey (NHAPS): A Resource For Assessing Exposure to Environmental Pollutants. J Exposure Anal Environ Epidemiol 11(3):231–252
2. Sarigiannis D (2013) Combined or Multiple Exposure to Health Stressors in Indoor Built Environments. Tech. Rep, World Health Organization, Regional Office for Europe
3. Constable G, Somerville B (2003) A century of innovation: twenty engineering achievements that transformed our lives. Joseph Henry Press
4. Stoecker WF (2000) Industrial refrigeration handbook. McGraw-Hill, New York
5. Bourdouxhe J-P, Grodent M, Lebrun J (1998) Reference guide for dynamic models of HVAC equipment, ASHRAE, Tech. Rep. Project 738-TRP
6. Bendapudi S, Braun J (2002) A review of literature on dynamic models of vapor compression equipment ASHRAE, Tech. Rep. Report #4036-5
7. Ding G-L (2007) Recent developments in simulation techniques for vapour-compression refrigeration systems. Int J Refrig 30(7):1119–1133
8. Rasmussen B (2011) Dynamic modeling for vapor compression systems - Part I: literature review. HVAC &R Res 18(5):934–955
9. Li P, Qiao H, Li Y, Seem JE, Winkler J, Li X (2014) Recent advances in dynamic modeling of hvac equipment. part 1: equipment modeling. HVAC&R Res 20(1):136–149
10. Eborn J (2001) On model libraries for thermo-hydraulic applications, Ph.D. dissertation. Lund University
11. Rasmussen B, Alleyne A, Musser A (2005) Model-driven system identification of transcritical vapor compression systems. IEEE Trans Control Syst Technol 13(3):444–451
12. Gruhle WD, Isermann R (1985) Modeling and control of a refrigerant evaporator. ASME J Dyn Syst Meas Control 107(4):235–240
13. MacArthur JW (1984) Transient heat pump behaviour: a theoretical investigation. Int J Refrig 7(2):123–132
14. MacArthur JW, Grald EW (1989) Unsteady compressible two-phase flow model for predicting cyclic heat pump performance and a comparison with experimental data. Int J Refrig 12(1):29–41
15. Bendapudi S, Braun J (2002) Development and validation of a mechanistic, Dynamic model for a vapor compression centrifugal chiller, ASHRAE, Tech. Rep. Report #4036-4
16. Anand G, Mahajan M, Jain N, Maniam B, Tumas T (2004) e-Thermal: automobile air conditioning module. In: SAE international Society of automotive engineers 2004 world congress, Paper 2004-01-1509
17. Skaugen G, Svensson M (1998) Dynamic modeling and simulation of a transcritical $CO_2$ heat pump unit, pp 230–239

18. Pfafferott T, Schmitz G (2000) Numeric simulation of an integrated $CO_2$ cooling system. In: Proceedings of the 2000 modelica workshop. Lund, Sweden, pp 89–92
19. Rigola J, Raush G, Prez-Segarra CD, Oliva A (2005) Numerical simulation and experimental validation of vapour compression refrigeration systems. special emphasis on $CO_2$ trans-critical cycles. Int J Refrig 28(8):1225–1237
20. Kapadia RG, Jain S, Agarwal RS (2009) Transient characteristics of split air-conditioning systems using R-22 and R-410A as refrigerants. pp 617–649
21. Beghi A, Cecchinato L (2009) A simulation environment for dry-expansion evaporators with application to the design of autotuning control algorithms for electronic expansion valves. Int J Refrig 32(7):1765–1775
22. Kohlenbach P, Ziegler F (2008) A dynamic simulation model for transient absorption chiller performance. Part I: the model. Int J Refrig 31(2):217–225
23. Kohlenbach P, Ziegler F (2008) A dynamic simulation model for transient absorption chiller performance. Part II: numerical results and experimental verification. Int J Refrig 31(2):226–233
24. Matsushima H, Fujii T, Komatsu T, Nishiguchi A (2010) Dynamic simulation program with object-oriented formulation for absorption chillers (modelling, verification, and application to triple-effect absorption chiller). Int J Refrig 33(2):259–268
25. Schicktanz M, Núñez T (2009) Modelling of an adsorption chiller for dynamic system simulation. Int J Refrig 32(4):588–595
26. Zhou X, Braun JE (2007) A simplified dynamic model for chilled-water cooling and dehumidifying coils-Part 1: development (RP-1194). pp 785–804
27. Zhou X, Braun JE (2007) A simplified dynamic model for chilled-water cooling and dehumidifying coils-Part 2: experimental validation (RP-1194). pp 805–817
28. Li P, Li Y, Seem JE (2010) Modelica-based dynamic modeling of a chilled-water cooling coil. HVAC&R Res 16:35–58
29. Grald EW, MacArthur JW (1992) A moving-boundary formulation for modeling time-dependent two-phase flows. Int J Heat Fluid Flow 13(3):266–272
30. He XD, Liu S, Asada H (1997) Modeling of vapor compression cycles for multivariable feedback control of HVAC systems. ASME J Dyn Syst Meas Control 119(2):183–191
31. Rasmussen BP, Alleyne A (2004) Control-oriented modeling of transcritical vapor compression systems. ASME J Dyn Syst Meas Control 126(1):54–64
32. Rice C (1987) Effect of void fraction correlation and heat flux assumption on refrigerant charge inventory. ASHRAE Trans 93(1):341–367
33. Wilson M, Newell T, Chato J (1998) Experimental investigation of void fraction during horizontal flow in larger diater refrigeration applications, Air conditioning and refrigeration center, vol. ACRC TR-140. University of Illinois, Urbana
34. Wedekind GL, Bhatt BL, Beck BT (1978) A system mean void fraction model for predicting various transient phenomena associated with two-phase evaporating and condensing flows. Int J Multiph Flow 4(1):97–114
35. Beck BT, Wedekind GL (1981) A generalization of the system mean void fraction model for transient two-phase evaporating flows. ASME J Heat Transf 103(1):81–85
36. He XD, Asada H, Liu S, Itoh H (1998) Multivariable control of vapor compression systems. HVAC&R Res 4(3):205–230
37. Bendapudi S, Braun JE, Groll EA (2008) A comparison of moving-boundary and finite-volume formulations for transients in centrifugal chillers. Int J Refrig 31(8):1437–1452
38. Rasmussen BP, Shenoy B (2011) Dynamic modeling for vapor compression systems, Part II: Simulation tutorial. HVAC&R Res 18(5):956–973
39. Pangborn H, Alleyne AG, Wu N (2015) A comparison between finite volume and switched moving boundary approaches for dynamic vapor compression system modeling. Int J Refrig 53:101–114
40. Pettit NBOL, Willatzen M, Ploug-Sorensen L (1998) General dynamic simulation model for evaporators and condensers in refrigeration. part ii: simulation and control of an evaporator. Int J Refrig 21(5):404–414

41. Willatzen M, Pettit NBOL, Ploug-Sorensen L (1998) General dynamic simulation model for evaporators and condensers in refrigeration. part i: moving-boundary formulation of two-phase flows with heat exchange. Int J Refrig 21(5):398–403
42. Zhang W-J, Zhang C-L (2006) A generalized moving-boundary model for transient simulation of dry-expansion evaporators under larger disturbances. Int J Refrig 29(7):1119–1127
43. McKinley T, Alleyne A (2008) An advanced nonlinear switched heat exchanger model for vapor compression cycles using the moving-boundary method. Int J Refrig 31(7):1253–1264
44. Li B, Alleyne A (2010) A dynamic model of a vapor compression cycle with shut-down and start-up operations. Int J Refrig 33(3):538–552
45. Cecchinato L, Mancini F (2012) An intrinsically mass conservative switched evaporator model adopting the moving-boundary method. Int J Refrig 35(2):349–364
46. Bonilla J, Dormido S, Cellier FE (2015) Switching moving boundary models for two-phase flow evaporators and condensers. Commun Nonlinear Sci Numer Simul 20(3):743–768
47. Qiao H, Laughman CR, Aute V, Radermacher R (2016) An advanced switching moving boundary heat exchanger model with pressure drop. Int J Refrig 65:154–171
48. Gordon BW, Liu S, Asada HH (1999) Dynamic modeling of multiple-zone vapor compression cycles using variable order representations. In: Proceedings of the American control conference. IEEE, pp 3765–3769
49. Eldredge B, Rasmussen B, Alleyne A (2008) Moving-boundary heat exchanger models with variable outlet phase. J Dyn Syst Meas Control 130: 61 003–1–61 003–12
50. Zhang W-J, Zhang C-L, Ding G-L (2009) On three forms of momentum equation in transient modeling of residential refrigeration systems. Int J Refrig 32(5):938–944
51. Sheng L, Xiangdong H, Asada H, Itoh H, Nakagawa K (1996) Multivariable control of vapor compression cycles: coordination of compressor and expansion valve. In: Proceedings of the 13th world congress, International federation of automatic control. Chemical process control, mineral, mining, metals, Vol. M. Pergamon. 1997. Oxford, UK, pp 193–8
52. Jensen JM, Tummescheit H (2002) Moving boundary models for dynamic simulations of two-phase flows. In: Proceedings of the 2nd international modelica conference. pp 235–244
53. Leducq D, Guilpart J, Trystram G (2003) Low order dynamic model of a vapor compression cycle for process control design. J Food Process Eng 26(1):67–91
54. Eisenhower B, Runolfsson T (2009) System level modeling of a transcritical vapor compression system for bistability analysis. Nonlinear Dyn 55(1):13–30
55. Rodriguez E, Rasmussen B (2015) A nonlinear reduced order modeling method for dynamic two-phase flow heat exchanger simulations. Science and technology for the built environment
56. Murphy W, Goldschmid V (1985) Cyclic characteristics of a residential air-conditioner - modeling of start-up transient. ASHRAE Trans 91(2A):427–444
57. García-Valladares O (2007) Numerical simulation of non-adiabatic capillary tubes considering metastable region. part i: mathematical formulation and numerical model. Int J Refrig 30(4):642–653
58. García-Valladares O (2007) Numerical simulation of non-adiabatic capillary tubes considering metastable region. Part II: experimental validation. Int J Refrig 30(4):654–663
59. Stoecker WF (1996) Stability of an evaporator-expansion valve control loop. ASHRAE Trans 72(2):4.3.1–4.3.8
60. Broersen PMT, van der Jagt MFG (1980) Hunting of evaporators controlled by a thermostatic expansion valve. ASME J DynSyst Meas Control 102(2):130–135
61. Higuchi K, Hayano M (1982) Dynamic characteristics of thermostatic expansion valves. Int J Refrig 5(4):216–220
62. Yasuda H, Touber S, Machielsen C (1983) Simulation model of a vapor compression refrigeration system. ASHRAE Trans 89(2787):408–42
63. Najork H (1973) Investigations on the dynamical behavior of evaporators with thermostatic expansion valve. In: Proceedings of the 13th international congress of refrigeration. pp 759–769
64. Broersen PMT, ten Napel J (1983) Identification of a thermostatic expansion valve. In: Proceedings of the sixth IFAC symposium, vol. 1. Pergamon, pp 415–20

65. James K, James R (1987) Transient Analysis of Thermostatic Expansion Valves for Refrigeration System Evaporators Using Mathematical Models. Trans Instit Meas Control 9(4):198–205
66. Ibrahim G (2001) Effect of sudden changes in evaporator external parameters on a refrigeration system with an evaporator controlled by a thermostatic expansion valve. In J Refrig 24(6):566–576
67. Mithraratne P, Wijeysundera N (2002) An experimental and numerical study of hunting in thermostatic-expansion-valve-controlled evaporators. Int J Refrig 25(7):992–998
68. Hariharan N, Rasmussen B (2010) Parameter estimation of hvac model with limited sensors. In: Proceedings of the American control conference
69. Outtagarts A, Haberschill P, Lallemand M (1997) Transient response of an evaporator fed through an electronic expansion valve. Int J Energy Res 21(9):793–807
70. Finn D, Doyle C (2000) Control and optimization issues associated with algorithm-controlled refrigerant throttling devices. ASHRAE Trans 106(1):524–533
71. Jain N, Li B, Keir M, Hencey B, Alleyne A (2010) Decentralized feedback structures of a vapor compression cycle system. IEEE Trans Control Syst Technol 18(1):185–193
72. Pollock D, Yang Z, Wen J (2015) Dryout avoidance control for multi-evaporator vapor compression cycle cooling. Appl Energy 160:266–285
73. Estrada-Flores S, Cleland D, Cleland A, James R (2003) Simulation of transient behavior in refrigeration plant pressure vessels: mathematical models and experimental validation. Int J Refrig 26(2):170–179
74. Qiao H, Aute V, Radermacher R (2014) Transient modeling of a flash tank vapor injection heat pump system - part i: model development. Int J Refrig 49:169–182
75. Rasmussen B, Alleyne A, Musser A (2004) Iterative modeling and identification of a $CO_2$ air conditioning system. In: Proceedings of the international mechanical engineering congress and exposition, American society of mechanical engineers, Dynamic systems and control division (Publication) DSC, vol. 73. American Society of Mechanical Engineers, New York, NY 10016-5990, United States, pp 813–820
76. Bendapudi S (2015) Development and evaluation of modeling approaches for transients in centrifugal chillers, Ph.D. dissertation. Purdue University
77. Haines RW, Hittle DC (2006) Control systems for heating, ventilating, and air conditioning, 6th edn. Springer Science & Business Media
78. Yamakawa Y, Yamazaki T, Kamimura K, Kurosu S (2010) Compensation of manual reset to offset thermal loads change for PID controller. ASHRAE Trans 116(1):303–315
79. Price C, Liang S, Rasmussen B (2015) HVAC nonlinearity compensation using cascaded control architectures. ASHRAE Trans 121:217–231
80. Wisniewski R, Chen L, Larsen LF (2009) Synchronization analysis of the supermarket refrigeration system. In: Proceedings of the 48th conference on decision and control, 2009 held jointly with the 2009 28th Chinese control conference (CDC/CCC). IEEE pp 5562–5567
81. Wisniewski R, Leth J, Rasmussen JG (2014) Analysis of synchronization in a supermarket refrigeration system. Control Theory Technol 12(2):154–162
82. N, A, (2010) ASHRAE Handbook - Refrigeration (SI Edition). American Society of Heating, Refrigerating and Air-Conditioning Engineers Inc, p 2010
83. Energy Star (2009) A guide to energy-efficient heating and cooling
84. Ulpiani G, Borgognoni M, Romagnoli A, Di Perna C (2016) Comparing the performance of On/Off, PID, and fuzzy controllers applied to the heating system of an energy-efficient building. Energy Build 116:1–17
85. Ziegler JG, Nichols NB (1942) Optimum settings for automatic controllers. ASME Trans 64:759–768
86. Astrom K, Hagglund T (2004) Revisiting the Ziegler-Nichols step response method for PID control. J Process Control 14:635–650
87. Visioli A (2006) Practical PID control. Springer, Berlin
88. Lim D, Rasmussen BP, Swaroop D (2009) Selecting PID control gains for nonlinear HVAC&R systems. HVAC&R Res 15(6):991–1019

89. Kasahara M, Matsuba T, Kuzuu Y, Yamazaki T, Hashimoto Y, Kamimura K, Kurosu S (1999) Design and tuning of robust PID controller for HVAC systems. ASHRAE Trans 105:154–156
90. Chintala R, Price C, Liang S, Rasmussen B (2015) Identification and elimination of hunting behavior in HVAC systems. ASHRAE Trans 121:294–305
91. Bi Q, Cai W-J, Wang Q-G, Hang C-C, Lee E-L, Sun Y, Liu K-D, Zhang Y, Zou B (2000) Advanced controller auto-tuning and its application in HVAC systems. Control Eng Pract 8(6):633–644
92. Bai J, Zhang X (2007) A new adaptive PI controller and its application in HVAC systems. Energy Convers Manag 48(4):1043–1054
93. Tahersima F, Stoustrup J, Rasmussen H (2013) An analytical solution for stability-performance dilemma of hydronic radiators. Energy Build 64:439–446
94. Rasmussen B, Alleyne A (2010) Gain scheduled control of an air conditioning system using the Youla parameterization. IEEE Trans Control Syst Technol 18(5):1216–1225
95. Rentel-Gomez C, Velez-Reyes M (2001) Decoupled control of temperature and relative humidity using a variable-air-volume HVAC system and non-interacting control. In: Proceedings of the 2001 International Conference on Control Applications (CCA). IEEE, pp 1147–1151
96. Lee J, Hyun Kim D, Edgar T (2005) Static decouplers for control of multivariable processes. AIChE J 51(10):2712–2720
97. Elliott M, Estrada C, Rasmussen B (2011) Cascaded superheat control with a multiple evaporator refrigeration system. In: Proceedings of the American control conference (ACC). pp 2065–2070
98. Camacho E, Bordons Alba C (2004) Model predictive control. Advanced textbooks in control and signal processing. Springer, New York
99. Afram A, Janabi-Sharifi F (2014) Theory and applications of HVAC control systems - a review of model predictive control (MPC). Build Environ 72:343–355
100. Siroky J, Oldewurtel F, Cigler J, Privara S (2011) Experimental analysis of model predictive control for an energy efficient building heating system. Appl Energy 88(9):3079–3087
101. Ma Y, Borrelli F, Hencey B, Coffey B, Bengea S, Haves P (2012) Model predictive control for the operation of building cooling systems. IEEE Trans Control Syst Technol 20(3):796–803
102. Xu M, Li S, Cai W-J, Lu L (2006) Effects of a GPC-PID control strategy with hierarchical structure for a cooling coil unit. Energy Convers Manag 47(1):132–145
103. Elliott M, Rasmussen B (2015) Optimal setpoints for HVAC systems via iterative cooperative neighbor communication. J Dyn Syst Meas Control 137(1):011006
104. Elliott M, Rasmussen B (2013) Decentralized model predictive control of a multi-evaporator air conditioning system. Control Eng Pract 21(12):1665–1677
105. Jalal R, Rasmussen B (2015) Neighbor-communication distributed model predictive control for coupled and constrained subsystems in networks. In: 2015 American Control Conference (ACC). IEEE, pp 743–749
106. Monsen J (2015) Part 1: an insider's guide to valve sizing & selection. Flow Control Mag 21(2)
107. Headley M (2003) Guidelines for selecting the proper valve characteristics. Valve Mag 15(2)
108. Singhal A, Salsbury T (2007) Characterization and cancellation of static nonlinearity in HVAC systems. ASHRAE Trans 113(1):391–399
109. He X-D, Asada HH (2003) A new feedback linearization approach to advanced control of multi-unit HVAC systems. Proc Am Control Conf 3:2311–2316
110. Guardabassi G, Savaresi S (2001) Approximate linearization via feedback - an overview. Automatica 37(1):1–15
111. Price C, Rasmussen B (2016) Optimal tuning of casacaded control architectures for nonlinear HVAC systems. Sci Technol Built Environ. https://doi.org/10.1080/23744731.2016.1262663
112. Siemens Global (2011) Texas A&M University: a detailed account of how one university is improving its energy efficiency through effective management and performance contracting, White Paper
113. Price C, Rasmussen B (2015) Effective tuning of cascaded control loops for nonlinear HVAC systems. In: Proceedings of the 2015 dynamic systems and control conference. ASME

# Chapter 5
# Model Predictive Control of Multi-zone Vapor Compression Systems

**Daniel J. Burns, Claus Danielson, Stefano Di Cairano, Christopher R. Laughman and Scott A. Bortoff**

## Notation

$\mathbb{R}$ and $\mathbb{Z}$ denote the set of real and integer numbers, respectively. The vector formed by concatenating $x \in \mathbb{R}^n$ and $y \in \mathbb{R}^m$ is denoted by $\mathrm{col}(x, y) \in \mathbb{R}^{n+m}$. $I \in \mathbb{R}^{n \times n}$ and $0 \in \mathbb{R}^{n \times m}$ denote the identity and the all-zero matrices of appropriate dimension $n$ and $m$, respectively. Inequalities are component-wise, and $\prec$ and $\succ$ ($\preceq$ and $\succeq$) indicate positive and negative (semi)definiteness. A continuous-time signal $x(\tau)$ sampled with period $T_s$ is denoted by the discrete-time signal $x(t) = x(tT_s)$ where $t \in \mathbb{Z}$. $x_{k|t}$ denotes the $k$-steps predicted value of $x$ at time $t$.

## 5.1 Introduction

After more than 100 years of design evolution, vapor compression systems (VCS) are now the most common means for commercial and residential space cooling [1], often employed for space or water heating [2], and extensively used in refrigeration (both stationary and mobile [3, 4]), desalination [5, 6], and cryogenic applications [7]. As discussed in the previous chapter, while early systems typically used fixed-speed

D.J. Burns (✉) · C. Danielson · S. Di Cairano · C.R. Laughman · S.A. Bortoff
Mitsubishi Electric Research Laboratories, 201 Broadway, Cambridge, MA, USA
e-mail: burns@merl.com

C. Danielson
e-mail: danielson@merl.com

S. Di Cairano
e-mail: dicairano@merl.com

C.R. Laughman
e-mail: laughman@merl.com

S.A. Bortoff
e-mail: bortoff@merl.com



J.T. Wen and S. Mishra (eds.), *Intelligent Building Control Systems*,
Advances in Industrial Control, https://doi.org/10.1007/978-3-319-68462-8_5

compressors and fans, capillary tubes, and single condensers and evaporators, modern systems use state-of-the-art technology such as variable-speed compressors and fans, electronically positioned expansion valves, and multiple heat exchangers operating in independent zones [8, 9]. These machines must operate reliably in heating or cooling modes, over a broad range of operating conditions and climates for many years.

One effect of the advances in cycle technology is the increasing sophistication of the control systems [10–12]. Control strategies in early products were simple because of on/off actuation, limited sensing, and limited performance requirements such as zone temperature regulation [13, 14]. Efficiency requirements were met by designing the refrigeration cycle for a narrow set of conditions, and could be accomplished outside the scope of the control system design. However, modern VCS control systems must accommodate a broader set of requirements, including (1) maximizing energy efficiency over a broad range of operating conditions, (2) enforcing equipment protection constraints such as ensuring critical temperatures and pressures remain within permissible operating limits, (3) providing rapid transient response, (4) activation or deactivation of individual zones, and (5) integrating with intelligent building control strategies that periodically demand reduced energy consumption.

Model predictive control (MPC) is a strategy for the control of multi-input multi-output (MIMO) plants that satisfies many of these requirements explicitly, especially enforcement of plant input (actuator) and output (measurement) constraints with guarantees on stability and performance [15–17]. In MPC, control variables are computed by solving a constrained optimization problem in real time that includes the predicted response of the system, resulting in a controller that achieves optimal performance according to specified objective function and enforces constraints on actuators and plant outputs. Furthermore, the MPC approach offers advantages related to the requirement of turning zones on and off, which can be met in a straightforward and scalable manner: the objective function and constraints can be changed online when a zone is activated, for example, altering the number of control objectives and decision variables in the associated optimal control problem [18–20]. Moreover, tuning the terms of the objective function is intuitive and leads to predictable performance and constraints can be modified or added during the design process while preserving system stability properties, eliminating the need for extensive revalidation during the product development. For the manufacturer, MPC thus provides an attractive strategy to meet a diverse set of requirements in a rigorous manner.

In this chapter, an MPC design for a VCS is presented. The design is offset-free to account for low-frequency model uncertainty and a state estimator in order to estimate variables not directly measured [21, 22]. In contrast with other offset-free methods [23], this approach does not require fitting an input disturbance model—a task that is difficult because of disturbances, specifically the heat load, are unmeasured and possess significant uncertainty especially at low frequencies. The design enforces hard constraints associated with actuator limits and rate limits, and also enforces several output limits including a maximum compressor temperature, a minimum compressor discharge superheat temperature, and maximum and minimum heat exchanger temperatures. In addition, the control architecture enables a separation of design concerns: distinct tuning parameters are provided to adjust the system for energy efficiency, constraint enforcement and transient performance, and these parameters are not coupled—enabling intuitive tuning.

The chapter is organized as follows. Section 5.2 describes the vapor compression system under consideration, derives a linear model, and introduces pertinent constraints. Section 5.3 discusses the prioritized set of control system requirements and the strategy employed to meet them. In Sect. 5.4, an MPC design is presented including prediction model augmentations and Kalman filter derivation. Also in this section, an HVAC laboratory and validation experiments are described. Section 5.5 describes how the proposed control strategy meets extended requirements by describing a method that permits individual zones to be turned on or off within the MPC framework, and discusses optimization algorithms appropriate for low power embedded processors. Finally, concluding remarks are offered in Sect. 5.6.

## 5.2 Multi-zone Vapor Compression System Description

This section describes the dynamics and constraints of the multi-evaporator vapor compression system (ME-VCS) shown in Fig. 5.1. We treat the single evaporator system as a special case of the ME-VCS. The specific vapor compression system considered here is an air conditioner operating in cooling mode, and therefore the objects of conditioning are indoor volumes of air, although other applications of vapor compression systems can be considered (heat pumps, chillers, refrigeration systems, etc.) with straightforward substitutions of machine endpoints.

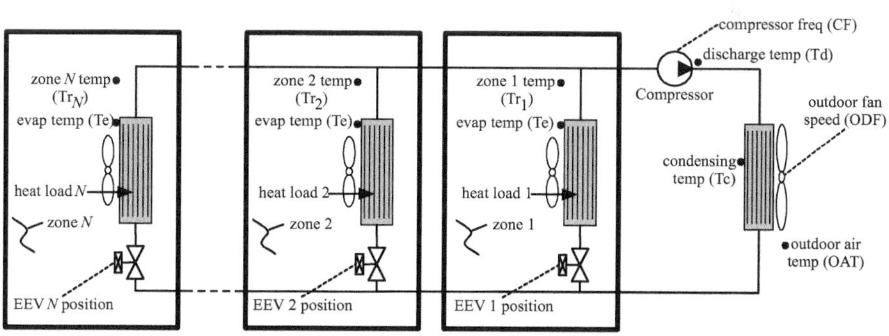


**Fig. 5.1** Refrigerant piping arrangement of a multi-evaporator vapor compression system. The main actuators in the system are (i) the compressor, (ii) the outdoor (condenser) fan (ODF), and (iii) $N$ electronic expansion valves (EEV). Sensors provide measurements of the compressor discharge temperature (Td), the evaporating temperature (Te), the condensing temperature (Tc) and the zone temperatures (Tr$i$)

### 5.2.1 ME-VCS Dynamics

Consider the multi-zone vapor compression system operating in cooling mode shown schematically in Fig. 5.1. The ME-VCS is comprised of a single outdoor unit and $N$ indoor units. The purpose of this system is to move both sensible and latent heat from the indoor zones to the outdoor air. When operating in cooling mode, the outdoor unit receives low pressure, low temperature refrigerant in the vapor state from the indoor units. The compressor performs work to increase the pressure and temperature of the refrigerant. The amount of work done is controlled by the compressor rotational frequency CF. A sensor measures the discharge temperature Td of the refrigerant leaving the compressor. The refrigerant then flows through the outdoor heat exchanger across which a fan forces air. Heat is transferred from the refrigerant causing it to condense from a vapor to a saturated liquid where the amount of heat removed from the refrigerant depends on the outdoor air temperature OAT and outdoor fan speed ODF. In cooling mode, the outdoor unit heat exchanger acts as a condenser, and the small pressure drop over the length of the heat exchanger enables the modeling assumption that the phase change of the refrigerant is approximately isobaric and occurs at a nearly constant condensing temperature Tc, which is measured by a sensor on the heat exchanger.

High pressure liquid refrigerant from the outdoor unit is then routed to the indoor units, indexed by $i \in \mathcal{I} = \{1, \ldots, N\}$. The amount of refrigerant that enters the indoor unit is controlled by the opening position $\text{EEV}_i$ of an electronic expansion valve. If the electronic expansion valve is open ($\text{EEV}_i 0$), then refrigerant flows into the $i$-th indoor unit. As the refrigerant flows through the expansion valve, it undergoes a isenthapic drop in pressure and temperature and changes state into a two-phase mixture of liquid and vapor. The low-temperature two-phase refrigerant then flows through the indoor unit heat exchanger. An indoor fan forces air from the zone across the heat exchanger, which absorbs heat from the zone. An unmeasured heat load acts in each zone. The temperature of each zone $\text{Tr}_i$ is measured in the return air stream. The heat absorbed by each indoor unit causes the refrigerant to evaporate from a two-phase mixture to a saturated vapor, so that these heat exchangers function as evaporators. As was also the case with the condenser, the small pressure drop over the length of the heat exchanger enables the modeling assumption of isobaric phase change that occurs at a nearly constant evaporating temperature Te and is measured by sensors on the indoor unit heat exchangers. There are no measurements of refrigerant pressures or flow rates [24].

As discussed in Chap. 4, the dynamics of the ME-VCS can be modeled by a set of nonlinear differential-algebraic equations (DAEs)

$$0 = f(\dot{\phi}, \phi, u, v, d), \tag{5.1a}$$

$$y = h(\phi, u, v), \tag{5.1b}$$

where $\phi \in \mathbb{R}^n$ are the states representing thermofluid variables, $u \in \mathbb{R}^m$ are the control inputs, $v \in \mathbb{R}^l$ are the measured disturbances, $d \in \mathbb{R}^q$ are the unmeasured disturbances, and $y \in \mathbb{R}^p$ are the measured outputs. These equations can be derived from the principles of fluid mechanics and thermodynamics, and typically involve spatial discretization of one-dimensional Navier–Stokes equations and empirically determined closure relationships, including heat transfer correlations and frictional pressure drops [25, 26].

Generally speaking, Eq. (5.1) is nonlinear, high-order ($n \approx O(100 - 1000)$), stiff, and marginally stable. Nonlinearity arises from thermofluid properties, the closure relations, and the mass, momentum and energy balance equations governing fluid dynamics and heat transfer [27]. The order can be high because of finite element or finite volume modeling methodologies which are required for practical reasons to accurately capture the spatial and temporal dynamics of the heat exchangers over a broad range of operating conditions [28]. Timescales can range from approximately 1 second for the fastest dynamics associated with pressure within the ME-VCS to as much as 1 week for the slowest timescale associated with heat transfer to the building envelope. This slowest time scale is inherent in (5.1) because we consider the zone air temperatures $\mathrm{Tr}_i$ to be among the measured outputs.

Idealized energy balance models of zone air thermodynamics assume an adiabatic process (no thermal losses to the environment), leading to an integrator in the equation for zone temperature dynamics. Physically, this marginal instability manifests as a diverging zone temperature whenever the cooling provided by the ME-VCS is not matched to the load in steady state. However, real systems are characterized by finite heat loads and nonadiabatic volumes and therefore the zone air temperature will ultimately reach some equilibrium, but this equilibrium is unknown and with the unmeasured load, contributes to uncertainty at low frequencies. Uncertainty in this context is used informally to mean that is difficult to experimentally calibrate or validate an instance of the model, and also that (5.1) is assumed to contain uncertainty in order to cover a diverse range of buildings in which a ME-VCS is installed and operated. The low-frequency uncertainty implies that it is practically impossible to precisely compute the equilibrium $0 = f(0, \bar{\phi}, \bar{u}, \bar{v}, \bar{d})$ and the algebraic relationship among $\bar{\phi}, \bar{u}, \bar{v}$ and $\bar{d}$ for any model instance. This recognition is important to the architecture of any feedback compensator that aims to regulate the zone temperature to a setpoint with zero steady state error.

Fortunately, the dynamic response of the system in the "medium frequency" range of interest to the control system designer is remarkably linear and low order. To support this assertion, consider the data shown in Fig. 5.2 where a vapor compression system is perturbed from three equilibrium points in order to characterize the sensitivity of nonlinearities to operating condition. The range of heat loads are selected to cover from 50 to 85% of the rated capacity of this particular system. For each heat load, the compressor speed input is increased by 8 Hz, and the resulting responses of the compressor discharge temperature and the zone air temperature are shown. Despite settling to different values at steady state, the time constants for both outputs at each heat load are similar. For the systems considered, a medium frequency range of approximately 0.001–0.1 rad/s (corresponding to time scales of approximately

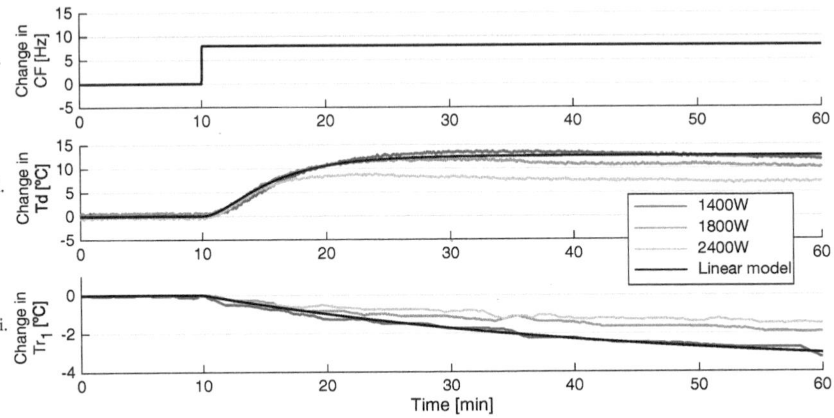


**Fig. 5.2** Experiments in which compressor steps are applied at three different equilibrium points demonstrate that ME-VCS dynamics are well approximated by linear low-order models

1–60 min) have been shown to be well approximated by low-order linear models. Feedback is employed to compensate the low-frequency uncertainty. This empirical observation enables the use of linear MPC.

Accordingly, the ME-VCS outdoor unit is modeled by

$$x_0(t+1) = A_{00}x_0(t) + \sum_{j=0}^{N} B_{0j}u_j(t) \tag{5.2a}$$

$$y_0(t) = C_{00}x_0(t) + w_0(t) \tag{5.2b}$$

where the inputs $u_0 = \text{col}(\text{CF}, \text{ODF}) \in \mathbb{R}^{m_0}$ are the compressor frequency CF and outdoor fan speed ODF, and the outputs $y_0 = \text{col}(\text{Td}, \text{Te}, \text{Tc})$ are the discharge Td, evaporator Te, and condenser Tc temperatures. The inputs $u_j$, $j = 1, \ldots, N$ pertain to the individual indoor units and are described below. The model is fit to experimentally obtained input–output data, and thus the state of the outdoor unit $x_0(t) \in \mathbb{R}^{n_0}$ is nonphysical. The discrete-time model (5.2) describes the ME-VCS system when sampled with a period of 1 min.

An additive output disturbance $w_0(t)$ is used to capture the effects of outdoor air temperature OAT on the measured outputs $y_0(t)$. Since the outdoor air temperature varies slowly with a diurnal period, we treat the disturbance $w_0(t)$ as constant on the timescale of the model (5.2),

$$w_0(t+1) = w_0(t). \tag{5.2c}$$

*Remark 5.1* The entire ME-VCS is characterized by a single evaporating temperature due to the arrangement of valves shown in Fig. 5.1. In particular, the ME-VCS considered does not have valves at the outlet of each indoor unit. Thus, all indoor units with open valves operate at the same pressure and therefore the same evaporating temperature. As a consequence, the evaporator temperature Te is modeled as an output of the outdoor unit (5.2b) and depends only on the state $x_0$ of the outdoor unit. □

Over its range operation, the effect of the electronic expansion valve position $\text{EEV}_i$ on the room temperature $\text{Tr}_i$ is nonlinear. Therefore, each indoor unit has an inner feedback loop that manipulates the expansion valve position $\text{EEV}_i$ to achieve a desired cooling capacity $\text{CC}_i$. The cooling capacity is the amount of heat removed from the zone by the corresponding evaporator per unit time. The cooling capacity controllers linearize the response from the reference cooling capacity command of each zone $\text{CC}_i$ to the associated zone temperature $\text{Tr}_i$. Additional details on the inner feedback loops are provided in [29].

The $i$-th indoor unit, $i = 1, \ldots, N$, is modeled by

$$x_i(t+1) = A_{ii}x_i(t) + A_{i0}x_0(t) + \sum\nolimits_{j=0}^{N} B_{ij}u_i(t) \tag{5.2d}$$

$$y_i(t) = C_{ii}x_i(t) + w_i(t) \tag{5.2e}$$

where the input $u_i = \text{CC}_i \in \mathbb{R}^{m_i}$ is the cooling capacity command (which also appears in (5.2)) and the output $y_i = \text{Tr}_i \in \mathbb{R}$ is the zone temperature. The state of the $i$-th indoor unit $x_i \in \mathbb{R}^{n_i}$ is nonphysical. The dynamics of the indoor unit depend on the state $x_0$ and input $u_0$ of the outdoor unit, as well as the inputs $u_i$ for $i = 1, \ldots, N$ for each of the indoor units.

Note that in some formulations, the indoor fan speed $\text{IDF}_i$ may be considered as a control input and included in the definition of $u_i(t)$. However, in this application we permit the occupants to set the IDF for comfort reasons and therefore we treat the IDF as a measured disturbance. Because the IDF is measured, its effect on the zone temperature $y_i(t)$ can either be modeled as an input disturbance (in which case a dynamic model must be created that describes how changes in this input influence the measurements) or as an output disturbance (in which case only the steady state offset in zone temperature due to this disturbance is captured). Because the IDF is anticipated to change infrequently and therefore not excite the system dynamics, the output disturbance method is chosen. Additionally, the effects of the unmeasured heat load $\text{Q}_i$ on the zone temperature are also modeled as an output disturbance. These are lumped into the term $w_i(t)$, assumed constant, and added to the output in (5.2e),

$$w_i(t+1) = w_i(t). \tag{5.2f}$$

It is convenient to gather both the outdoor (5.2a)–(5.2c) and indoor (5.2d)–(5.2f) models into a single representation

$$x(t+1) = Ax(t) + Bu(t) \tag{5.3a}$$
$$w(t+1) = Iw(t) \tag{5.3b}$$
$$y(t) = Cx(t) + Iw(t) \tag{5.3c}$$

where $x = \text{col}(x_0, x_1, \ldots, x_N)$, $w = \text{col}(w_0, w_1, \ldots, w_N)$, $u = \text{col}(\text{CF, ODF, CC}_i)$, $y = \text{col}(\text{Td, Te, Tc, Tr}_i)$, $I$ is the identity matrix, and $(A, B, C)$ are assembled as indicated in (5.2).

The models (5.2) and (5.3) are experimentally identified with the structure described above for the ME-VCS operating under typical conditions. The models (5.2) and (5.3) are a minimal realization of the dynamics of the ME-VCS, and the pairs $(A_{ii}, B_{ii})$ and $(A_{ii}, C_{ii})$ are controllable and observable, respectively, for $i = 0, \ldots, N$. The signals $u_i(t)$, $y_i(t)$, and $w_i(t)$ for $i \in \mathcal{I}_0 = \{0, \ldots, N\}$ are the deviations of the inputs, outputs, and disturbance from their nominal values, respectively. The physical meaning of the control inputs, measured outputs, and disturbances is summarized in Table 5.1.

### 5.2.2 ME-VCS Constraints

This section describes the constraints on outputs, states, and inputs of the multi-evaporator vapor compression system.

The outdoor unit measurements $y_0(t)$ of the discharge, evaporating, and condensing temperatures are constrained to protect the equipment. Physical damage to the compressor motor can occur when its internal temperature exceeds some critical value, thus we have an upper bound $\text{Td}_{\max}$ on the compressor discharge temperature $\text{Td} \leq \text{Td}_{\max}$. If the indoor unit evaporators become too cold, frost can accumulate on the heat exchanger inhibiting heat transfer, thus we have a lower bound $\text{Te}_{\min}$ on the evaporating temperature $\text{Te} \geq \text{Te}_{\min}$. Additionally, excessive condenser-side

**Table 5.1** Definition of physical signals

| Type | Symbol | Description | Units |
|---|---|---|---|
| Control Inputs | $u_0$ | Compressor Frequency (CF) | Hz |
| | | Outdoor Fan Speed (ODF) | rpm |
| | $u_i$ | Cooling Capacity for $i$-th Zone ($\text{CC}_i$) | % |
| Measured Outputs | $y_0$ | Discharge Temp (Td) | °C |
| | | Evaporating Temp (Te) | °C |
| | | Condensing Temp (Tc) | °C |
| | $y_i$ | $i$-th Zone Temp ($\text{Tr}_i$) | °C |
| Disturbances | $w_0$ | Temp offset due to Outdoor Air Temp (OAT) | °C |
| | $w_i$ | Temp offset due to heat load ($\text{Q}_i$) and indoor fan speed ($\text{IDF}_i$) in $i$-th Zone | °C |

pressures (as measured through the surrogate condensing temperature) can rupture components of the equipment, and thus we have an upper bound $\text{Tc}_{\text{max}}$ on the condenser temperature $\text{Tc} \leq \text{Tc}_{\text{max}}$. Finally, to prevent the ingestion of damaging liquid refrigerant into the compressor, we define the refrigerant superheat temperature *at the compressor discharge port* Tdsh as an algebraic combination of other measurements $\text{Tdsh} = \text{Td} - \text{Tc}$ and impose a lower bound $\text{Tdsh}_{\text{min}}$ on the discharge superheat temperature $\text{Tdsh} \geq \text{Tdsh}_{\text{min}}$.

Note that the output constraints only apply to the outputs (or algebraic combinations thereof) (5.2b) of the outdoor unit. The outputs (5.2e) of the indoor units are unconstrained. In particular, the evaporating temperature constraint, which physically relates to the indoor units, is modeled as an output of the outdoor unit for reasons described in Remark 5.1.

The outdoor unit has lower and upper bounds on the compressor frequency $\text{CF}_{\text{min}} \leq \text{CF} \leq \text{CF}_{\text{max}}(t)$ and outdoor fan speed $\text{ODF}_{\text{min}} \leq \text{ODF} \leq \text{ODF}_{\text{max}}$. Furthermore, we allow $\text{CF}_{\text{max}}(t)$ to be time-varying to accommodate the action of external protection logic, or as a means to implement building-level demand response.

Since the cooling capacity commands represent a fraction of the total rated capacity of each evaporator [29], the inputs to the indoor unit capacity controllers have lower and upper bounds $\text{CC}_{i,\text{min}} \leq \text{CC}_i \leq \text{CC}_{i,\text{max}}$. The actuator constraints associated with the electronic expansion valves $\text{EEV}_i$ are enforced within the indoor unit capacity controllers.

The system also has constraints on the amount the inputs can change during each sample period. The outdoor unit has constraints on the change in compressor frequency $\Delta\text{CF}_{\text{min}} \leq \Delta\text{CF} \leq \Delta\text{CF}_{\text{max}}$ and outdoor fan speed $\Delta\text{ODF}_{\text{min}} \leq \Delta\text{ODF} \leq \Delta\text{ODF}_{\text{max}}$.

The indoor units have inner feedback loops that control the position of the electronic expansion valves to track the cooling capacity command. The change in cooling capacity command is bounded $\Delta\text{CC}_{i,\text{min}} \leq \Delta\text{CC}_i \leq \Delta\text{CC}_{i,\text{max}}$ to ensure that the transient response of the inner loop controllers settles during the sample period.

## 5.3 Control Requirements

The control requirements for a production-grade vapor compression system extend beyond temperature regulation. The ME-VCS must also protect itself from damage, permit some zones to shut off while others remain in service, minimize energy consumption to the extent possible and accomplish these objectives while operating over a wide and variable range of outdoor and load conditions. Additionally, emerging trends in building efficiency standards, smart grid compatibility and tighter integration with other HVAC equipment require flexibility in the control method to, for example, adapt the operation of the machine to a demand response event or cooperate with other equipment such as dedicated ventilation systems to efficiently manage sensible and latent loads. Finally, economic and reliability pressures demand the use of a limited number of low-cost sensors and low-end embedded processors.

The following is a prioritized list of control objectives that is divided into two parts: baseline requirements and extended requirements. The baseline requirements are control objectives for the system during nominal operating modes and include constraint enforcement, disturbance rejection, and transient response metrics. The extended requirements address practical concerns often neglected in the literature such as compatibility with startup, shutdown, equipment protection and defrost logic, response to building management system requests for energy reduction, and the ability to shut off some zones.

### *5.3.1 Baseline Controller Requirements*

This section lists control requirements for baseline operation of the ME-VCS. Here, "baseline operation" refers to the system operating around an equilibrium state.

**B.1 Constraint Enforcement** The controller should enforce the input (actuator), input rate and output constraints described in Sect. 5.2.2.

**B.2 Setpoint Regulation** If B.1 is satisfied, the controller should drive each zone air temperature to its setpoint with zero steady-state error.

**B.3 Disturbance Rejection** The controller should achieve B.2 given changes in heat loads and outdoor air temperatures.

**B.4 Energy Minimization** If B.1–B.3 are satisfied, the controller should drive the system to energy-optimal operating points in steady state. Note that energy performance is not defined during transients.

**B.5 Transient Performance** While ensuring B.1 is satisfied, the controller should minimize the time required to adjust to a new setpoint with an overshoot of less than 1 °C. The response time is left unspecified because it depends on (i) whether the setpoint change is a step-up or step-down, (ii) the unknown volume of air in the zone, and (iii) the unmeasured load.

### *5.3.2 Extended Controller Requirements*

In addition to the baseline requirements, the controller must also meet the following extended requirements:

**E.1 Permit individual zones to be shut off** Often low heat loads in some zones or occupant desires are such that some evaporators must be turned off while others remain in service. The control system should enable this type of behavior.

**E.2 Computationally tractable** Embedded processors are often limited in processing ability, memory availability, and communication speeds. The algorithmic realization of the controller should be conformable to such computational platforms.

**E.3 Compatible with Demand Response** The controller should adapt its energy consumption in response to signals received from a Building Management System during a Demand Response event.
**E.4 Independent parameter-to-performance metric relationship** The parameters that adjust the behavior of the controller should independently and predictably do so. For example, a parameter that specifies an output constraint should not affect energy performance and the parameter that specifies energy performance should not affect the output constraints. Parameter independence enables late-stage modification of specifications without extensive revalidation, and therefore speeds product development.

### *5.3.3 Control Design Strategy*

Our strategy to meet these requirements is outlined in this section, with additional details and controller derivations provided in Sects. 5.4 and 5.5. Model Predictive Control (MPC) is selected as the main approach because (i) constraint enforcement is critical (B.1), (ii) a model (5.3) of the process dynamics is available and a prediction model for the references can be created, and (iii) control invariant sets and online updates to the cost function enable satisfaction of the extended requirements.

An outline of the design process for model predictive control is shown in Fig. 5.3. Designing a model predictive controller requires specifying a receding horizon optimal control problem in a series of offline steps. Briefly, starting from a model of the ME-VCS dynamics, augmentations to the model are performed to meet controller requirements, resulting in a prediction model that achieves desired closed loop dynamics. The prediction model is used in the specification of the optimal control problem and as the basis of a online state estimator. The controller requirements

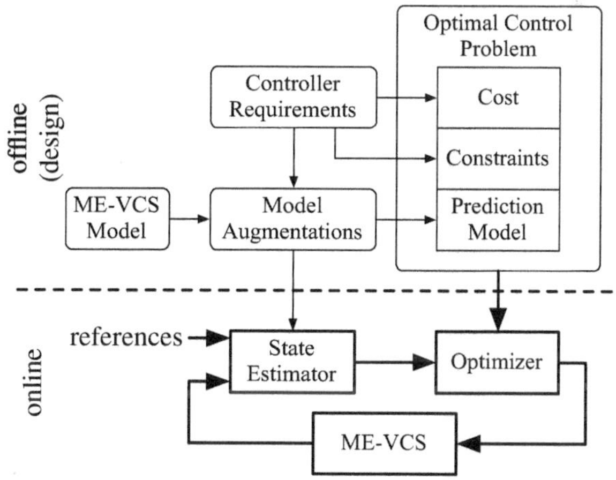


**Fig. 5.3** A model predictive controller is designed in a series of offline steps to specify an optimal control problem. Controller requirements inform the creation of the cost function, constraints, and the model augmentations needed create the prediction model. The optimal control problem is then converted into a representation suitable for online calculation of the control action

are also used to specify the cost function and constraint components of the optimal control problem. A control problem with quadratic cost and linear constraints is equivalent to a constrained quadratic program that is solved online by an optimization algorithm [30].

The following describes how we apply the MPC design framework of Fig. 5.3 to meet the control requirements. Specifically, and in order of priority, the constraint enforcement requirement (B.1) is satisfied through appropriate specification of the constraints in the optimal control problem. Zone temperature regulation (B.2) is addressed by specifying a setpoint temperature for each zone and penalizing tracking errors in the cost function. Disturbance rejection (B.3) is addressed with output disturbance models incorporated and the use of integral action on the zone temperature tracking errors. A Kalman filter is used to estimate unmeasured states that are used to initialize the prediction model.

To address the requirement to minimize energy consumption (B.4), we note that the number of actuators is greater than the number of zones, which implies that sufficient degrees of freedom exist to regulate an additional variable while simultaneously meeting zone temperature setpoint objectives, and we therefore introduce another reference correlated to steady state thermodynamic efficiency. While many control approaches for vapor compression systems regulate evaporator superheat as a surrogate for efficiency, we instead select the compressor discharge temperature Td because it is measured and because the steady-state power consumption at any operating point is convex with respect to it [31, 32]. Therefore, energy-optimal values of Td are used to create a reference, and tracking errors of this reference are penalized in the cost function.

After introducing the Td reference there still remains an additional degree of freedom. However, it is more important to enforce constraints, especially through transients, than to regulate an additional process variable. We therefore choose to use the additional degree of freedom to provide flexibility to the optimizer to meet the constraints. The transient performance objective (B.5) is addressed through the selection of appropriate penalty terms in the cost function and the reliance on the actuator and output constraint enforcement to safely extract the high performance only obtained near equipment limits.

The extended controller requirements are addressed by exploiting properties of the MPC method. Selectively turning zones on or off (E.1) is accomplished with online modification to the penalty terms in the MPC cost function, effectively removing zone-level process variables and control inputs from the optimization problem. The manipulation of the controller cost function is performed when either an occupant specifies that a zone is to be shut off or supervisory logic determines that a zone should be deactivated. A detailed derivation of this approach and validation experiments are provided in Sect. 5.5.

By requirement (E.2), the realization of the online calculations method must be computationally tractable and fit within the limited memory available on low-cost embedded systems. The reconfigurable MPC strategy for shutting off zones with online modifications avoids the need to store individual controllers for every possible on/off configuration, and the relatively slow dynamics of (5.2) provide ample time between controller updates for advanced algorithms [16] to reliably solve the associated quadratic program.

The requirement of adapting to a demand response request (E.3) is enabled through the parameter independence property of our MPC design and therefore (E.3) is satisfied when (E.4) is satisfied. The controller derived in Sects. 5.4 and 5.5 is shown to decouple the effect of one parameter from another. Not only does this property simplify tuning and validation, but it also provides tunable parameters that can be exposed to higher level supervisory control systems such as with a building management system. For example, zone temperature set points, or the controller discharge reference or the maximum compressor frequency can each be dynamically adjusted by external logic to alter the energy consumption behavior without modifying, for example, the controller's ability to enforce constraints.

As a result of the controller structure presented in the next section, the ability to meet the baseline requirements degrades gracefully and in accordance with the listed priority. We show this in experiments in the following section.

## 5.4 Model Predictive Control for Baseline Requirements

In this section, we demonstrate how a model predictive control strategy can be applied to meet the baseline requirements described in Sect. 5.3.1. The key elements of the control algorithm are described, including the prediction model used to define performance and constrained outputs, the Kalman Filter and the optimal control problem formulation. Finally, experiments are conducted to validate the approach.

A block diagram of the closed loop system is shown in Fig. 5.4. A state estimator which includes a Kalman Filter (plus additional functions for generating other states) receives zone temperature references $r_\iota(t)$, a signal $u_{0,\max}(t)$ indicating the time-varying maximum compressor speed from external control logic, the outputs $y(t)$ measured from the ME-VCS, and the control inputs $u(t)$ supplied to it, and computes estimated states $\bar{x}(t)$ of a prediction model. The model predictive controller uses $\bar{x}(t)$ to define and to solve a constrained optimal control problem to generate the control inputs $u(t)$.

### 5.4.1 State Estimation Using a Kalman Filter

The structure of the Kalman Filter is designed to satisfy the setpoint regulation requirement (B.2) and the disturbance rejection requirement (B.3), which demands that the states be estimated such that the outputs of the estimator $\hat{y}$ match the measurements $y$ at steady state when the disturbances are constant.

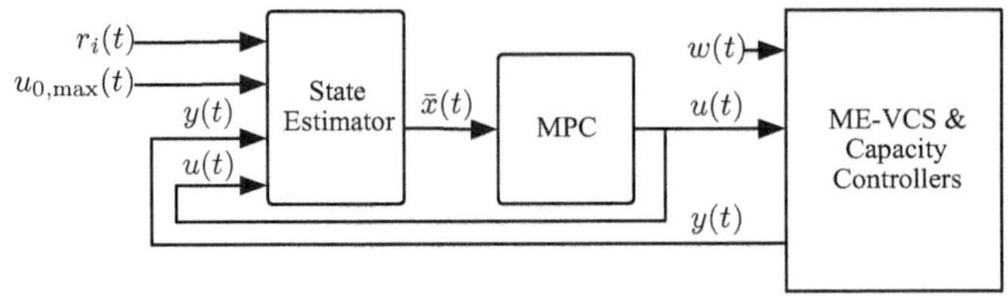


**Fig. 5.4** Block diagram of the closed loop system

Starting from model (5.3), we construct the Kalman Filter for the ME-VCS plant dynamics

$$\hat{x}(k+1) = A\hat{x}(k) + Bu(k) + L_x\left(y(k) - \hat{y}(k)\right) \tag{5.4a}$$

$$\hat{w}(k+1) = \hat{w}(k) + L_w\left(y(k) - \hat{y}(k)\right) \tag{5.4b}$$

$$\hat{y}(k) = C\hat{x}(k) + \hat{w}. \tag{5.4c}$$

The estimator gain matrix $L = [L_x\ L_w]$ can be computed using conventional methods such as minimizing an $H_2$ norm.

In addition to the estimated states $[\hat{x}^T\, \hat{w}^T]^T$, there are additional variables required by the MPC that are added to the state vector, resulting in the augmented state vector $\bar{x}$. The block labeled "State Estimator" in Fig. 5.4 contains the Kalman Filter (5.4) which computes $[\hat{x}^T\, \hat{w}^T]^T$, and also other equations for computing the remaining components of $\bar{x}$, as described in the next section.

### 5.4.2 *Prediction Model Augmentations*

In this section, we begin with the plant model (5.3) and augment the dynamics to define performance outputs and constrained outputs for the optimal control problem. Each augmentation is performed to meet one or more control requirements, as will be indicated.

**Augmentation 1** (Incremental Inputs) *Because there are constraints on both the magnitude of the inputs and the amount they can change at each sample period, we redefine the inputs of (5.3) to explicitly track the incremental control action [33]. Let* $u(t) = u(t-1) + \Delta u(t)$*, define the actuator state* $x_u(t) := u(t-1)$ *and augment the state in (5.3) with* $x_u$*. Finally, augment the constrained outputs with this new state so that the magnitude of the control inputs can be constrained, satisfying B.1.*

**Augmentation 2** (Time-varying Constraints) *In order to be compatible with commands (E.4), we allow the maximum compressor maximum frequency constraint to be time-varying* $u_{0,max}(t) = CF_{max}(t)$*. Therefore, we want to enforce* $CF(t) \leq CF_{max}(t)$

*so we rewrite the constraint as* $CF(k) - CF_{max}(t) \leq 0$ *and define the new constrained output* $y_{tv}(t) := CF(t) - CF_{max}(t)$ *where the subscript is meant to indicate a time-varying constraint. This new constraint will have a maximum value of 0, which ensures that CF remains less than the constrained value.*

*Because the value of* $CF_{max}$ *comes from an external source, we define a new state* $x_{tv}(t) := CF_{max}(t)$ *and include it in the definition of the prediction model state. The dynamics of this exogenous constraint cannot be predicted, so we assume that it is constant* $x_{tv}(t+1) = x_{tv}(t)$.

*Note that allowing a building management system to specify the maximum compressor speed permits a more direct influence over the vapor compression system's energy consumption than traditional means where zone temperature setpoints are adjusted with offsets.*

**Augmentation 3** (References and Performance Outputs) *The baseline requirements include setpoint regulation (B.1) of the zone air temperatures and energy minimization (B.4) in steady state. To achieve both requirements, we first introduce the references into the prediction model, then define the performance outputs as the tracking errors and integrals thereof.*

*The energy-optimal discharge temperature reference is a linear function of the compressor frequency CF (control input) and the outdoor air temperature, OAT (disturbance input) [34] given by*

$$r_0 = F_{00}u_0 + G_{00}w_0 \tag{5.5}$$

*where the coefficients* $F_{00}, G_{00}$ *are experimentally characterized to minimize power consumption over a range of operating conditions. Since* $w_0$ *is assumed constant over the timescale of the model (5.3), the prediction model of the reference discharge temperature is given by*

$$r_0(t+1) = r_0(t) + F_{00}\Delta u_0(t). \tag{5.6}$$

*The reference zone temperatures are assumed constant and modeled by*

$$r_i(t+1) = r_i(t), \tag{5.7}$$

*for* $i = 1, \ldots, N$. *For consistency we define* $F_{ii} = 0$ *and* $G_{ii} = 0$ *for* $i = 1, \ldots, N$. *Thus, the composite reference* $r = \mathrm{col}(r_0, r_1, \ldots, r_N)$ *is modeled by*

$$r(t+1) = r(t) + F\Delta u_0(t)$$

*where* $F = \mathrm{col}(F_{00}, F_{11}, \ldots, F_{NN})$ *and* $G = \mathrm{col}(G_{00}, G_{11}, \ldots, G_{NN})$. *The prediction model state vector is then augmented to include these references.*

*Using these references, tracking errors are computed for the compressor discharge temperature and the zone temperatures*

$$e_0(t) = Td - r_0(t) \tag{5.8}$$

$$e_i(t) = Tr_i(t) - r_i(t) \quad i = 1, \ldots, N \tag{5.9}$$

*Additionally, the zone temperature tracking errors are integrated*

$$\xi_i(t+1) = \xi_i(t) + e_i(t) \quad i = 1, \ldots, N \tag{5.10}$$

*The zone temperature error integrators (5.10) provide integral action and hence zero steady-state tracking error in the presence of uncertainties in zone air volume and heat loads. Although the auxiliary output offset (5.2c) and (5.2f) provide offset-free tracking [35, 36], our experimental results have shown that combining both integral action and output offsets provides improved transient performance.*

*Finally, the performance outputs $\bar{z}$ are defined as the tracking errors and the integrals of the zone temperature tracking errors $\bar{z} := \mathrm{col}(e_o, e_i, \xi_i)$, for $i = 1, \ldots, N$.*

The augmentations result in the prediction model

$$\bar{x}(t+1) = \bar{A}\bar{x}(t) + \bar{B}\Delta u(t) \tag{5.11}$$

$$\bar{y}(t) = \bar{C}\bar{x}(t) + \bar{D}\Delta u(t) \tag{5.12}$$

$$\bar{z}(t) = \bar{E}\bar{x}(t) \tag{5.13}$$

where $\bar{x} = \mathrm{col}(\hat{x}(t), \hat{w}(t), r(t), \xi_i(t), x_{tv}(t), x_u(t))$, $i = 1, \ldots, N$ are the augmented states, $\bar{y} = \mathrm{col}(\mathrm{Td}, \mathrm{Te}, \mathrm{Tc}, \mathrm{Tdsh}, \mathrm{CF}, \mathrm{ODF}, \mathrm{CC}_i, \mathrm{CF} - \mathrm{CF}_{max})$, $i = 1, \ldots, N$ are the constrained outputs and $\bar{z} = \mathrm{col}(\mathrm{Td} - r_0, \mathrm{Tr}_i - r_i, \xi_i)$, $i = 1, \ldots, N$ are the performance outputs and $\Delta u = \mathrm{col}(\Delta\mathrm{CF}, \Delta\mathrm{ODF}, \Delta\mathrm{CC}_i)$, $i = 1, \ldots, N$ are the inputs.

### 5.4.3 Optimal Control Problem Formulation

As shown in Fig. 5.5, at each sample time the estimator computes the augmented state $\bar{x}(t)$, and an optimizer computes the control input by solving the following constrained finite-time optimal control problem

$$\min_{\Delta \mathbf{u}} \quad \bar{x}'_{T|t} P \bar{x}_{T|t} + \sum_{k=0}^{T-1} \bar{z}'_{k|t} Q \bar{z}_{k|t} + \Delta u'_{k|t} R \Delta u_{k|t} \tag{5.14a}$$

$$\text{s.t.} \quad \bar{x}_{k+1|t} = \bar{A}\bar{x}_{k|t} + \bar{B}\Delta u_{k|t} \tag{5.14b}$$

$$\bar{y}_{k|t} = \bar{C}\bar{x}_{k|t} + \bar{D}\Delta u_{k|t} \tag{5.14c}$$

$$\bar{z}_{k|t} = \bar{E}\bar{x}_{k|t} \tag{5.14d}$$

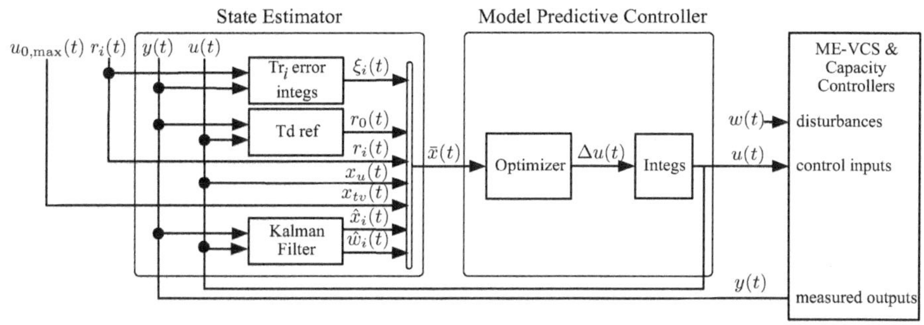


**Fig. 5.5** Controller block diagram. A state estimator computes the states defined by the prediction model from zone integrators, the Td reference function and the Kalman Filter. An optimizer solves the constrained optimal control problem to determine incremental input vales that are integrated to generate the control inputs

$$\Delta u_{\min} \le \Delta u_{k|t} \le \Delta u_{\max} \tag{5.14e}$$

$$y_{\min} \le \bar{y}_{k|t} \le y_{\max} \tag{5.14f}$$

$$\bar{x}_{0|t} = \bar{x}(t). \tag{5.14g}$$

where $\bar{x}_{k|t}$ is the predicted augmented state under the incremental input $\Delta u_{k|t}$ over the finite prediction horizon of $T$-steps, and $\Delta \mathbf{u} = \text{col}(\Delta u_{0|t}, \ldots, \Delta u_{T-1|t})$ is the optimization variable. The first element from $\Delta \mathbf{u}$ is selected from the solution to the optimization problem, $\Delta u(t) = \Delta u^{\star}_{0|t}$ and because this variable is an incremental input (by Augmentation 1), the model predictive controller integrates the incremental input $\Delta u(t)$ (see Fig. 5.5) to obtain the implemented input

$$u(t) = u(t-1) + \Delta u_t \tag{5.15}$$

which is then applied the ME-VCS.

The cost function (5.14a) penalizes the predicted performance outputs $\bar{z}_{k|t}$ and incremental inputs $\Delta u_{k|t}$ over the prediction horizon, and the states at the end of the prediction horizon $\bar{x} + T|t$. The performance output $\bar{z}_0 = z_0 - r_0$ is the difference between the discharge temperature $z_0 = \text{Td}$ and the reference discharge temperature $r_0 = \text{Td}_{\text{ref}}$, which maximizes energy efficiency at steady state. Additional performance outputs $\bar{z}_i = \text{col}(z_i - r_i, \xi), i = 1 \ldots N$ relate to the $N$ indoor units and include the room temperature tracking errors $z_i - r_i$, and their integrals $\xi$. The matrix $Q \geq 0$ is used to weigh the relative importance of the performance outputs. Changes in control inputs $\Delta u_{k|t}$ are also penalized, and the matrix $R > 0$ specifies their relative importance. The terminal state matrix $P > 0$ is designed to ensure that tracking errors are locally asymptotically stable, and may be computed by solving the Ricatti equation with slight modifications as discussed in [37].

The optimal control problem (5.14) is converted to a quadratic program in an offline step [30] and parameters representing the quadratic program are stored for online calculation of the control action. An algorithm for solving the quadratic program (labeled "Optimizer" in Fig. 5.5) uses these parameters with the current instance of the augmented state $\bar{x}$ to solve the control problem (5.14) on the embedded ME-VCS processor to compute $\Delta u$.

### 5.4.4 Experimental Validation

In this section, we present experimental results that demonstrate that the MPC design satisfies the baseline requirements. We begin by briefly describing the experimental testing facility, then show two experimental results aimed at validating the controller design.

The ME-VCS system is a commercially available two-zone unit installed in a test facility that includes a balance-of-plant system to supply heating and cooling loads, as shown schematically in Fig. 5.6. The ME-VCS outdoor unit (consisting of the compressor, outdoor heat exchanger and fan, and EEVs) is installed in a 6.3 $m^3$ insulated test chamber and is connected via refrigerant lines to two indoor units (each consisting of indoor heat exchangers and fans), which are installed in separate 9.9 $m^3$ insulated test chambers. The balance-of-plant system consists of variable power heaters and a variable power chiller, hydronic fan coils and associated controllers and is configured to regulate the heat loads in the indoor unit test chambers and the air temperature in the outdoor unit test chamber.

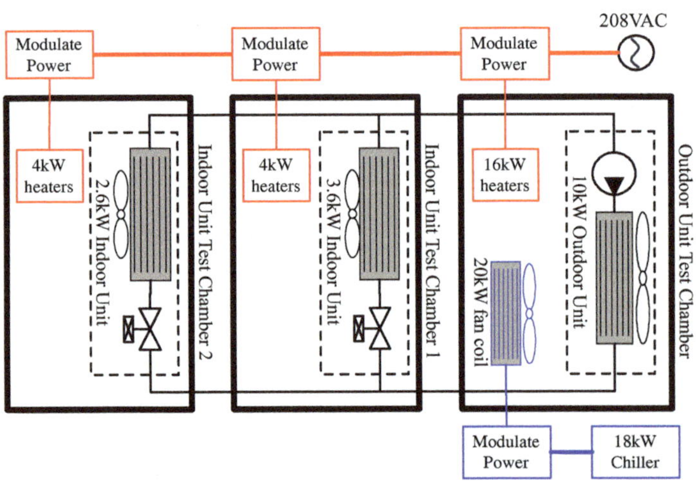


**Fig. 5.6** The outdoor unit and two indoor units of a split-ductless style vapor compression system is installed in three test chambers. A balance-of-plant system consists of a set of adjustable power heaters (red) and an adjustable power hydronic system (blue), and is used to set the environmental conditions of the vapor compression system under test

The plant model (5.2) is derived from experimental data collected in this test facility as follows: With 1.6 kW fixed heat loads applied in the indoor unit test chambers, the outdoor test chamber regulated to 35° C, and the vapor compression system operating at steady state, steps are separately applied to each control input $u$ (CF, ODF, $CC_1$ and $CC_2$). Measurements of the system outputs $y$ are collected the data is used to fit a model of the plant in the form of Eq. (5.2).

#### 5.4.4.1 Experiment 1: Enforcing Multiple Simultaneous Constraints

This section describes an experiment where a step change to a setpoint is applied in order to induce a transient response where an output constraint becomes active and thereby demonstrate enforcement (B.1).

Referring to Fig. 5.7, the vapor compression system is operated until steady-state conditions are present with a zone setpoint $r_1 = 25°$ C and the boundary conditions $Q_1 = 2200$ W, and OAT = 35° C. At $t = 5$ min, the setpoint is lowered to $r_1 = 22°$ C (Fig. 5.7B-ii), and the controller increases the CF and saturates $CC_1$ to reduce the temperature in that zone. As a result of the increased CF, the discharge temperature increases until it reaches its constraint at $t = 8$ min (Fig. 5.7B-i). The CF commands (Fig. 5.7A-i) computed by the MPC maintain Td below its constraint. From $t = 20$ to 35 min, Td largely follows the constraint until at $t = 35$ min a constraint violation

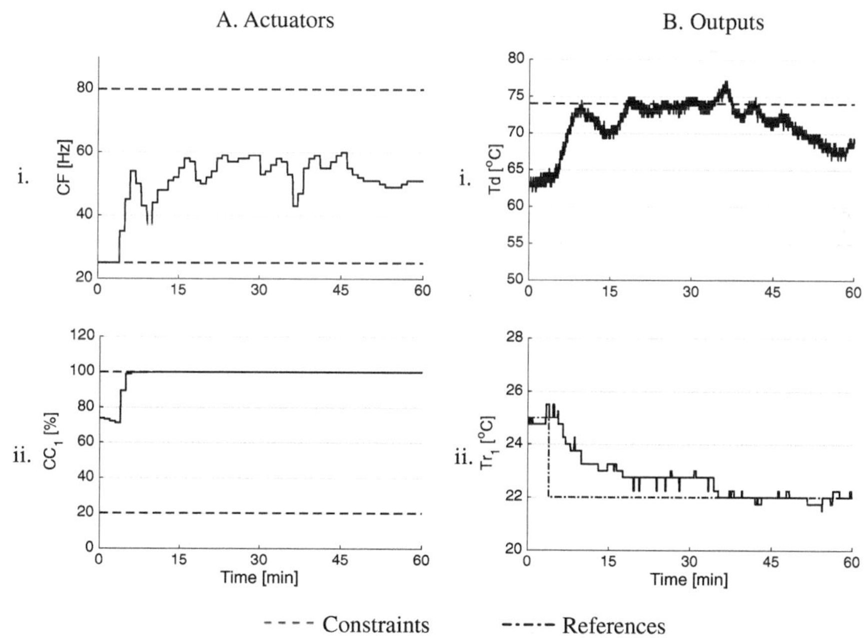


**Fig. 5.7** A step down in the zone temperature setpoint causes the Td and capacity command constraints to become active

of about 2° C occurs (Fig. 5.7B-i), which is attributed to modeling errors. Additionally, the $CC_1$ input is also saturated at its maximum value (Fig. 5.7A-ii), making two constraints active during this transient. The Td constraint violation causes an immediate reduction in CF at $t = 35$ min, and Td is decreased accordingly, relieving the violation. Finally, zone 1 achieves the setpoint with zero steady state error at $t = 37$ min.

This experiment demonstrates that both input (actuator) and output (sensor) constraints can be simultaneously enforced by the MPC during transients, satisfying the constraint enforcement requirement (B.1). Additionally, because the maximum cooling capacity occurs at the constraints, the zone temperature response in Fig. 5.7B-ii represents the fastest pull-down in zone temperature possible from this system at these conditions, satisfying the transient performance requirement (B.5).

#### 5.4.4.2 Experiment 2: Recovery from Constraint Violation

This section describes an experiment showing the controller meeting requirement priority. Specifically, transient performance (B.5) is reduced when a constraint becomes violated demonstrating that the controller alters behavior to prioritize constraint enforcement (B.1).

Referring to Fig. 5.8, two experiments are performed wherein an MPC that includes a $Tdsh_{min}$ constraint (black lines) is compared to the same controller where Tdsh is unconstrained (gray lines). A zone temperature reference $r_1$ step up from 24 to 27° C is applied at $t = 5$ min (Fig. 5.8B.-ii) to cause a reduction in both the CF and $CC_1$ commands. Note that the capacity command can be intuitively understood as proportional to the associated expansion valve (i.e., when $CC_1$ is reduced, $EEV_1$ becomes more closed). This simultaneous reduction in both actuators causes the discharge superheat temperature Tdsh to fall rapidly, resulting in a constraint violation due to modeling errors. For the case where Tdsh is unconstrained (gray lines), the MPC acts to raise the zone temperature (Fig. 5.8B.-ii) as quickly as possible. The CF is reduced to its minimum constraint for the duration of the transient, and the capacity command is smoothly reduced in an effort for Td to track its reference $r_0$ (not shown for clarity). Under these actuator commands, the room temperature first reaches the setpoint at about $t = 25$ min. But the violation for Tdsh is larger in this case.

In this experiment, the MPC controller with a $Tdsh_{min} = 10°$ C constraint (black lines) selects actuator commands that reflect the prioritized baseline requirement list. In particular, actuator commands and the resulting room temperature response initially follow the same trajectories, but when the Tdsh constraint is violated at $t = 8$ min, the controller prioritizes recovery from the constraint violation at the penalty of increased time in meeting the setpoint. In particular, when Tdsh falls below the minimum value, the compressor frequency is rapidly increased and the expansion valve is simultaneously closed down, where both actions serve to increase Tdsh, but at a penalty of slower warming of the room. In this case, the room temperature first reaches the setpoint at $t = 31$ min, but Tdsh recovers quickly from the constraint violation.

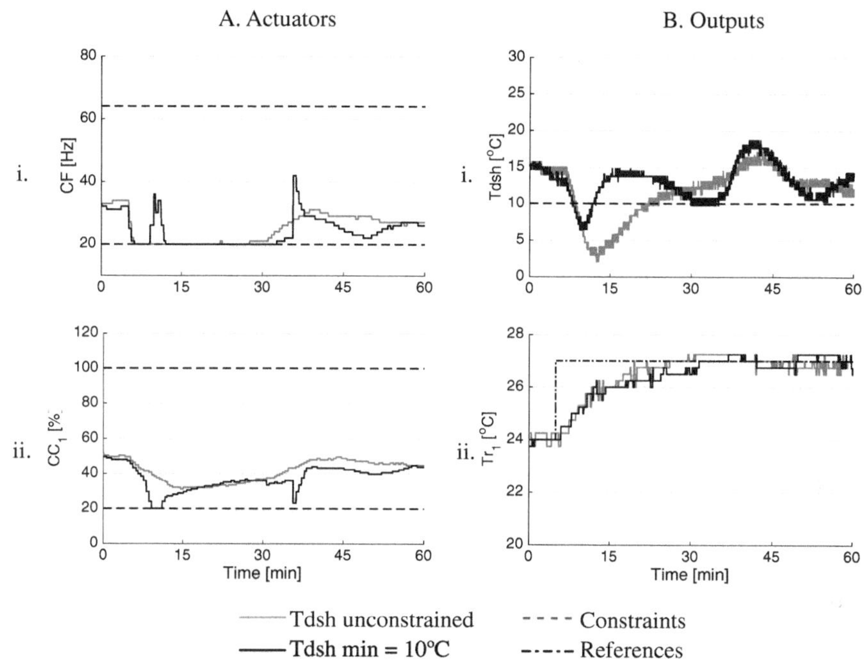


**Fig. 5.8** The MPC controller raises the room temperature as quickly as possible, despite a constraint violation that slows the zone temperature response

## 5.5 Model Predictive Control for Extended Requirements

In this section, we expand on the MPC design in order to meet the extended controller requirements described in Sect. 5.3.2. We present a new method that permits reconfiguration of the model predictive controller to accommodate the changing number of control objectives and inputs when individual zones are turned on or off, and demonstrate the method in experiments. We also provide an overview of two optimization algorithms that are suitable for the embedded processors typically used in commercial HVAC equipment.

### *5.5.1 Selectively Deactivating Zones with Reconfigurable MPC*

In practice, many multi-evaporator systems often experience low heat loads in zones such that the corresponding evaporator no longer needs to provide cooling and should be shut off while the remaining evaporators continue to operate. This function is captured in the extended controller requirement (E.1). Turning an indoor unit on or off alters the model of the plant dynamics, and therefore induces changes in the prediction model and number of regulated variables, actuators, sensors, and constraints. A structural change of this nature conventionally would require a separate controller for each machine configuration, where at each control cycle the appropriate controller is switched in at runtime [19, 20]. However, this standard approach is too memory intensive for the available embedded processor and violates requirement (E.2).

The reconfigurable MPC approach described here features a single "master" controller designed and tuned for the configuration where all subsystems are active, and enables automatic reconfiguration of the controller by simple operations for configurations when any number of evaporators are deactivated. In this way, the master controller is designed and tuned using a single appropriately partitioned prediction model. The method scales to any combination of active evaporators in a multi-evaporator vapor compression system.

In this section, we briefly describe how the configuration-dependent prediction model is obtained and used to specify a reconfigurable optimal control problem. The terms in the cost function are then manipulated online based on the ME-VCS configuration to obtain an controller specific to the machine configuration. Finally, we show an experiment where the reconfigurable MPC autonomously activates and deactivates a zone due to low heat load conditions.

#### 5.5.1.1 Configuration-Dependent Model

An indoor unit is said to be active when its associated expansion valve is open allowing refrigerant to flow through the evaporator providing cooling. Conversely, an indoor unit is said to be inactive when its associated expansion valve is closed and no cooling occurs. A configuration of the ME-VCS is a combination of active and inactive indoor units.

To each indoor unit $i \in \mathcal{I}$ we assign a configuration variable $\varsigma_i \in \{0, 1\}$ where $\varsigma_i = 1$ if the unit is active and $\varsigma_i = 0$ if the unit is inactive. The configuration of the entire ME-VCS is given by the vector $\varsigma = \text{col}(\varsigma_0, \varsigma_1, \ldots, \varsigma_N)$ where $\varsigma_0 = 1$ since the outdoor unit is always active. The configuration $\varsigma \in \{1\} \times \{0, 1\} \times \cdots \times \{0, 1\}$ is used to obtain a parameter-dependent controller [38] which operates for any ME-VCS configuration. In the subsequent analysis, the configuration is assumed to be kept constant, at least for a sufficiently long dwell time.

For configuration $\varsigma$, the dynamics of the outdoor unit are modeled by

$$x_0(t+1) = A_{00}x_0(t) + \sum\nolimits_{j=0}^{N} \varsigma_j B_{0j} u_j(t) \tag{5.16a}$$

where the future state $x_0(t+1)$ of the outdoor unit is only affected by the inputs $u_j(t)$ of active units $\varsigma_j = 1$. The dynamics of the $i$-th indoor unit are modeled by

$$x_i(t+1) = A_{ii}x_i(t) + \varsigma_i A_{i0}x_0(t) + \sum_{j=0}^{N} \varsigma_i \varsigma_j B_{ij} u_j(t) \tag{5.16b}$$

If the $i$-th indoor unit is inactive $\varsigma_i = 0$, then its future state $x_i(t+1)$ depends only on its current state $x_i(t)$. If the $i$-th indoor unit is active $\varsigma_i = 1$, then its future state $x_i(t+1)$ also depends on the outdoor unit state $x_0(t)$ and the inputs $u_j(t)$ to the other active indoor units $\varsigma_j = 1$.

The configuration-dependent model (5.16) can be written more compactly in the form

$$x(t+1) = A^{\varsigma}x(t) + B^{\varsigma}u(t) \tag{5.17}$$

where $x = \mathrm{col}(x_0, x_1, \dots, x_N)$ and $u = \mathrm{col}(u_0, u_1, \dots, u_N)$ are the composite state and input respectively. The state update matrix $A^{\varsigma}$ is block lower arrowhead and the input matrix $B^{\varsigma}$ is dense.

#### 5.5.1.2 Configuration-Dependent Augmentations

In the design of the reconfigurable MPC, similar augmentations are made as described in Sect. 5.4.2 to create the prediction model. Here, we highlight the configuration-dependent augmentations. Specifically, the integrators of the zone temperature tracking errors are expressed in configuration-dependent form as

$$\xi_i(t+1) = \varsigma_i\xi_i(t) + \varsigma_i\big(e_i(t)\big). \tag{5.18}$$

And we define the state $x_{u,i}(t) = u_i(t-1)$. Note that when the $i$-th indoor unit is inactive $\varsigma_i = 0$ the corresponding integrator is shut-off $\xi_i(t+1) = 0$. The incremental inputs are configuration-dependent

$$u_i(t) = \varsigma_i u_i(t-1) + \varsigma_i \Delta u_i(t) \tag{5.19}$$

Note that when the $i$-th indoor unit is inactive $\varsigma_i = 0$ Eq. (5.19) sets the input $u_i(t)$ to zero, and the associated inner feedback loop fully closes the expansion valve.

The remaining augmentations are applied as in Sect. 5.4.2. The configuration-dependent prediction model then becomes

$$\bar{x}(t+1) = \bar{A}^{\varsigma}\bar{x}(t) + \bar{B}^{\varsigma}\Delta u(t) \tag{5.20a}$$

$$\bar{y}(t) = \bar{C}^{\varsigma}\bar{x}(t) + \bar{D}^{\varsigma}\Delta u(t) \tag{5.20b}$$

$$\bar{z}(t) = \bar{E}^{\varsigma}\hat{x}(t) \tag{5.20c}$$

where $\bar{x}=\mathrm{col}(\hat{x}_i(t), \hat{w}_i(t), r(t), \xi_i(t), x_{tv}(t), x_{u,i}(t))$, $i=1, \dots, N$ are the augmented states, $\bar{y} = \mathrm{col}(\mathrm{Td}, \mathrm{Te}, \mathrm{Tc}, \mathrm{Tdsh}, \mathrm{CF}, \mathrm{ODF}, \mathrm{CC}_i, \mathrm{CF} - \mathrm{CF}_{max})$, $i = 1, \dots, N$ are the constrained outputs and $\bar{z} = \mathrm{col}(\mathrm{Td} - r_0, \mathrm{Tr}_i - r_i, \xi_i)$, $i = 1, \dots, N$ are the performance outputs and $\Delta u = \mathrm{col}(\Delta\mathrm{CF}, \Delta\mathrm{ODF}, \Delta\mathrm{CC}_i)$, $i = 1, \dots, N$ are the inputs.

#### 5.5.1.3 Configuration-Dependent Optimal Control Problem

The reconfigurable model predictive controller computes the control input by solving the following configuration-dependent finite-time optimal control problem

$$\min_{\Delta \mathbf{u}} \quad \sum\nolimits_{i=0}^{N} \left( \varsigma_i \bar{x}'_{i,T|t} P_i \bar{x}_{i,T|t} + \sum\nolimits_{k=0}^{T-1} \varsigma_i \bar{z}'_{i,k|t} Q_i \bar{z}_{i,k|t} + \lambda_i \Delta u'_{i,k|t} R_i \Delta u_{i,k|t} \right) \tag{5.21a}$$

$$\text{s.t.} \quad \bar{x}_{k+1|t} = \bar{A}^{\mathbf{1}} \bar{x}_{k|t} + \bar{B}^{\mathbf{1}} \Delta u_{k|t} \tag{5.21b}$$

$$\bar{y}_{k|t} = \bar{C}^{\mathbf{1}} \bar{x}_{k|t} + \bar{D}^{\mathbf{1}} \Delta u_{k|t} \tag{5.21c}$$

$$\bar{z}_{k|t} = \bar{E}^{\mathbf{1}} \bar{x}_{k|t} \tag{5.21d}$$

$$\Delta u_{\min} \leq \Delta u_{k|t} \leq \Delta u_{\max} \tag{5.21e}$$

$$y_{\min} \leq \bar{y}_{k|t} \leq y_{\max} \tag{5.21f}$$

$$\bar{x}_{0|t} = \bar{x}(t), u_{-1|t} = \varsigma_i u(t-1). \tag{5.21g}$$

where $\bar{x}_{k|t}$ is the predicted augmented state under the incremental input $\Delta u_{k|t}$ over the horizon $T$, and $\Delta \mathbf{u} = \text{col}(\Delta u_{0,0|t}, \ldots, \Delta u_{N,0|t}, \ldots, \Delta u_{N,T-1|t}, \ldots, \Delta u_{N,T-1|t})$ is the optimization variable. Note that the prediction model used here is for the nominal configuration $\varsigma = \mathbf{1}$ where all units are active $\varsigma_i = 1$ for $i = 0, \ldots, N$. The reconfigurable model predictive controller integrates the optimal incremental input $\Delta u^\star_{i,k|t}$ to obtain the implemented input

$$u_i(t) = \varsigma_i u_i(t-1) + \Delta u^\star_{i,0|t} \tag{5.22}$$

for each unit $i = 0, \ldots, N$.

The cost function (5.21a) of the (5.21) is configuration-dependent. The term $\varsigma_i \bar{z}'_{i,k|t} Q_i \bar{z}_{i,k|t}$ penalizes the performance outputs of the configuration-dependent prediction model (5.20). For the outdoor unit $i = 0$, the performance output $\bar{z}_0 = z_0 - r_0$ is the difference between the discharge temperature $z_0 = \text{Td}$ and the reference discharge temperature $r_0 = \text{Td}_{\text{ref}}$, which maximizes energy efficiency. Since the outdoor unit is always active $\varsigma_0 = 1$, the model predictive control always regulates the discharge temperature. For the indoor units $i = 1, \ldots, N$, the performance outputs $\bar{z}_i = \text{col}(z_i - r_i, \xi_i)$ include the zone temperature tracking errors $z_i - r_i$, and their integrals $\xi_i$. If an indoor unit is inactive $\varsigma_i = 0$, then the zone temperature tracking error and integrated error for that indoor unit do not appear in the cost $\varsigma_i \bar{z}'_{i,k|t} Q_i \bar{z}_{i,k|t} = 0$. Thus, the reconfigurable MPC does not regulate the zone temperature of inactive zones.

The term $\lambda_i \Delta u'_{i,k|t} R_i \Delta u_{i,k|t}$ in the cost function (5.21a) penalizes changes $\Delta u_{i,k|t}$ to the inputs $u_{i,k|t}$. For the outdoor unit $i = 0$, this term penalizes changing the compressor frequency and outdoor fan speed. For the indoor units, $i = 1, \ldots, N$ this term penalizes changing the capacity command. The scalar $\lambda_i$ is defined as

$$\lambda_i = \begin{cases} 1 & \text{if } \varsigma_i = 1 \\ M & \text{if } \varsigma_i = 0 \end{cases} \tag{5.23}$$

where the "big-M" scalar $M$ is chosen to be large compared to the eigenvalues of the matrices $Q_i$, $R_i$, and $P_i$. The scalar $\lambda_i$ ensures the optimal incremental input $\Delta u^{\star}_{i,k|t} = 0$ is zero for inactive indoor units $\varsigma_i = 0$ [39]. Thus, the capacity command $u_{i,k|t} = u_{i,k-1|t} + \Delta u_{i,k|t} = 0$ to an inactive room $\varsigma_i = 0$ is zero since it is initially zero $u_{i,-1|t} = \varsigma_i u_i(t-1)$ and does not change $\Delta u_{i,k|t} = 0$ for $k = 0, \ldots, T-1$.

The terminal cost term $\varsigma_i \bar{x}'_{i,T|t} P_i \bar{x}_{i,T|t}$ penalizes the deviation of the augmented state $\bar{x}_{i,N|t}$ for $i = 0, \ldots, N$. When the $i$-th indoor unit is inactive $\varsigma_i = 0$, its terminal cost is zero $\varsigma_i \bar{x}'_{i,T|t} P_i \bar{x}_{i,T|t} = 0$. The terminal cost matrices $P_i$ are the Lyapunov matrices for a linear controller designed using the configuration-dependent model (5.20). The prediction model (5.21b)–(5.21d) used by the model predictive controller does not depend on the configuration $\varsigma$ of the ME-VCS. Instead, the optimal control problem (5.21) uses the model (5.20) for the nominal configuration $\varsigma = \mathbf{1}$. Due to the structure of the cost (5.21a), solving the optimal control problem (5.21) with the configuration-independent prediction model (5.21b)–(5.21d) is equivalent to solving the problem with the correct configuration-dependent model (5.20). For more details see [18].

A block diagram summarizing the online calculations performed for the reconfigurable MPC is shown in Fig. 5.9. The controller is similar to the non-reconfigurable controller (in particular, note that the same estimator is used in Fig. 5.5), with the exception of a new configuration signal $\varsigma(t)$ supplied by external logic. This configuration signal could originate from an occupant indicating that a particular zone is to be turned on ($\varsigma_i = 1$), or from a higher level controller that determines whether an indoor unit should become activated based, for example, on the magnitude of the zone error signal $e_i$. This configuration signal is provided to the optimizer, which is used to modify the cost function (5.21a) as previously described.

The reconfigurable MPC approach enables zones to be turned on or off, satisfying requirement (E.1). Additionally, because only one set of controller parameters needs

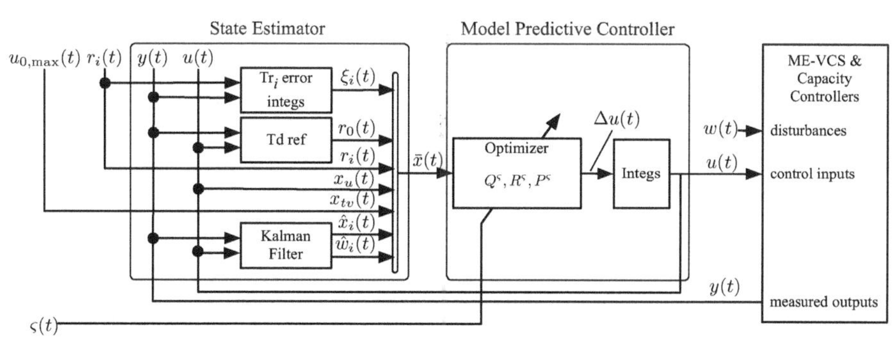


**Fig. 5.9** Block diagram for the reconfigurable MPC. A configuration signal indicating the zones to be turned on modifies the parameterized optimizer

to be stored, the memory usage requirements are reduced, which satisfies requirement (E.2). In order to validate the method, experiments are shown wherein external logic automatically deactivates a zone based on the zone temperature error.

#### 5.5.1.4 Experiment 3: Validation of Reconfigurable MPC

The control system must turn zones on or off automatically in order to regulate zone temperatures when the heat loads are lower than the minimum continuously available cooling capacity. This requires supervisory logic to compute the system configuration $\varsigma(t)$. In this section, a state machine is designed to detect low heat load conditions and reconfigure the ME-VCS automatically, and experiments are presented to validate its performance. The objective of this experiment is to empirically demonstrate reference tracking and constraint enforcement of the reconfigurable MPC controlling a switching system where the dwell times for a particular configuration are determined by a supervisory state machine [40]. The state machine logic considers the sign and magnitude of the zone temperature error signal and associated cooling command to determine when to switch a zone on or off. Specifically, if a zone is off and has become overheated by 1° C, then the state machine will activate that zone. If a zone is active and either the zone has become overcooled by 2.5° C, or the cooling capacity command for that zone has been low enough for long enough, then the zone is deactivated.

Regarding the latter condition, an integrator is used to determine the low actuator condition as follows: If $CC_i$ is less than 40%, then an integrator state $e_i(t)$ increases according to

$$e_i(t+1) = e_i(t) + (40 - CC_i) \tag{5.24}$$

Once the integrator has reached a predetermined value, then the zone is deactivated. This predetermined value is chosen so that if $CC_i$ has been at its low constraint of 20% for about 5 min, then condition (2) becomes true. This test on the capacity command will cause a zone to be deactivated even if good setpoint tracking is achieved, but requires a low capacity command to do so, which is the intended behavior. The parameters used in the state machine have been determined heuristically.

Using this state machine to automatically determine the configuration signal $\varsigma(t)$, an experiment is conducted wherein the heat load in zone 2 is reduced. Referring to Fig. 5.10, the vapor compression system is brought into steady-state operation with setpoints $r_1 = 21$, $r_2 = 25$° C and the boundary conditions are set to $Q_1 = Q_2 = 1800\ W$, and OAT = 35° C. At this initial condition, both zone loads are met in steady state with a compressor frequency of CF = 40 Hz, and capacity commands of $CC_1 = 100$ and $CC_2 = 60\%$. At $t = 10$ min, the heat load in zone 2 is decreased to 500 W (Fig. 5.10B-iii, red), which is a load that is about equal to the cooling provided when operated at $CC_{2,min} = 20\%$. As a result of the load step, the temperature in zone 2 is reduced, and the MPC decreases the associated actuator command to its minimum value in an attempt to raise the zone temperature back to the setpoint.

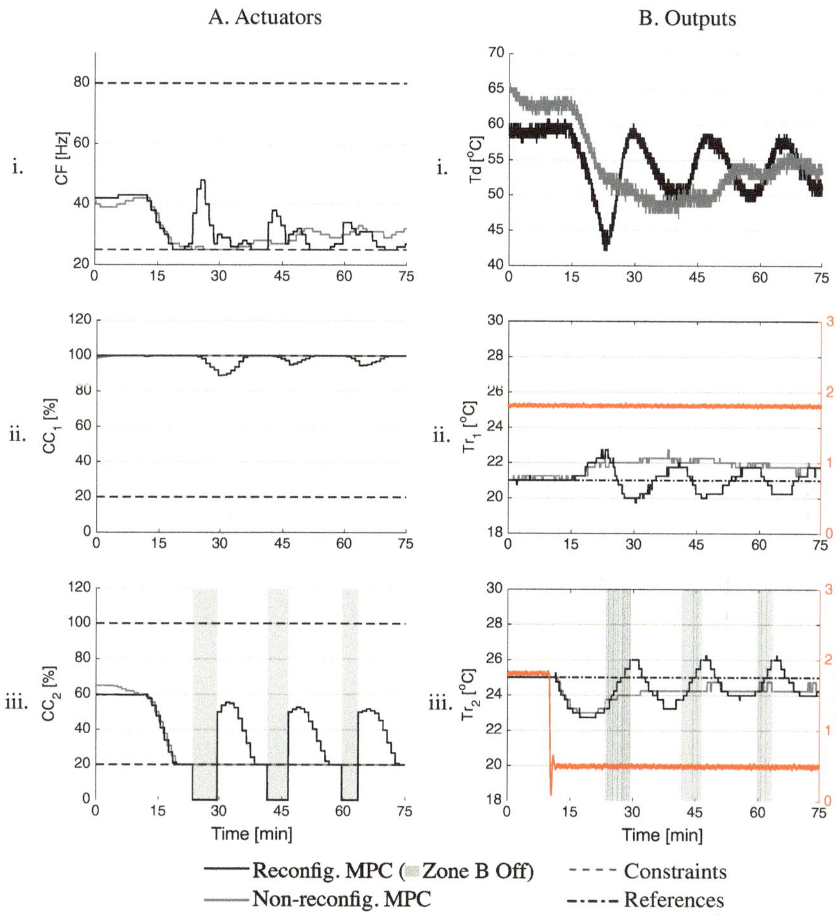


**Fig. 5.10** Reconfig MPC shuts off zones

After about 10 min, the integrator in the supervisory state machine has reached its predetermined value, and $\varsigma_2$ is set to 0, triggering reconfiguration of the controller. Zone 2 is automatically switched off, and the $CC_2$ command is set to 0 (Fig. 5.10A-iii, black), which closes the associated EEV. The zone temperature subsequently increases under the influence of the 500 W load until it becomes overheated by 1° C (Fig. 5.10B-iii, black), at which point the state machine sets $\varsigma_2$ to 1 and the controller is again reconfigured to turn zone 2 back on. This pattern is repeated under automatic control, establishing a cyclic response.

Note that the coupling inherent in the vapor compression system induces a periodic disturbance in zone 1 (Fig. 5.10B-ii, black) as zone 2 is switched on and off, which can only be partially rejected since the associated actuator $CC_1$ is saturated at its maximum value for large periods of the disturbance cycle (Fig. 5.10A-ii, black).

Despite the large imbalance in loads between the two zones, both zone temperatures cycle around their respective setpoints, and when averaged over multiple periods, both zones are shown to achieve their setpoints. The experiment demonstrates that the reconfigurable MPC can operate in combination with supervisory logic determining the on/off conditions.

Figure 5.10 also shows the same experiment for the case where the controller is not permitted to reconfigure. This non-reconfigurable MPC is shown in gray for the same test conditions. Since the controller cannot turn zone 2 off, $CC_2$ is driven to its lower limit (Fig. 5.10A-iii, gray) while $CC_1$ remains saturated at its upper limit (Fig. 5.10A-ii, gray). The corresponding temperatures for these zones settle to a nonzero steady-state error (about 1°C overheated in zone 1 (Fig. 5.10B-ii, gray), and about 1°C overcooled in zone 2 (Fig. 5.10B-iii, gray)). Note that since both zone temperature tracking errors are equally penalized in the cost function, the MPC controller selects compressor frequency commands that equally distribute the zone tracking errors despite both zone capacity commands being saturated.

The experiment presented in this section demonstrates that reconfigurable MPC permits individual zones to be selectively deactivated, satisfying extended requirement (E.1). Further, despite similar transient performance, reconfigurable MPC is much more efficient in memory storage requirements and tuning effort than traditional approaches where an MPC must be created and stored for each configuration and this satisfies extended requirement (E.2).

### 5.5.2 *Optimization Algorithms for Embedded Platforms*

While in general, building control systems may be supported by a fairly powerful computing infrastructure, at the equipment level, the computing platforms are significantly more limited. Thus, in order to solve the finite horizon optimal control problem (5.14) and (5.21), HVAC equipment control needs optimization algorithms that are fast, but also simple to allow for implementation in low computational power embedded platforms and for rapid validation.

In general, (5.14) and (5.21) can be formulated as convex parametric Quadratic Programs (pQPs) with linear equalities and bound constraints

$$\min_{\varphi} \frac{1}{2}\varphi' Q_u \varphi \tag{5.25a}$$

$$\text{s.t.}\ G_u\varphi = W_u + S_u\bar{x}(t) \tag{5.25b}$$

$$\underline{\varphi} \leq \varphi \leq \overline{\varphi} \tag{5.25c}$$

where $\varphi = \text{col}(\mathbf{x}_t, \Delta\mathbf{u}_t, \mathbf{y}_t)$, $\mathbf{x}_t = \text{col}(\bar{x}_{1|t}, \ldots, \bar{x}_{T|t})$, $\mathbf{y}_t = \text{col}(\bar{y}_{0|t}, \ldots, \bar{y}_{T-1|t})$, and $\underline{\varphi}, \overline{\varphi}$ are constructed from lower and upper bounds on inputs and outputs. Furthermore, the equality constraints in pQP (5.25) can be eliminated by applying basic concepts in linear systems theory [41] since the real degrees of freedom are only the inputs in $\Delta\mathbf{u}$. The result is the pQP

$$\min_{\Delta \mathbf{u}} \frac{1}{2}\Delta \mathbf{u}' Q_p \Delta \mathbf{u} + x(t)' C_c' \Delta \mathbf{u} \tag{5.26a}$$

$$\text{s.t. } G_c \Delta \mathbf{u} \leq W_c + S_c \bar{x}(t) \tag{5.26b}$$

where all the constraints are general linear inequalities.

Over the last few years, there has been a large activity in developing simple and fast optimization algorithms for solving (5.25), (5.26) as required for MPC [16, 42–49]. In particular, the results in Sects. 5.4 and 5.5 are obtained using the PQP method [16] that solves (5.26) by solving the dual problem

$$\min_{\varphi} \frac{1}{2}\varphi' Q_d \varphi + F_d' \varphi \tag{5.27a}$$

$$\text{s.t. } \varphi \geq 0 \tag{5.27b}$$

where $Q_d = G_p Q_P^{-1} G_p'$, $F_d = (G_p Q_p^{-1} C_p + S_p)\bar{x}(t) + W_p$, and then computing $\Delta \mathbf{u}^* = \Gamma_d \theta + \Upsilon_d \varphi^*$, where $\Upsilon_d = -Q_c^{-1} G_c'$, $\Gamma_d = -Q_c^{-1} C_c$. The PQP method solves (5.27) by iterating

$$[\varphi_{(\ell+1)}]_i = \frac{[(Q_d^- + \phi)\varphi_{(\ell)} + F_d^-]_i}{[(Q_d^+ + \phi)\varphi_{(\ell)} + F_d^+]_i}[\varphi_{(\ell)}]_i \tag{5.28}$$

where $[a]_i$ is the $i^{th}$ component of vector $a$, and $A^+$, $A^-$ are the positive and negative parts of a matrix $A$, respectively, until the convergence conditions are reached.

Another algorithm that is most commonly applied to (5.25) is based on the alternating direction method of multipliers (ADMM), where in order to obtain a simple iteration, a "copy" $\zeta$ of the optimization vector $\varphi$ is used to enforce bound constraints, and the equality between $\zeta$ and $\varphi$ is dualized in the augmented Lagrangian,

$$\min_{\varphi,\zeta} \frac{1}{2}\varphi' Q_u \varphi + \frac{\beta}{2}\|\varphi - \zeta - \lambda\|^2 \tag{5.29a}$$

$$\text{s.t. } G_p \varphi = K_p \tag{5.29b}$$

$$\underline{\varphi} \leq \zeta \leq \overline{\varphi} \tag{5.29c}$$

where $\lambda$ is the Lagrange multiplier vector and $\beta$ is a stepsize parameter that can be optimally determined [48, 49]. The ADMM algorithm iteratively adjusts $\lambda$ to seek the values of $\varphi$, $\zeta$ that solve (5.29) and such that at optimum $\zeta = \varphi$. In [50], the solution is achieved by the iterations

$$\varphi^{(k+1)} = \mathcal{M}(\zeta^{(k)} + \lambda^{(k)}) + \mathcal{N} K_E \tag{5.30a}$$

$$\zeta^{(k+1)} = \text{proj}_{\zeta \in [\underline{\varphi}, \overline{\varphi}]}(\varphi^{(k+1)} - \lambda^{(k)}) \tag{5.30b}$$

$$\lambda^{(k+1)} = \lambda^{(k)} + \zeta^{(k+1)} - \varphi^{(k+1)} \tag{5.30c}$$

where $\mathcal{M}, \mathcal{N}$ are matrices computed from the matrices in (5.29), and proj denotes the projection, i.e., in this case the clipping within the box determined by $\underline{\varphi}, \overline{\varphi}$.

Note that both iterations (5.28) and (5.30) involve only basic operations and no additional libraries for algebra operations are required. Hence, both methods are simple to code and to verify and do not impose stringent requirements on the embedded platform, thus enabling the usage of the algorithms in low-cost embedded platforms that are suitable for HVAC equipment. Despite their simplicity, it is shown in [16, 48] that both methods are relatively fast, and hence allow for solving the finite horizon optimal control problems (5.14), (5.21) at rates that are suitable for the target HVAC applications, where the control sampling period is usually on the order of 10–100 s.

### *5.5.3 Reducing Power Consumption for Demand Response*

In the demand response (DR) framework, utilities send signals to consumers to indicate a planned short-term increase in energy rates to incentivize reduced consumption during anticipated periods of expected peak demand. Because HVAC equipment typically consumes significant amounts of energy and users can tolerate a range of temperatures, a small reduction in individual cooling capacity can meaningfully reduce electrical consumption for the utility with minimal impact to zone occupants. Therefore, it is important for HVAC control systems to be responsive to DR requests for reduced consumption.

DR is often realized through zone setbacks. The reduction in energy consumption is an indirect consequence of the setpoint change, and predicting the exact energy reduction depends on equipment details, load conditions, and building construction. It is thus difficult to use this approach to reduce energy consumption by a known amount.

However, the model predictive control architecture presented includes a time-varying maximum constraint on the compressor frequency $\mathrm{CF}(t) \leq \mathrm{CF}_{\max}(t)$, which can be used to directly limit power consumption, provided a performance map (similar to Eqs. (4.18)–(4.19) in Chap. 4) is available relating compressor speed to power consumption. Suppose a DR event is initiated with the ME-VCS operating in steady state with a nonzero compressor frequency and the heat loads are such that the system does not duty cycle. The supervisory logic receives the DR signal, evaluates the current power consumption, and uses the compressor map to determine the new maximum compressor frequency that results in the desired reduction of energy consumption. This new $\mathrm{CF}_{\max}$ is provided to the model predictive controller and the compressor speed is subsequently limited. When a DR event limits compressor speed, the baseline control requirements will continue to be enforced, potentially giving up on setpoint regulation in order to enforce constraints. This strategy has the same effect of allowing zone temperatures to increase as in the zone setback method, but in the case of MPC, the reduction in power consumption is directly specified and reliably achieved.

## 5.6 Conclusion

A controller for a commercial vapor compression system must be capable of meeting an extensive set of control objectives. MPC is a design framework that enables satisfaction of both conventional controller requirements (constraint satisfaction, setpoint regulation, etc.), and also requirements not typically discussed in academic treatments of vapor compression system control, including prioritization, design separation of concerns and scalability. Through careful architecture considerations, MPC can satisfy requirements for set point regulation, disturbance rejection, energy efficiency, zone on/off, scalability, and ease of tuning. MPC also enables integration with higher level functionality at the building level. Finally, efficient numerical optimization algorithms have been developed that enable this technology to be deployed into commercial products. We expect this technology to see increased application to this industry in the coming years.

## References

1. Chua KJ, Chou SK, Yang WM (2010) Advances in heat pump systems: a review. Appl Energy 87(12):3611–3624. https://doi.org/10.1016/j.apenergy.2010.06.014
2. Hepbasli A, Kalinci Y (2009) A review of heat pump water heating systems. Renew Sustain Energy Rev 13(6–7):1211–1229
3. Liu M, Saman W, Bruno F (2012) Development of a novel refrigeration system for refrigerated trucks incorporating phase change material. Appl Energy 92:336–342
4. Tassou SA, De-Lille G, Ge YT (2009) Food transport refrigeration - approaches to reduce energy consumption and environmental impacts of road transport. Appl Therm Eng 29(8–9):1467–1477
5. Hawlader M, Dey PK, Diab S, Chung CY (2004) Solar assisted heat pump desalination system. Desalination 168:49–54
6. Slesarenko V (2001) Heat pumps as a source of heat energy for desalination of seawater. Desalination 139(1–3):405–410
7. Burger J, Holland H, Berenschot E, Seppenwodde J, ter Brake M, Gardeniers H, Elwenspoek M (2001) 169 kelvin cryogenic microcooler employing a condenser, evaporator, flow restriction and counterflow heat exchangers. In: 14th IEEE international conference on micro electro mechanical systems, pp 418–421
8. Aynur T (2010) Variable refrigerant flow systems: a review. Energy Build 42:1106–1112
9. Goetzler W (2007) Variable refrigerant flow systems. ASHRAE J 49(4):24–31
10. Qureshi T, Tassou S (1996) Variable-speed capacity control in refrigeration systems. Appl Therm Eng 16(2):103–113
11. He X-D, Liu S, Asada HH (1997) Modeling of vapor compression cycles for multivariable feedback control of HVAC systems. J Dyn Syst Meas Contr 119(2):183–191
12. He X-D, Liu S, Asada HH, Itoh H (1998) Multivariable control of vapor compression systems. HVAC&R Res 4(3):205–230
13. Amano K (2012) Heat pump apparatus and control method thereof, June 2012, eP Patent App. EP20,110,007,843. https://www.google.com/patents/EP2469201A2
14. Glattfelder AH, Schaufelberger W (2003), Control systems with input and output constraints. Springer
15. Qin S, Badgwell T (2003) A survey of industrial model predictive control technology. Control Eng Pract 11(7):733–764

16. Di Cairano S, Brand M, Bortoff SA (2013) Projection-free parallel quadratic programming for linear model predictive control. Int J Control 86(8):1367–1385
17. Maciejowski, JM (2002) Predictive control with constraints. Pearson Education Limited
18. Burns DJ, Danielson C, Zhou J, Di Cairano S, Reconfigurable model predictive control for multi-evaporator vapor compression systems. IEEE Trans Control Syst Technol, under review
19. Bendtsen J, Trangbaek K, Stoustrup J (2013) Plug-and-play control: modifying control systems online. IEEE Trans Control Syst Technol 21(1):79–93
20. Stoustrup J (2009) Plug & play control: control technology towards new challenges. Eur J Control 15(3):311–330
21. Pannocchia G, Rawlings JB (2003) Disturbance models for offset-free model-predictive control. AIChE J 49(2):426–437
22. Maeder U, Borrelli F, Morari M (2009) Linear offset-free model predictive control. Automatica 45(10):2214–2222
23. Wallace M, Das B, Mhaskar P, House J, Salsbury T (2012) Offset-free model predictive control of a vapor compression cycle. J Process Control 22(7):1374–1386
24. Bejan A (2006) Advanced engineering thermodynamics, 3rd edn. Wiley
25. Li P, Qiao H, Li Y, Seem J, Winkler J, Li X (2014) Recent advances in dynamic modeling of HVAC equipment. Part 1: equipment modeling. HVAC&R Res 20(1):136–149
26. Li P, Li Y, Seem J, Qiao H, Li X, Winkler J (2014) Recent advances in dynamic modeling of HVAC equipment. Part 2: modelica-based modeling. HVAC&R Res 20(1):150–161
27. Ishii M, Hibiki T (2011) Thermo-fluid dynamics of two-phase flow, 2nd edn. Springer
28. Laughman C, Qiao H, Aute V, Radermacher R (2015) A comparison of transient heat pump cycle models using alternative flow descriptions. Sci Technol Built Environment 21(5):666–680
29. Burns DJ, Bortoff SA (2016) Cooling capacity control for multi-evaporator vapor compression systems. In: 16th international refrigeration and air conditioning conference at Purdue
30. Borrelli F, Bemporad A, Morari M (2015) Predictive control for linear and hybrid systems. Springer
31. Burns DJ, Laughman CR, Guay M (2016) Proportional-integral extremum seeking for vapor compression systems. In: Proceedings of the 2016 American controls conference
32. Burns DJ, Laughman CR (2012) Extremum seeking control for energy optimization of vapor compression systems. In: International refrigeration and air conditioning conference at Purdue
33. Di Cairano S, Tseng HE, Bernardini D, Bemporad A (2013) Vehicle yaw stability control by coordinated active front steering and differential braking in the tire sideslip angles domain. IEEE Trans Control Syst Technol 21(4):1236–1248
34. Burns DJ, Weiss WK, Guay M (2015) Realtime setpoint optimization with time-varying extremum seeking for vapor compression systems. In: Proceedings of the 2015 American controls conference
35. Pannocchia G, Rawlings JB (2003) Disturbance models for offset-free model-predictive control. AIChE J 49(2):426–437
36. Maeder U, Borrelli F, Morari M (2009) Linear offset-free model predictive control. Automatica 45(10):2214–2222
37. Di Cairano S, Pascucci CA, Bemporad A (2012) The rendezvous dynamics under linear quadratic optimal control. In: Proceedings of the 51st IEEE conference on decision and control. IEEE, pp 6554–6559
38. Di Cairano S (2015) Model adjustable predictive control with stability guarantees. In: American control conference (ACC), July 2015, pp 226–231
39. Floudas CA (1995) Nonlinear and mixed-integer optimization. Oxford University Press
40. Bridgeman L, Danielson C, Di Cairano S (2016) Stability and feasibility of mpc for switched linear systems with dwell-time constraints. In: American control conference (ACC)
41. Di Cairano S, Bemporad A (2010) Model predictive control tuning by controller matching. IEEE Trans Autom Control 55(1):185–190
42. Necoara I, Suykens J (2008) Application of a smoothing technique to decomposition in convex optimization. IEEE Trans Autom Control 53(11):2674–2679

43. Giselsson P (2012) Execution time certification for gradient-based optimization in model predictive control. In: Proceedings of the 51st IEEE conference on decision and control, Maui, HI, pp 3165–3170
44. Kögel M, Findeisen R (2011) Fast predictive control of linear systems combining Nesterov's gradient method and the method of multipliers. In: Proceedings of the 50th IEEE conference on decision and control, Orlando, FL, pp 501–506
45. Richter S, Jones CN, Morari M (2012) Computational complexity certification for real-time mpc with input constraints based on the fast gradient method. IEEE Trans Autom Control 57(6):1391–1403
46. Patrinos P, Bemporad A (2014) An accelerated dual gradient-projection algorithm for embedded linear model predictive control. IEEE Trans Autom Control 59(1):18–33
47. Ghadimi E, Teixeira A, Shames I, Johansson M (2015) Optimal parameter selection for the alternating direction method of multipliers (admm): quadratic problems. IEEE Trans Autom Control 60(3):644–658
48. Raghunathan AU, Di Cairano S (2014) Optimal step-size selection in alternating direction method of multipliers for convex quadratic programs and model predictive control. In: Proceedings of the symposium on mathematical theory of networks and systems, pp 807–814
49. Raghunathan AU, Di Cairano S (2014) Infeasibility detection in alternating direction method of multipliers for convex quadratic programs. In: Proceedings of the 53rd IEEE conference on decision and control, pp 5819–5824
50. Raghunathan AU, Di Cairano S (2014) ADMM for convex quadratic programs: Q-linear convergence and infeasibility detection. arXiv:1411.7288

# Chapter 6
# Multi-zone Temperature Modeling and Control

**Justin Koeln, Bryan Keating, Andrew Alleyne, Christopher Price and Bryan P. Rasmussen**

## 6.1 Motivation

This chapter addresses a central obstacle to achieving higher efficiency, enhanced performance, and greater occupant comfort using advanced control approaches: *building zone temperature model development* [1]. With the proliferation of model-based control research and variable speed actuator components, such as variable air volumes (VAVs) and variable speed pumps and fans, the major obstruction to implementation of a diverse set of control architectures and predictive control strategies is the identification of a system model. In Chap. 7, zone temperature models are used to minimize energy consumption of air handling units (AHUs) using a distributed predictive control approach; in Chap. 8, a consensus-based algorithm for choosing optimal temperature set points is proposed. The success of both approaches hinges on development of control-oriented models that balance complexity and accuracy.

Depending on the application, building modeling approaches lie on a spectrum from general/data-based to specific/physics based. For models that require data for identification, the building dynamics must be sufficiently excited by system input signals. For models that employ physical parameter data, there are weak guarantees that the model will yield good predictions [1]. A complete physical model would include the effect of the HVAC system on temperature, relative humidity, and carbon dioxide. A model of the HVAC equipment dynamics must be generated as well, and its interaction with the building system may be difficult to capture because of the coupling between zone temperature, relative humidity, and heat transfer. Ideally, the entire system model would be easily reconfigurable in order to account for changes

J. Koeln (✉) · B. Keating · A. Alleyne
University of Illinois at Urbana-Champaign, Champaign, IL, USA
e-mail: justin.koeln@gmail.com

C. Price · B.P. Rasmussen
Texas A&M University, College Station, TX, USA



J.T. Wen and S. Mishra (eds.), *Intelligent Building Control Systems*,
Advances in Industrial Control, https://doi.org/10.1007/978-3-319-68462-8_6

to the building geometry and the effects of switching off and replacing components for routine maintenance. These requirements motivate a system modeling framework that is rooted in physical understanding of the system, but is general enough to account for unique features of buildings and flexible enough to be easily reconfigurable. In the next section, a single zone temperature modeling example is used to discuss three categories of modeling approaches: black-box, gray-box, and white-box.

## 6.2 Review of Modeling Approaches

Modeling approaches are often classified by their dependence on knowledge of governing physics and knowledge of system input/output data. White-box modeling constitutes gathering geometrical and material properties and using first-principles conservation equations to derive the dynamic equations that govern air flow and heat transfer. Gray-box modeling also uses these dynamic equations but relies on system input/output data to identify the parameters. Black-box modeling uses a combination of model structure, model order, and system identification methods to generate predictions based exclusively on a model constructed using knowledge of input/output data generated by an experiment.

To illustrate the basic ideas behind these modeling approaches and the basic challenges to system identification, Fig. 6.1 is used as a starting point for each modeling procedure. Figure 6.1 represents a simplified single-zone building with air temperature, $T_{air}$, and wall temperature, $T_{wall}$. These temperatures are coupled by a convective heat transfer coefficient, $h_i$. The wall temperature is affected by both convective heat

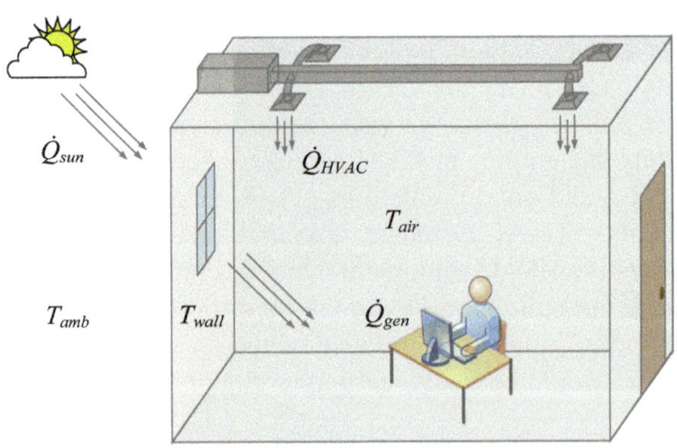


**Fig. 6.1** Simplified single-zone building model with well-mixed air at temperature, $T_{air}$, a thin wall with surface area, $A_{wall}$, at temperature, $T_{wall}$, an internal heat load, $\dot{Q}_{gen}$, a solar heat load $\dot{Q}_{sun}$, a heat flow representing HVAC heat removal or input, $\dot{Q}_{HVAC}$, and an outdoor air temperature, $T_{amb}$. Heat transfer between the wall and the room air is influenced by the inner wall convection coefficient, $h_i$, while heat transfer between the wall and the ambient is influenced by the outer wall convection coefficient, $h_o$

transfer from its inner surface to the building environment and from its outer surface to the ambient air at temperature $T_{amb}$. It is assumed that zone air is well mixed and that the wall's temperature profile is uniformly distributed across its thickness.

Starting with a white-box first-principles approach, the governing equations for this simplified model can be derived using an energy balance for the air and wall temperature states, $T_{air}$ and $T_{wall}$, where

$$(\rho V c_p)_{air}\dot{T}_{air} = h_i A_{wall}(T_{wall} - T_{air}) + \dot{Q}_{gen} - \dot{Q}_{HVAC}, \tag{6.1}$$

$$\begin{aligned}(\rho V c_p)_{wall}\dot{T}_{wall} = &- h_i A_{wall}(T_{wall} - T_{air}) \\ &- h_o A_{wall}(T_{wall} - T_{amb}) + \dot{Q}_{sun},\end{aligned} \tag{6.2}$$

and $(\rho V c_p)_{air}$ denotes the thermal mass of the air with density $\rho$, volume $V$, and specific heat $c_p$. The thermal mass of the walls is defined similarly with $(\rho V c_p)_{wall}$.

With a white-box approach, nominal geometry, specific heat, and heat transfer parameters are calculated and used to generate coefficients for a linear model. The white-box approach has several benefits: (1) models can be constructed from physical properties alone and render a time-consuming identification experiment unnecessary, (2) models of varying fidelity levels including PDEs or nonlinear/linear ODEs can be generated and validated against each other in software, and (3) the model is fully illustrative of the system's physical structure. The most significant drawback of the white-box approach is that using physical quantities to choose simplified parameter values may not result in good prediction performance. Additionally, for white-box models that are higher fidelity, such as discretized PDEs or nonlinear ODEs, the resulting model may be too complex for control design.

The modeling problem in Fig. 6.1 could alternatively be tackled using a gray-box approach where the system is described by the differential equations

$$\begin{bmatrix}\dot{T}_{air}\\ \dot{T}_{wall}\end{bmatrix} = \begin{bmatrix}a_{11} & a_{12}\\ a_{21} & a_{22}\end{bmatrix}\begin{bmatrix}T_{air}\\ T_{wall}\end{bmatrix} + \begin{bmatrix}b\\ 0\end{bmatrix}\dot{Q}_{HVAC} + \begin{bmatrix}v_{11} & 0 & 0\\ v_{21} & v_{22} & v_{23}\end{bmatrix}\begin{bmatrix}\dot{Q}_{gen}\\ \dot{Q}_{sun}\\ T_{amb}\end{bmatrix}, \tag{6.3}$$

where the parameter set, $(a_{11}, a_{12}, a_{21}, a_{22}, b, v_{11}, v_{21}, v_{22}, v_{23})$, can be identified by discretizing (6.3) and fitting the model parameters to experimental data. A key advantage of the gray-box approach is the combination of the constructive nature of the white-box model with specific data collected from the building. Additionally, the parameter set could take values that are more representative of the true system behavior given the uncertainties incurred by simplifying assumptions. Meanwhile, system structure is preserved with the states and parameter values having physical meaning that can be checked against the white-box calculations. A disadvantage of this approach is that system disturbance effects may not be known and may corrupt the parameter estimates. As with white-box modeling, the model structure may be too restrictive to be sufficiently representative of the system's dynamics. Additionally, achieving a sufficiently rich input signal to identify the many parameters for a multi-zone building may be difficult. Note that Chaps. 6 and 8 employ gray-box approaches to modeling.

A black-box approach to modeling the system in Fig. 6.1 would involve performing evaluation of candidate model structures using input/output data. Both linear (state space or ARX) or nonlinear (neural network) model structures could be tested to determine the simplest model with the strongest predictive capability. A significant benefit of black-box modeling is that the model order can be increased to improve prediction quality and ultimately compensate for the simplifying assumptions made during the white-box or gray-box modeling processes. However, black-box modeling does not easily accommodate changes to the building architecture; any significant change made to the building mandates a recalibration of the black-box model or a sacrifice of the predictive capability of the model. Black-box approaches are most appropriate when the plant is too complicated to represent with physics-based models. Chapter 7 applies black-box modeling to vapor compression systems, which are often difficult to model for control design.

While there are clear benefits and drawbacks to each class of models, the gray-box modeling has the distinct advantage of preserving system structure while allowing sufficient flexibility for identification. As will be shown in the sections that follow, resistance–capacitance (RC) modeling provides an intuitive parameterization of building dynamics that can be constructed using fundamental heat transfer relationships. The RC network has two additional interesting aspects for discussion: (1) it is shown to be a special case of a more generalized graph-based conservation model and (2) it can model not only the building zones, but the HVAC equipment as well. Thus, an entire HVAC system can potentially be constructed in a modular fashion, assembled into a state space parameterization, and identified using experimental data.

## 6.3 RC Network Thermal Modeling

Based on the trade-offs discussed in the previous section, the RC network electrical analogy modeling approach is the focus of the remainder of the chapter. When modeling the dynamic thermal behavior of a building, a capacitor represents the thermal capacitance of the air within a zone or the material of the building elements that constitute the walls, floors, ceilings, furniture, etc. The capacitor voltage represents the temperature of the associated thermal capacitance. The major underlying assumption is that the single temperature sufficiently represents the thermal state of the air or building elements and can be used to accurately capture the heat transfer associated with this element of the building. Resistors in the RC network represent the possible paths for heat to flow among the thermal capacitances. The location of the resistors in the network defines the assumed structure of heat transfer throughout the building and the associated resistance value governs the heat transfer rate between neighboring thermal elements.

In electrical systems, capacitor voltage is governed by the differential equation

$$C\frac{dV(t)}{dt} = I_{in}(t) - I_{out}(t), \tag{6.4}$$

where $C$ is the capacitance, $V(t)$ is the time-varying voltage, and $I_{in}(t)$ and $I_{out}(t)$ are the total time-varying currents going into and out of the capacitor. Following the analogy, the corresponding differential equation for the temperature of a thermal element is

$$C\frac{dT(t)}{dt} = Q_{in}(t) - Q_{out}(t), \tag{6.5}$$

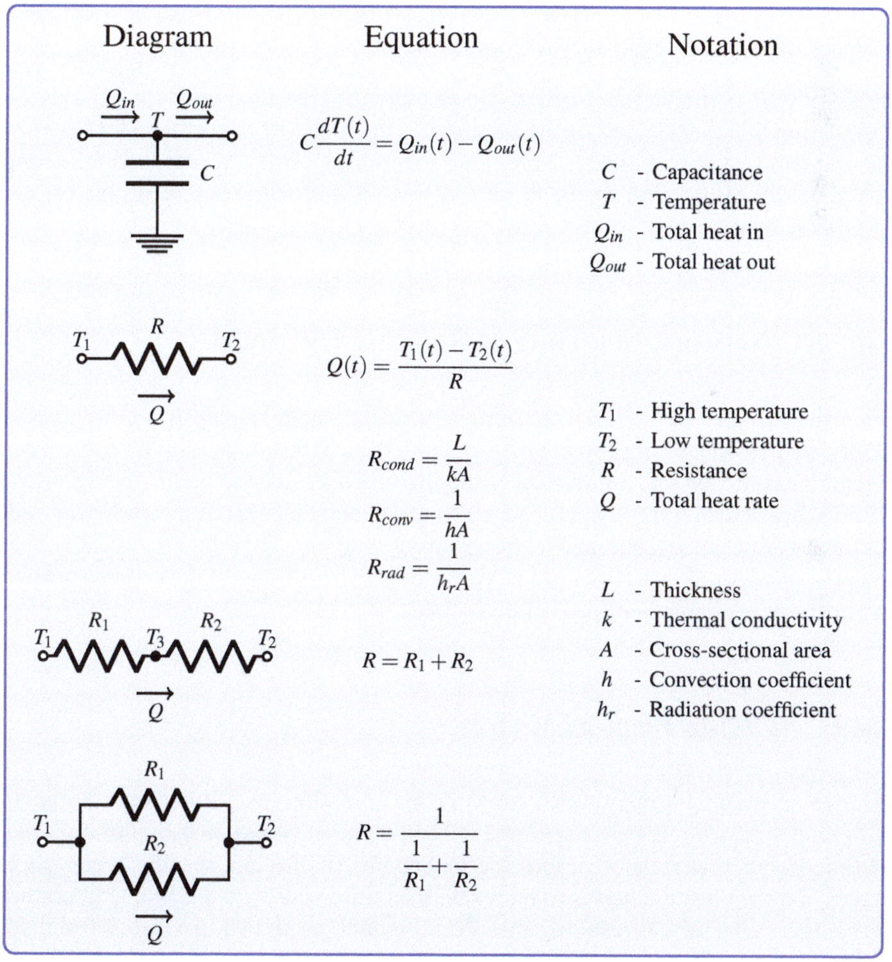


**Fig. 6.2** Basic RC elements and corresponding governing equations

where the temperature $T$ is analogous to voltage and the heat rate $Q$ is analogous to current. The capacitance $C$ now represents the thermal capacitance, typically defined as $C = \rho V c_p$, where $\rho$ is the material density, $V$ is the material volume, and $c_p$ is the material specific heat capacity.

Similarly, in electrical systems, the current flow through a resistor is governed by the static equation

$$I(t) = \frac{V_1(t) - V_2(t)}{R}, \tag{6.6}$$

where $R$ is the resistance and $V_1$ and $V_2$ are the voltages on either side of the resistor. Following the analogy, the corresponding static equation for the heat rate between thermal elements is

$$Q(t) = \frac{T_1(t) - T_2(t)}{R}, \tag{6.7}$$

where now $R$ is the thermal resistance separating the two temperatures $T_1$ and $T_2$. Calculating this thermal resistance depends on the mode(s) of heat transfer between the two temperatures. Figure 6.2 provides a summary of the RC network modeling fundamentals along with methods for calculating thermal resistances (see [2] for additional details).

## 6.4 RC Network Modeling of Building Dynamics

Extensive research has focused on capturing the thermal dynamics of buildings using the RC network modeling approach. In general, these efforts can be divided into three categories based on the complexity of the thermal dynamics to be modeled starting with (1) heat transfer through individual building elements, (2) heat transfer through the walls of a single zone to the surrounding ambient environment, and (3) heat transfer among multiple zones in a large building. The following subsections present the major developments in each of these areas.

### *6.4.1 Building Element Models*

Prior to developing a thermal RC network for an entire zone, it is natural to begin with modeling a single building element. Despite most walls consisting of multiple, heterogeneous materials of various thermal conductivities and specific heat capacitances, the simplest method is to represent the entire wall with a single temperature and single capacitance, resulting in a 2R1C model, as shown in Fig. 6.3 from [3]. Using slightly different notation from that used in [3], the value of the single capacitance, $C_{total}$, for a wall with $N$ layers, is simply the sum of the capacitances for each layer,

$$C_{total} = \sum_{i=1}^{N} C_i = \sum_{i=1}^{N} \rho_i V_i c_{p,i}. \tag{6.8}$$

Similarly, the total resistance, $R_{total}$, is the sum of the resistances for each layer,

$$R_{total} = \sum_{i=1}^{N} R_i = \sum_{i=1}^{N} \frac{L_i}{k_i A_i}, \tag{6.9}$$

where $L_i$ is the thickness, $A_i$ is the cross-sectional area, and $k_i$ is the material thermal conductivity of the $i^{th}$ layer. The two resistances, $R_{in}$ and $R_{out}$, are calculated as complementary fractions of the total resistance, where

$$R_{in} = \alpha R_{total}, \quad R_{out} = (1-\alpha) R_{total}, \tag{6.10}$$

where $\alpha$ is known as the "accessibility factor" and is calculated as

$$\alpha = 1 - \left( \frac{\sum_{i=1}^{N} R_i^* C_i}{R_{total} C_{total}} \right), \quad R_i^* = \sum_{j=1}^{N-1} R_j + \frac{R_j}{2}. \tag{6.11}$$

However, a series of simulation tests from [3] that compared low order models to a 21R20C benchmark model demonstrated that a slightly higher order model, 3R2C, provided significant model accuracy improvements over the 2R1C model with a tolerable increase in computational effort. Diagrams of the 2R1C, 3R2C, and 21R20C networks are shown in Fig. 6.4. Resistance and capacitance fractions were used to determine the distribution of the total resistance and capacitance among the various resistors and capacitors of the RC network. For the 3R2C model, these fractions are defined as $R_1 = f_1 R_{total}$, $R_2 = f_2 R_{total}$, $R_3 = f_3 R_{total}$, $C_1 = g_1 C_{total}$, and $C_2 = g_2 C_{total}$, where a subscript of 1 denotes the outermost element. Based on tests with rooms of both high and low relative thermal capacities, resistance fractions of $f_1 = 0.5$, $f_2 = 0.4$, and $f_3 = 0.1$ and capacitance fractions of $g_1 = 0.85$

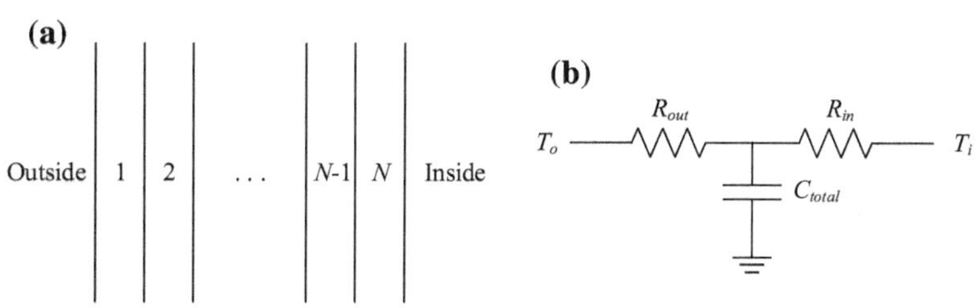


**Fig. 6.3** **a** Building element with $N$ layers, **b** simplest RC model with single capacitor and inner and outer resistors. Figure adapted from [3], copyright (2002), with permission from Elsevier

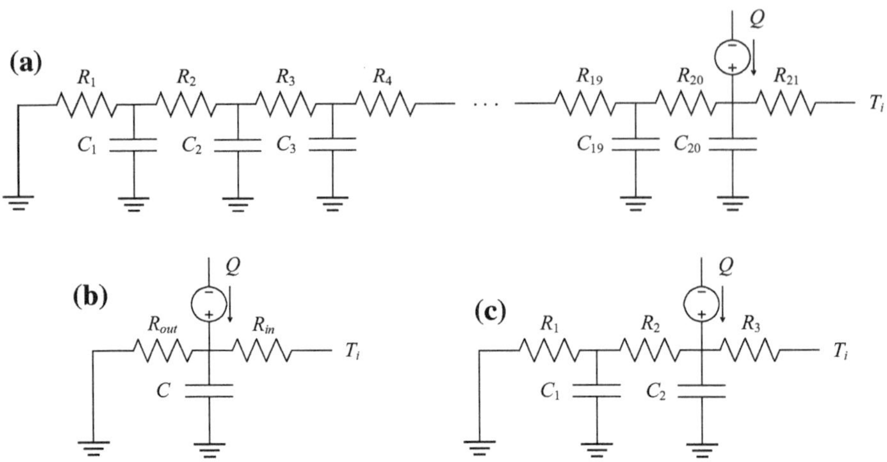


**Fig. 6.4** RC network models based on **a** 21R20C, **b** 2R1C, and **c** 3R2C model fidelities. Figure reprinted from [3], copyright (2002), with permission from Elsevier

and $g_2 = 0.15$, where found to be near optimal in most cases. This 3R2C model for building elements has become widely adopted. In fact, [4] has performed an extensive validation of this model fidelity, on a set of 45 different construction element types including internal partitions, floors, roofs, and external walls.

**Parameter Identification**

There are numerous ways to identifying the parameters of the resistors and capacitors based on the resources available. A common approach is to develop a high fidelity model of the building element using a finite-difference modeling approach [4] or a high-order RC network model [3] based on the material properties of the individual layers of the building element. This model serves as the true model from which the R and C values of the 3R2C network are identified. In [3], a sequential quadratic programming (SQP) approach is used to find the optimal model parameters. The square root of the sum-square error between the internal air temperature of the high fidelity and 3R2C model was minimized for step changes in external temperature and internal surface heat rate. In [4], an active-set approach is used to solve the nonlinear program for identifying the optimal values for resistance and capacitance fractions. The root-mean-square error of the inner and outer surface temperatures was minimized for sinusoidal excitation of internal and external air temperatures. A very similar approach is used in [5].

Additional approaches include identifying the parameters based on either experimental data [6] or data from detailed modeling software like EnergyPlus or TRNSYS [7]. Alternatively, [8] suggests a genetic algorithm approach to better handle the non-smooth and multimodal characteristics associated with nonlinear optimization. Unlike the previous methods, which are all based on optimizing time series responses, [9] uses a genetic algorithm to determine optimal parameters to best match frequency domain characteristic between the 3R2C model and the theoretical frequency characteristics of heat transfer through a multilayer building element.

The appropriate method for determining the RC model parameters largely depends on the resources available. If measured building data is available, data-based parameter estimation methods are ideal because features of the true system often neglected in even a detailed model, such as air leaks and complex geometries, will be captured. Often experimental data is unavailable, but a detailed model developed in software such as EnergyPlus or TRNSYS can be readily used as a substitute. The fact that these models provide easy access to many temperatures and heat rates that could not be measured on a real building allows for even better identification of the reduced model parameters. Finally, if building data and detailed models are unavailable, then modeling building elements based on material properties and dimensions can be a valuable place to start with minimal effort. Details for modeling many types of building elements can be found in [10].

### 6.4.2 Zone Models

From the development of RC network models for individual building elements, there are two main approaches to modeling an entire thermal zone, referred to as: (1) the lumped approach and (2) the constructive approach. While a lumped approach can be used to model multi-zone buildings, this approach does not directly utilize the building elements models discussed previously and instead attempts to directly model the zone air and wall temperatures with an appropriate, typically very low, number of resistors and capacitors. For this reason, the lumped approach will not be covered in this chapter; details can be found in [11–14]. Alternatively, the constructive approach systematically builds up a zone model based on the individual building element models. Thus it is recommended that single zone, and multi-zone, building models follow a constructive approach where, in general, 3R2C models are developed for each building element and then assembled to generate the complete RC network model for the zone. This provides an advantageous balance between simplicity, physical insight (transparency), and computational demand. As an example, this approach has been used in [15], as shown in Fig. 6.5 for a single zone, as well as many of the multi-zone modeling efforts presented in the following subsection.

With the 3R2C model used to capture the thermal dynamics of building elements, the following presents the details of modeling additional features that influence heat transfer within a typical room.

- *Partitions, exterior walls, floors, and roofs*—Despite their difference in construction, internal partitions, external walls, floors, and roofs are often modeled using the 3R2C model. As observed in [4], single layer elements, or symmetric multilayer elements, have $R_1 = R_3$ and $C_1 = C_2 = C_{total}/2$. Additionally, $R_2$ tends to dominate for elements with many layers.

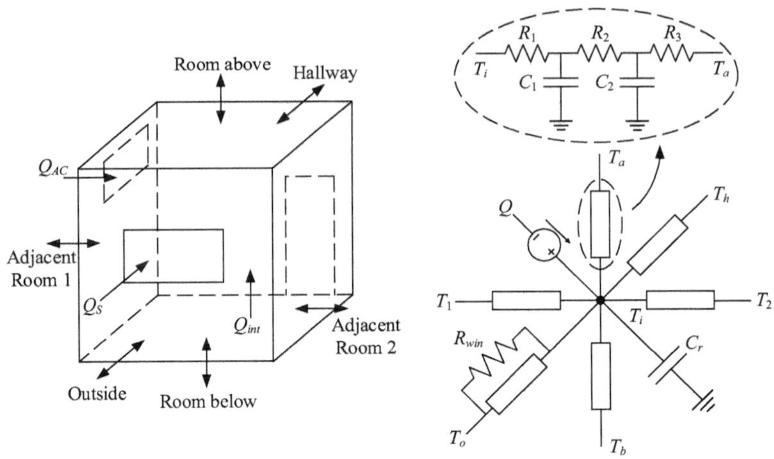


**Fig. 6.5** Example of constructive approach for developing single zone model from individual building element models. Figure reprinted, with permission, from [15], copyright (2002), IEEE.

- *Zone air* - Zone air is usually assumed to be well mixed and is represented by a single capacitor. The capacitance of the air is calculated as $C_{air} = \rho V c_p$, where air properties are usually taken at a nominal humidity.
- *Internal mass*—The internal mass refers to any internal structure, furniture, carpet, etc. that may be found within the zone. The need to specifically include the internal mass varies greatly depending on the items in the zone. Thus the internal mass can be assumed negligible, modeled with a single capacitance connected to the zone air capacitor via a single resistor, or even modeled as a pair of capacitances, as is done in [8]. In general, determining the values for these capacitances and resistances can be quite difficult and most authors suggest data-based system identification methods for determining these parameters [8, 9, 16, 17].
- *Windows*—Windows are assumed to have negligible thermal mass and thus are modeled as a resistance, in parallel with the resistance of the corresponding wall. This resistance is calculated as $R_{win} = \frac{1}{UA}$, where $A$ is the area of the window and $U$ is the overall heat transfer coefficient that includes the convective and conductive resistances to heat transfer. Values of $U$ are provided in [10] and generally range from $U = 0.7 - 4\ W/(m^2K)$, depending on the construction. A more detailed approach is provided in [18], which includes the effect of blinds.
- *Heat loads*—Heat loads are modeled as ideal, time-varying current sources. A typical zone has a number of heat loads including zone occupancy, electrical equipment, lighting, and solar radiation. Detailed methods for computing the heat gain due to solar radiation are provided in [13, 19]. Extensive reviews of heat load calculations for a wide variety of sources are presented for residential and nonresidential buildings in [20] and Chap. 30 of [10], respectively.

- *HVAC system*—From a building thermal zone modeling perspective, the HVAC system is typically modeled as an ideal current source/sink. Many building control studies aim to determine the optimal heat rates for the HVAC system in each zone of a building based on the thermal dynamics, heat loads, and ambient conditions [21]. More detailed models that capture the dynamics of various HVAC systems are presented in Sect. 6.6.
- *Infiltration/exfiltration*—Infiltration is the flow of outdoor air into a zone through cracks and other unintentional, often unmodeled, openings in the walls of the zone. Exfiltration denotes the same type of air flow but in the opposite direction. Chapter 29 of [10] presents a method, along with typical values, for calculating the thermal effect of infiltration for a range of general construction qualities.

**Parameter Identification**

There are numerous ways to determine the appropriate values for the R and C parameters of the RC network model. Clearly, if all the building element parameters have already been identified, then these values can be used together to model the zone dynamics. However, one must be aware of the possible error propagation that can occur when combining multiple imperfect models, as discussed in [1].

Alternatively, the parameters of an RC network model can be estimated online using data-driven parameter estimation. An Unscented Kalman Filter (UKF) approach is presented in [22] that demonstrates the ability to use less than 2 weeks of training date to determine parameter estimates capable of making reliable 48-hour predictions for a five-zone building model simulated using EnergyPlus. A very difficult and common problem when identifying these parameters is a lack of sufficient excitation of the system [14, 15, 23, 24], which resulted in very accurate estimation of some parameters and poor estimation of others. However, as a whole, the identified parameters led to excellent predictions that closely matched the results from the EnergyPlus simulation. Additionally, [8] presents a genetic algorithm-based method for determining the parameters of a 2R2C network model for the internal mass inside a zone including internal structures, furniture, and carpet. Finally, a recursive least squares approach is used in [25] to estimate the thermal load of a zone and the corresponding HVAC system.

Regardless of the algorithm used for identification, achieving persistent excitation of the building system is a significant challenge that directly affects the accuracy of the obtained parameters and the resulting system model. Highlighting this issue, [15] compared the predictive capability of an RC network model with parameters obtained from system identification based on three different sets of data from a physical building. A model obtained using Data Set A appeared to provide acceptable prediction accuracy (within about $1^o C$) when validated with Data Set B, but resulted in over a $10^o C$ prediction error when validated with Data Set C. The room temperatures in Data Set A were nearly the same throughout the entire data set, resulting in poor prediction in cases where the room temperatures differed. Because of the inability to sufficiently perturb the thermal state of the building, along with the occupant discomfort that can result for even mild perturbations, most model parameters for

RC network models are often identified using detailed models, like ones developed using EnergyPlus, as a surrogate for the physical building.

### 6.4.3 Multi-zone Modeling

With most buildings consisting of more than one zone, the RC network modeling framework must be capable of capturing multi-zone thermal dynamics. A key benefit of the RC network approach is that two RC networks, developed for neighboring zones, can be combined into a single, larger, network. Figure 6.6 provides an example from [26] of a four-zone building and its respective RC network.

While numerous modeling efforts have taken a similar approach, there is a rapid growth in the number of resistors and capacitors required to model an entire multi-zone building. In fact, for the four-zone building in Fig. 6.6, 37 capacitors and 51 resistors are needed to represent the thermal dynamics of the building. Moreover, the 15-room building in [27] required a 454 capacitor RC network. Thus, despite using a simplified model for each building element, the number of resistors and capacitors, and their corresponding parameters can grow very large for even moderately sized multi-zone buildings.

Numerous programs exist for high fidelity modeling and simulation of entire buildings, such as EnergyPlus, ESP-r, IDA ICE, and TRNSYS (see [28] for a complete review). However, the models used by these programs are highly nonlinear, require extensive user input, and are too complex for fast simulation and control development (EnergyPlus uses between 6 and 18 states per wall layer using its conduction transfer function algorithm [29]). While these modeling tools are certainly valuable, RC network models can extend their value by providing a simpler, more transparent, and more computationally efficient framework that captures the basic thermal behavior of the building. Thus, tools capable of semi-autonomously generating RC

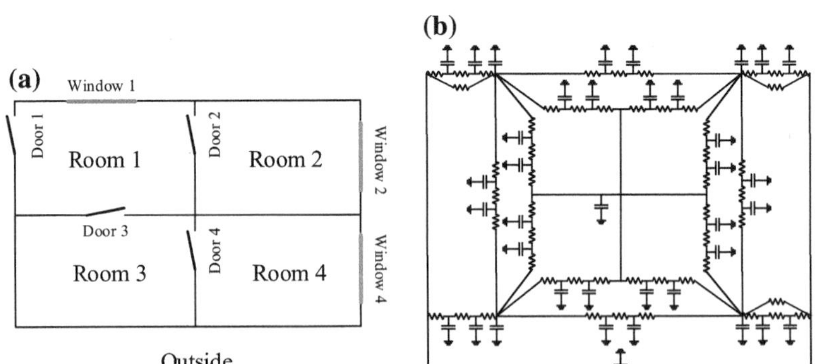


**Fig. 6.6** Example of multi-zone model developed using 3R2C building element models. Figure reprinted, with permission, from [26], copyright (2010), IEEE

network models based on these high fidelity models can expedite the development of RC network models, enabling the many control-oriented end uses for these models. The Building Resistance-Capacitance Modeling (BRCM) Toolbox [30] is a MATLAB toolbox specifically designed to convert detailed building models developed in EnergyPlus to reduced order RC network models to be used in model-based control. Additional details can be found in [30, 31].

While RC network models provide a significant reduction in model complexity compared to the high fidelity models developed used programs like EnergyPlus, RC networks can still be too complex for use in model-based control of large multi-zone buildings, especially when online optimization is required. Thus, prior to controller development, further model reduction steps are often taken as discussed in Sect. 6.7.1. These model reduction approaches utilize the detailed RC network to develop a reduced RC model that preserves structure and physical meaning. The detailed RC network models are also very useful for determining control architecture. As discussed in Sect. 6.7.2, a single centralized control is not always the best control architecture, and the structure of the detailed RC model can be used to determine appropriate ways of partitioning the building for a decentralized control approach.

**Parameter Identification**

For larger buildings with multiple dynamically coupled zones, parameter identification becomes a significant challenge. In [23], a Building Data-Dependent Identification (BDDI) algorithm is presented that dynamically improves the experimental design for system excitation to generate higher quality data and thus better parameter estimates. Along with the active system ID algorithm for generating the excitation signals, parameter estimation is performed online using prediction error identification methods. A 13-zone RC model with 52 capacitors and a total of 150 parameters provides a proof-of-concept, demonstrating the need for well-designed excitation signals to achieve proper parameter identification. This approach is extended in [32], where a large building is decomposed into smaller building zones allowing for decentralized parameter identification of each zone.

Approaches using Extended Kalman Filter (EKF) and Unscented Kalman Filter (UKF) techniques are developed in [33], which simultaneously estimate the parameters and states of an RC network model. These approaches are demonstrated using experimental data for a single zone, but can be scaled to a multi-zone building. The similarities in the physical properties such as wall materials and thickness were exploited in order to reduce the number of independent parameters to be estimated. While both approaches were able to sufficiently estimate the zone and wall temperature, the UKF approach proved to be more accurate.

## 6.5 Generic Graph-Based Modeling Framework

In the RC network modeling framework, an element with thermal capacitance is represented using electrical capacitors and the paths for heat transfer between these thermal elements are represented by resistors. A graph-based modeling framework is based on a very similar formulation. For a graph, a vertex (or node) is equivalent to a capacitor and represents an element capable of storing energy. An edge is similar to a resistor and denotes a path for energy transfer between vertices.

*The main advantage of the graph-based approach is the flexibility* introduced by the algebraic relationship that defines the heat rate, or in general, the power flow, along each edge. For a resistor, the heat rate is a function of the temperature difference and the resistance, as defined in (6.7). In the graph-based framework, the heat rate is still dependent on the neighboring temperatures, but now has the more general nonlinear form

$$Q(t) = f(T_1(t), T_2(t), u(t), \phi), \tag{6.12}$$

where $T_1$ and $T_2$ still denote the high and low temperatures driving the heat transfer, $u$ is a system input capable of augmenting the heat transfer (such as the speed of a fan), and $\phi$ denotes a vector of time-invariant parameters used to establish the relationships between these variables. The following section presents the generic formulation for graph-based model development followed by some application-specific extensions.

Following the general formulation outlined in [34], let the oriented graph $\mathcal{G} = (V, E)$ represent the structure of the system dynamics with the set of $N_v$ vertices $V = \{v_i\}$ and set of $N_e$ edges $E = \{e_j\}$. The orientation of each edge, representing the assumed direction of power flow (heat transfer), is used to formulate the differential equations for the system and does not prevent power from flowing in the opposite direction during simulation. Each of these oriented edges is defined by the tail and head vertices as $e_j = (v_j^{tail}, v_j^{head})$, where power flows from the tail vertex $v_j^{tail}$ to the head vertex $v_j^{head}$ along edge $e_j$. The associated power flow is denoted $P_j$. Each vertex $v_i$ has an associated state, $x_i$, which represents the amount of energy stored at that vertex. The dynamic of each vertex satisfies the energy conservation equation

$$C_i \dot{x}_i = \sum_{e_j \in E_i^{in}} P_j - \sum_{e_j \in E_i^{out}} P_j, \tag{6.13}$$

where $C_i > 0$ is the energy storage capacitance of the vertex and $E_i^{in}$ and $E_i^{out}$ represent the sets of edges oriented into and out of the vertex.

The most general form for the algebraic relationship between the power flow along edge $e_j$ and the states $x_j^{tail}$ and $x_j^{head}$ is

$$P_j = f_j(x_j^{tail}, x_j^{head}, u_j, \phi_j), \tag{6.14}$$

where $u_j$ is the actuator input associated with the edge and $\phi_j$ are the time-invariant parameters relating the variables. Common, increasingly simple, forms for this power flow relationship are the nonlinear, input affine form

$$P_j = f_j(x_j^{tail}, x_j^{head}, \phi_j) + g_j(x_j^{tail}, x_j^{head}, \phi_j)u_j, \tag{6.15}$$

where $f_j$ and $g_j$ are nonlinear, the bilinear form

$$P_j = \bar{f}_j(x_j^{tail}, x_j^{head}, u_j, \phi_j) + \bar{g}_j(x_j^{tail}, x_j^{head}, \phi_j)u_j, \tag{6.16}$$

where $\bar{f}_j$ and $\bar{g}_j$ are linear, and the linear form

$$P_j = a_j x_j^{tail} + b_j x_j^{head} + c_j u_j + d_j, \tag{6.17}$$

where $\phi_j = [a_j\ b_j\ c_j\ d_j]^T$. For a resistor $\phi_j = [\frac{1}{R_j}\ \frac{-1}{R_j}\ 0\ 0]^T$.

Figure 6.7 shows a graph of an example system used to identify key components. In addition to the capacitive vertices, governed by (6.13), ideal current sources and current sinks can be represented as vertices, as indicated by the vertices with dashed lines. Current sources can be used to represent heat loads inside a zone. The time-varying power flow along these edges act as disturbances to the system and are not considered part of the graph $\mathcal{G}$, denoted by the dashed edges. Additionally, current sinks can be used to represent the ambient environment surrounding a building, where the state of the vertex is a time-varying disturbance representing the ambient temperature and the edge entering this sink vertex can have any of the power flow relationships listed above.

With the graph $\mathcal{G}$ representing the structure of energy storage and power flow throughout the system, the matrix $M = [m_{i,j}]$, referred to as the incidence matrix of $\mathcal{G}$ [35], numerically captures this structure where

$$m_{i,j} = \begin{Bmatrix} +1 & v_i \text{ is the tail of } e_j \\ -1 & v_i \text{ is the head of } e_j \\ 0 & \text{else} \end{Bmatrix}. \tag{6.18}$$

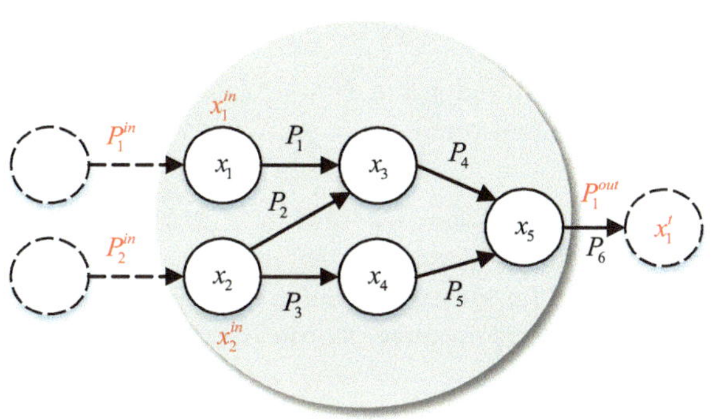


**Fig. 6.7** Notional system exemplifying the graph-based power flow representation with key power flows and states high-lighted in red. Dashed lines indicate elements that serve as disturbances to the system

Then, based on (6.13), the subsystem dynamics are captured by

$$\begin{bmatrix} C\dot{x} \\ \dot{x}^t \end{bmatrix} = -MP + \begin{bmatrix} D \\ 0 \end{bmatrix} P^{in}, \tag{6.19}$$

where $x$ are the states of the dynamic vertices, $x^t$ are the states of the sink vertices, $C = diag([C_i])$ is a diagonal matrix of the capacitances of the dynamic vertices, $P$ are the power flows along the edges of $\mathcal{G}$, $P^{in}$ are the source power flows entering the system, and $D = [d_{i,j}]$ is a matrix where

$$d_{i,j} = \begin{Bmatrix} 1 & v_i \text{ is the head of } P_j^{in} \\ 0 & \text{else} \end{Bmatrix}. \tag{6.20}$$

Since $x^t$ are disturbances to the subsystem, not states, $M$ is partitioned as $M = \begin{bmatrix} \bar{M} \\ \underline{M} \end{bmatrix}$, resulting in

$$C\dot{x} = -\bar{M}P + DP^{in}. \tag{6.21}$$

For the input affine power flow relationship from (6.15), the vector of power flows for the entire system is represented as

$$P = F\left(x, x^t\right) + G\left(x, x^t\right)u, \tag{6.22}$$

where $F\left(x, x^t\right) = \left[f_j\left(x_j^{tail}, x_j^{head}\right)\right]$ and $G\left(x, x^t\right) = diag\left(\left[g_j\left(x_j^{tail}, x_j^{head}\right)\right]\right)$. Thus the dynamics for the system are

$$C\dot{x} = -\bar{M}F\left(x, x^t\right) - \bar{M}G\left(x, x^t\right)u + DP^{in}. \tag{6.23}$$

Alternatively, for the linear power flow relationship from (6.17) when $a_j = -b_j$ and $d_j = 0$, the vector of power flows for the entire system is represented as

$$P = \mathcal{A}M^T \begin{bmatrix} x \\ x^t \end{bmatrix} + \mathcal{B}u, \tag{6.24}$$

where $\mathcal{A} = diag(a_j)$ and $\mathcal{B} = diag(c_j)$. This form can be used to represent RC network systems where $a_j = \frac{1}{R_j}$. The dynamics for the linear system are

$$C\dot{x} = -\bar{M}\mathcal{A}\bar{M}^T x - \bar{M}\mathcal{B}u + DP^{in} - \bar{M}\mathcal{A}\underline{M}^T x^t. \tag{6.25}$$

With $C_i > 0$ for all vertices and $a_j > 0$ for all edges, (6.25) is a stable system since $-C^{-1}\bar{M}\mathcal{A}\bar{M}^T$ is negative definite, i.e., $\text{Re}\lambda_i < 0$ for all eigenvalues $\lambda_i$ of $A$ [36].

## 6.6 Graph-Based Modeling of HVAC Equipment

### 6.6.1 *Forced-Air Systems*

Multi-zone buildings are often air-conditioned using variable air volume (VAV) systems. In [37], a bilinear model is developed for a VAV system and model predictive control is used to effectively control zone temperatures within desired upper and lower limits while minimizing electrical power consumption. Figure 6.8 shows the VAV system along with the corresponding graph model, which captures the structure of the system. Both the VAV system and building zone dynamics can be represented using the same modeling framework.

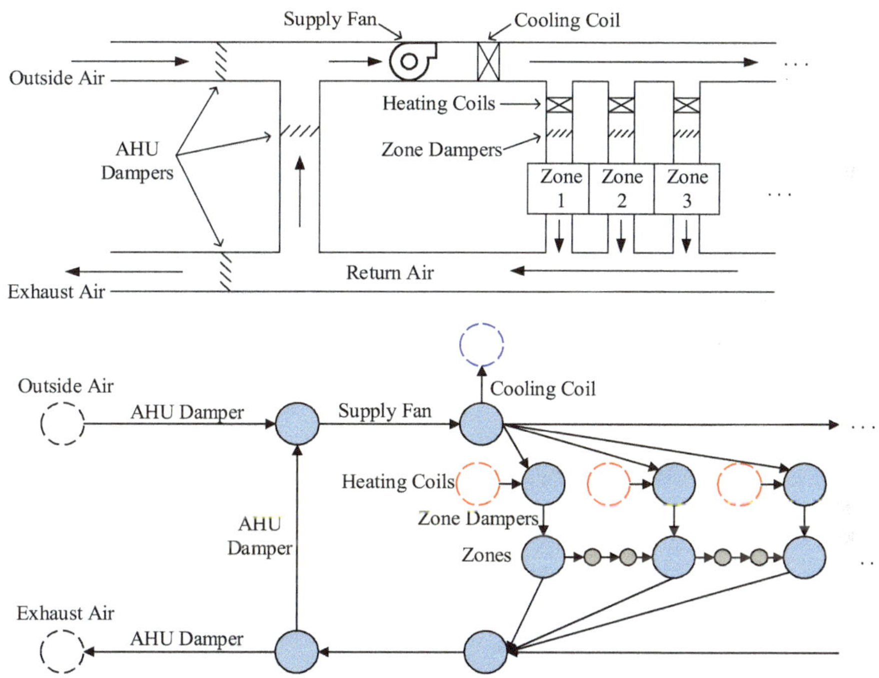


**Fig. 6.8** VAV air-conditioning system from [37] with the corresponding graph where blue vertices represent temperature states of the air, the gray vertices represents the temperature state of the walls between zones, and the dashed red and blue vertices represent the heat gain and loss to the air from the heating and cooling coils, respectively. Note that if the dynamics of the heating and cooling coils were also to be consider, these components can also be represented as a graph, typically with a vertex corresponding to the temperature of the wall of the heat exchanger, and easily integrated into the overall model. Figure reprinted, with permission, from [37], copyright (2011) IFAC

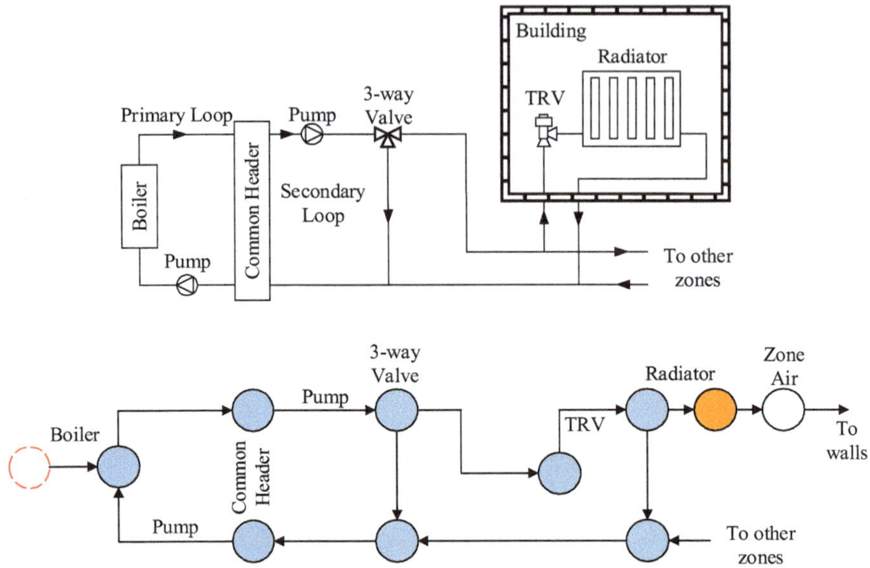


**Fig. 6.9** Radiator-based heating system from [38] with the corresponding graph where blue vertices represent temperature states of the fluid, the orange vertex represents the temperature state of radiator wall, the white vertex represents the temperature state of the zone air, and the dashed red vertex represents the heat gain to the fluid from the boiler. Figure reprinted from [38], copyright (2004), with permission from Elsevier

## 6.6.2 *Pumped-Liquid Systems*

In [38], a heating system consisting of boilers and radiators is developed for a multi-zone building. A schematic of the system is shown in Fig. 6.9. While an RC model is presented for an individual zone, based on a number of assumptions, the dynamics of the boilers, pumped water loops, and radiators are not represented in the RC network. However, following the same set of modeling equations presented in [38], a graph-based model can readily be constructed.

One of the key features of the RC network and graph-based modeling approaches is the scalability. Thus, complex multi-chiller, multi-zone hydronic systems, such as the chilled water plants presented in [39, 40], can be modeled using the same procedure. Accordingly, [40] develops a graph-based approach, slightly different from the one presented in this chapter, to capture the mass and energy flow throughout the system shown in Fig. 6.10. Using the approach outlined in Sect. 6.5, the graph for this system is also shown in Fig. 6.10. The corresponding equations to model this system can be found in [40]. Graph-based modeling of a similar thermal fluid system has been validated experimentally in [41].

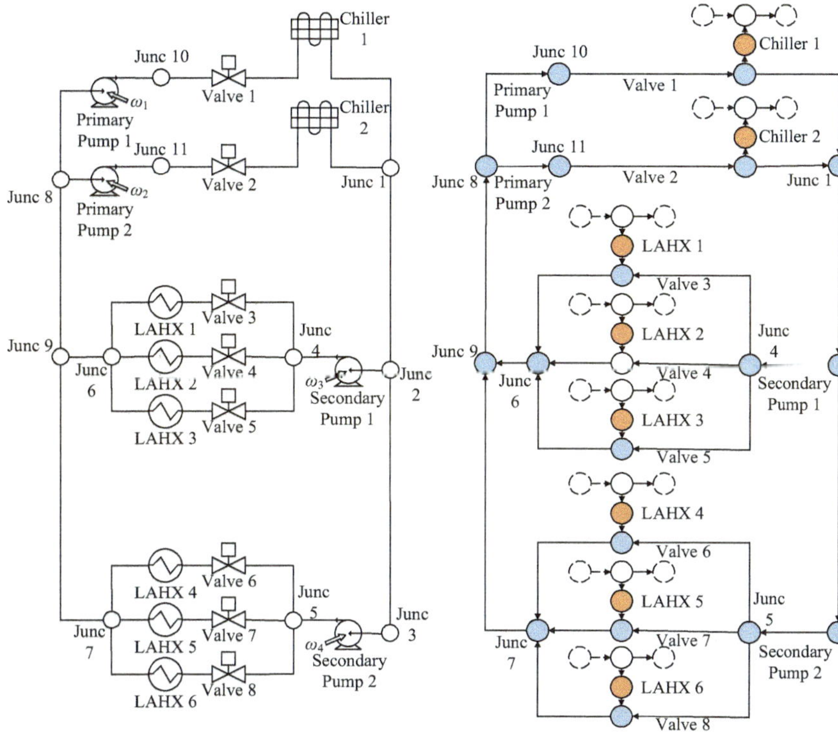


**Fig. 6.10** Chilled water system from [40] represented as a graph where blue vertices represent temperature states of the fluid, orange vertices represent temperature states of heat exchanger walls, and the white vertices represent fluid temperatures on the secondary sides of the heat exchangers. Figure reprinted, with permission, from [40], copyright (2009), ASME

## *6.6.3 Vapor Compression Systems*

Multi-zone vapor compression systems (VCSs), often referred to as multi-evaporator or variable refrigerant flow (VRF) systems [42], shown in Fig. 6.11, are gaining significant market share and are the focus of extensive research efforts due to their associated high efficiencies compared to conventional air duct systems [43–50]. A detailed review of modeling efforts, often used to support these control developments, is provided in [51].

Compared to single-phase pumped-liquid or forced-air systems, the phase change and highly nonlinear dynamics of VCSs provide an exacting test of the capabilities of graph-based modeling. As with the pumped-liquid and forced-air systems, capturing the mass and energy conservation through hydraulic and thermal graphs is required, but now there is a significantly higher degree of coupling between the two graphs due to the relationship between pressure and temperature in the two-phase fluid state. For brevity, the details of the graph model development are not included in this chapter, but can be found in [52], which are further developed and validated in [53].

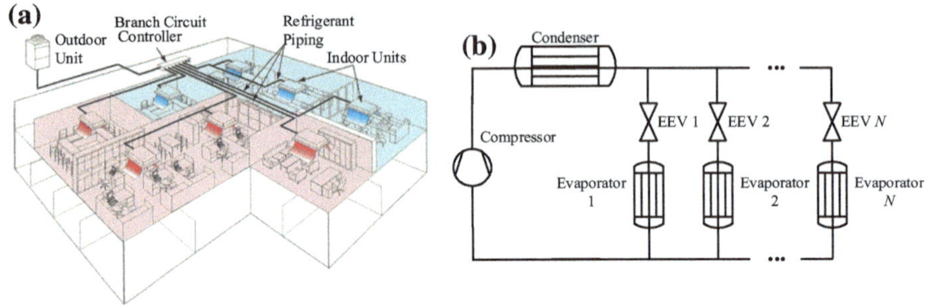


**Fig. 6.11** Schematic of a multi-evaporator vapor compressor system used to cool/heat a multi-zone building along with a simplified schematic of the system highlighting the key components. Figure **a** is reprinted from [54], copyright (2011) Mitsubishi, with permission from Mitsubishi Electric Cooling & Heating, and **b** is reprinted from [53] with permission from the author

### *6.6.4 A Unified Framework*

Even in one of the earliest works, [55] stresses the importance of understanding and modeling the dynamic interaction between the building and the HVAC system. Despite this, the majority of modeling and control efforts have decomposed the problem, with focus on either the building dynamics or the HVAC system dynamics. More recently, however, this has begun to change. A increasing number of studies have aimed to model buildings from a holistic multi-subsystem point of view [24, 56–58]. This is largely due to increases in computational power and highly capable modeling software such as EnergyPlus. However, as previously discussed, while these high fidelity models are extremely valuable, RC network and graph-based models proved additional, complementary capabilities. With the wide adoption of the RC network approach and the extended flexibility of the graph-based approach, these methods can be readily used to capture the complex dynamics of both the building and HVAC systems within a single unifying framework, as shown previously using several examples in this section. The following section discusses some of the many ways these models can be utilized, with an emphasis on control structure and control algorithm design.

## 6.7 Utility of RC Network and Graph Models

### *6.7.1 Model Reduction*

As discussed in the multi-zone modeling section, the number of resistors and capacitors, and the resulting number of states can grow very quickly even for relatively small buildings. Thus prior to control development, it is necessary to employ various model reduction techniques. A comparison of three such techniques is provided in [59], including state deletion, state aggregation, and structured balanced truncation. The state deletion approach, also known as residualization [60], identifies the states with the fastest dynamics and replaces them with their steady-state values. An algorithm for identifying the best states to remove is provided in [61]. While state deletion is the simplest of the three, it also produces the largest error between the full and reduced models. The state aggregation approach, while not as generic, applies to RC network models and maintains insight on the thermal structure of the building. The full RC network model is transformed into a corresponding Markov chain using an approach developed in [26] and used in [27, 59]. Finally, the structured balanced truncation approach from [62] maintains system structure by identifying and reducing subsystems while preserving the interconnection topology between these subsystem. While providing excellent reduced model accuracy, this approach is the most complex and requires solving linear matrix inequalities (LMIs).

With the extensive effort required to generate a full-order RC network model, a model reduction technique that outputs a reduced RC network is preferable. Thus the aggregation approach is recommended. With the original approach presented in [26], this method has been modified in [27] to produce a tree structure and included in the modeling software *Sustain* [63]. In [64], the same approach is extended to included nonlinear systems. Finally, [65] adapted the approach to produce observable reduced order models to allow development of extended state observers to estimate unmeasured states and disturbances.

Using this approach, the four-room building from [26] with 37 states is successfully reduced to a reduced order RC network model with only 6 states. The 15-room, 454-state, building model from [59] demonstrated the relationship between number of states in the reduced model and the weighted RMS error between the true and reduced models. This highlights a value of the approach, where the number of states in the reduced model can be an important decision variable used to find the appropriate balance between model accuracy and complexity.

### 6.7.2 Control Architecture Selection

While reduced models can be used to develop centralized controllers for the entire building, for large buildings it may be desirable to decompose the overall building into *partitions* and perform model reduction and control for each partition independently. In theory, a centralized controller (see Fig. 6.12), with access to the entire building model, all sensor data, and full authority over all actuators, provides the optimal control of the building by maintaining thermal comfort while minimizing energy consumption. However, in practice, buildings are complex systems with numerous sensors and actuators operating over a long life cycle, and thus are prone to faults and equipment failures. Due to the centralized decision-making of a centralized controller, the effects of these faults can be distributed throughout the systems via the controller, in addition to the natural coupling from the building dynamics.

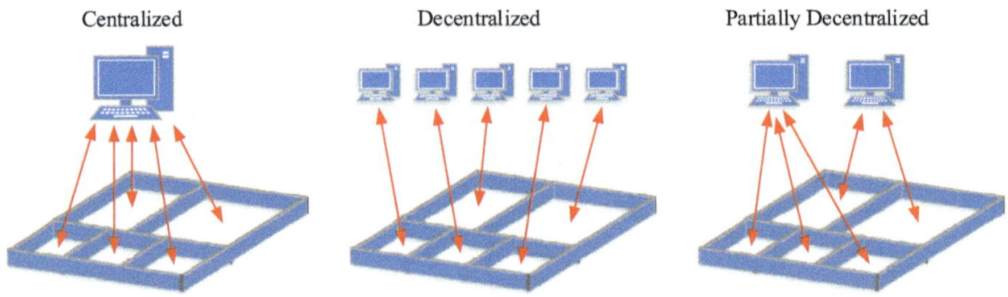


**Fig. 6.12** Examples of centralized, decentralized, and partially decentralized control architectures for a five zone building

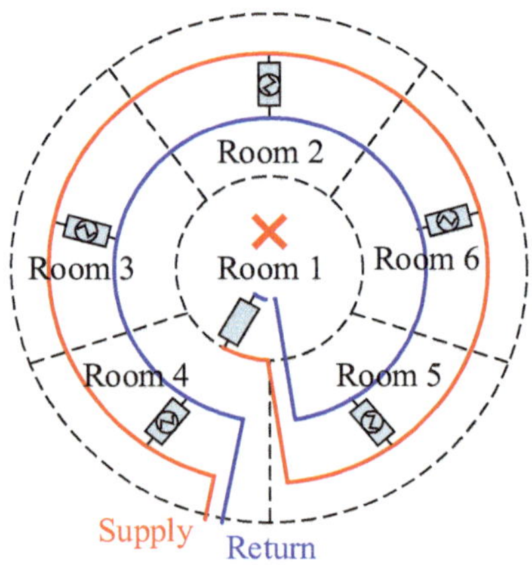


**Fig. 6.13** Six-zone building used to demonstrate the robustness of decentralized control to sensor faults compared to a centralized control approach. Figure reprinted, with permission, from [66], copyright (2013), IEEE

Alternatively, a decentralized controller (Fig. 6.12) consists of multiple controllers that independently control different parts of the building. Thus, a fault affects only the controller for that particular part of the building and the effect of this fault is propagated only through the building dynamics. This improved robustness is demonstrated in [66] for an example six-zone building, shown in Fig. 6.13, where a temperature sensor fault occurs in Room 1. Figure 6.14 compares the zone temperature regulation error for the centralized and decentralized control approaches. Since the decentralized controllers do not communicate, the effect of the temperature sensor fault is isolated to only Room 1. Under centralized control, the effect of the fault is propagated to all zones.

While a decentralized control approach benefits from robustness to faults, under nominal operation the lack of a complete system model and the coupling between adjacent zones of the building result in suboptimal control where temperature regulation performance and energy efficiency are reduced. Thus there is a trade-off

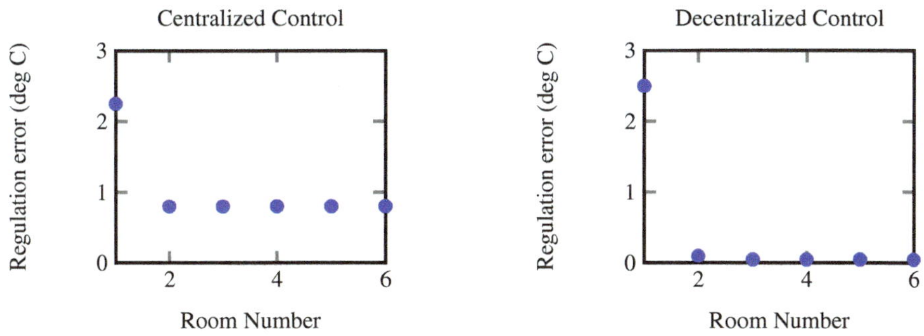


**Fig. 6.14** Effect of temperature sensor failure in Room 1 of building from Fig. 6.13 under centralized and decentralized control. Figure reprinted, with permission, from [66], copyright (2013), IEEE

between the two standard control objectives, performance and robustness, which influences the choice of control architecture.

In addition to completely centralized and completely decentralized control approaches, a partially decentralized approach (Fig. 6.12) can provide a balance between these competing control objectives. Instead of each room or zone having a dedicated controller, as with a completely decentralized approach, a partially decentralized approach decomposes the building into multiple partitions, each containing multiple rooms or zones and a corresponding controller. Thus zones with a strong degree of coupling can be contained within a partition and controlled by a single controller, improving the control performance. Meanwhile, faults within this partition are only weakly propagated to other partitions of the building.

Determining the optimal partitioning of a building is the focus of [66], which introduces the optimality loss factor (OLF) and the fault propagation metric (FPM) to qualify the two competing control objectives. As discussed in [67], the number of possible partitions, defined as the Bell number, grows exponentially with the number of zones in the building. For example, a building with nine zones has 4140 possible ways of partitioning. In order to analyze and identify optimal partitioning of large buildings, a less computationally intensive method is required and leading to the use of an agglomerative clustering approach, detailed in [66]. Using this approach, only 240 partitions need to be analyzed and compared to identify the desired trade-off between optimality and robustness.

### *6.7.3 Control Strategies*

With increasing demands for energy efficiency while maintaining occupant comfort, model-based control of building thermal systems has been researched extensively. However, practical adoption of these techniques has been slow and wide-scale market implementation has yet to be seen [68]. Developing simple yet accurate models is certainly a major hurdle and is considered the most challenging part of the controller design process. Accordingly, a number of papers have been published that discuss the current open challenges in model-based control design for buildings [69, 70].

A review of control strategies for thermal control of buildings found in [71] compared 46 different control strategies from the literature and identified three overall trends for controller development: improving "classical" control methods, use of predictive control based on models, and use of intelligent control methods. Improving classical control methods includes incorporating feedforward with feedback controllers [72, 73], optimizing On-Off control [74], and advanced tuning [75, 76] and input–output paring of PID controllers [77]. In general, predictive control strategies are all based on model predictive control (MPC) and use either linear [78, 79] or nonlinear models [37, 80]. Finally, intelligent control techniques typically refer to various forms of fuzzy logic [81–84]. Based on such an extensive review, the authors of [71] found that classical controllers, such as PID controllers, are still the most commonly implemented control solution in practice, despite many results demonstrating the superiority of model predictive and intelligent control strategies in terms of energy efficiency. There is a strong need to identify methods that are commercially viable and better address the variability of buildings and their surrounding environments. Chapters 4, 6, and 7 discuss specific control formulations for single-unit vapor compression systems, forced-air and radiant slab systems, and multiunit vapor compression systems, respectively, and the models presented in this chapter can be used to develop, implement, and test these controllers.

In addition to using the RC network and graph-based models for model predictive control, the structure of these models can be used explicitly in the control formulation. In fact, [85] utilizes a graph-based modeling framework to capture the thermal dynamics of a multi-zone building in order to develop a decentralized control framework. Due to the structure of the model, passivity of the open-loop building dynamics was readily established. Due to the fact that two passive systems in negative feedback is also passive, and thus stable [36], establishing the passivity of the system allows for a wide range of passive controllers to be designed and implemented with guaranteed stability. Future efforts should attempt to follow this example and utilize the structure of the system, captured by the graph-based model, to develop novel control strategies with guaranteed properties.

## 6.8 Conclusions and Future Work

While modeling the thermal dynamics of a multi-zone building often requires significant engineering effort, the RC network and graph-based modeling approaches divide this effort into smaller, more manageable, tasks. The first task is to understand the structure of the heat transfer throughout the building by identifying where energy is stored and the possible paths for energy transfer between these storage elements. Both the RC network and the more generic graph-based modeling procedures are designed to represent this structure. The next task is to identify the parameters of the model. While this task is significantly harder, this chapter has cited various parameter identification techniques that can be used on individual building elements, a single zone, or the entire multi-zone building. The appropriate technique depends on the information available about the building. Due to the structure and physical meaning of the RC network and graph-based models, the model development and parameter identification can be performed modularly, on different parts of a large building, further reducing the barrier to model generation. Once a model is obtained, there are many control-oriented uses that leverage the structure and simplicity of these models. Strategic control architecture design is critical for achieving the multiple demanding control objectives for buildings. The structure of these models can be readily used to evaluate and identify these strategic control architectures. Additionally, structure preserving model reduction enables these models to be used directly for model-based predictive control, which has been widely shown to be a promising control approach for buildings. With these valuable modeling frameworks identified, future efforts should focus on further utilizing the structure of these models in the control architecture design, to achieve industry-friendly, high-performance control formulations.

## References

1. Atam E (2016) Decentralized thermal modeling of multi-zone buildings for control applications and investigation of submodeling error propagation. Energy Build 126:384–395
2. Bergman TL, Incropera FP, DeWitt DP, Lavine AS (2011) Fundamentals of Heat and Mass Transfer. Wiley, New York
3. Gouda M, Danaher S, Underwood C (2002) Building thermal model reduction using nonlinear constrained optimization. Build. Environ. 37(12):1255–1265
4. Underwood C (2014) An improved lumped parameter method for building thermal modelling. Energy Build 79:191–201
5. Harish V, Kumar A (2016) Reduced order modeling and parameter identification of a building energy system model through an optimization routine. Appl Energy 162:1010–1023
6. McKinley TL, Alleyne AG (2008) Identification of building model parameters and loads using on-site data logs. IBPSA-USA J 3(1):9–16
7. Bouache T, Ginestet S, Limam K, Lindner G, Bosschaerts W (2013) Identification of thermal characteristics of a building. Energy Procedia 42:280–288
8. Wang S, Xu X (2006) Parameter estimation of internal thermal mass of building dynamic models using genetic algorithm. Energy Convers Manag 47(13):1927–1941
9. Xu X, Wang S (2007) Optimal simplified thermal models of building envelope based on frequency domain regression using genetic algorithm. Energy Build 39(5):525–536
10. Handbook A et al. (2001) Fundamentals. American Society of Heating, Refrigerating and Air Conditioning Engineers, vol 111. Atlanta
11. Crabb J, Murdoch N, Penman J (1987) A simplified thermal response model. Build Serv Eng Res Technol 8(1):13–19
12. Tindale A (1993) Third-order lumped-parameter simulation method. Build Serv Eng Res Technol 14(3):87–97
13. Berthou T, Stabat P, Salvazet R, Marchio D (2014) Development and validation of a gray box model to predict thermal behavior of occupied office buildings. Energy Build 74:91–100
14. Reynders G, Diriken J, Saelens D (2014) Quality of grey-box models and identified parameters as function of the accuracy of input and observation signals. Energy Build 82:263–274

15. Lin Y, Middelkoop T, Barooah P (2012) Issues in identification of control-oriented thermal models of zones in multi-zone buildings. In: Conference on Decision and Control (CDC), pp 6932–6937
16. Wang S, Xu X (2006) Simplified building model for transient thermal performance estimation using ga-based parameter identification. Int J Therm Sci 45(4):419–432
17. Zhou Q, Wang S, Xu X, Xiao F (2008) A grey-box model of next-day building thermal load prediction for energy-efficient control. Int J Energy Res 32(15):1418–1431
18. Lehmann B, Gyalistras D, Gwerder M, Wirth K, Carl S (2013) Intermediate complexity model for model predictive control of integrated room automation. Energy Build 58:250–262
19. Nielsen TR (2005) Simple tool to evaluate energy demand and indoor environment in the early stages of building design. Solar Energy 78(1):73–83
20. Barnaby CS, Spitler JD, Xiao D (2005) The residential heat balance method for heating and cooling load calculations. ASHRAE Transactions, vol 111, no 1
21. Shaikh PH, Nor NBM, Nallagownden P, Elamvazuthi I, Ibrahim T (2014) A review on optimized control systems for building energy and comfort management of smart sustainable buildings. Renew Sustain Energy Rev 34:409–429
22. Radecki P, Hencey B (2012) Online building thermal parameter estimation via unscented kalman filtering. In: American Control Conference (ACC), pp 3056–3062
23. Agbi C, Song Z, Krogh B (2012) Parameter identifiability for multi-zone building models. In: Conference on Decision and Control (CDC), pp 6951–6956
24. Cai J, Kim D, Jaramillo R, Braun JE, Hu J (2016) A general multi-agent control approach for building energy system optimization. Energy Build 127:337–351
25. Wen J, Smith TF (2002) Development and validation of online parameter estimation for HVAC systems. In: ASME 2002 International mechanical engineering congress and exposition. American Society of Mechanical Engineers, pp 211–217
26. Deng K, Barooah P, Mehta PG, Meyn SP (2010) Building thermal model reduction via aggregation of states. In: American Control Conference (ACC), pp 5118–5123
27. Dobbs JR, Hencey BM (2012) Automaticmodel reduction in architecture: Awindowinto building thermal structure. IBPSA-USA J 5(1):562–568
28. Crawley DB, Hand JW, Kummert M, Griffith BT (2008) Contrasting the capabilities of building energy performance simulation programs. Build Environ 43(4):661–673
29. U. D. of Energy (2010) Energyplus engineering reference: The reference to energyplus calculations
30. Sturzenegger D, Gyalistras D, Semeraro V, Morari M, Smith RS (2014) Brcm matlab toolbox: Model generation for model predictive building control. In: American Control Conference (ACC), pp 1063–1069
31. Sturzenegger D, Gyalistras D, Morari M, Smith RS (2012) Semi-automated modular modeling of buildings for model predictive control. In: Proceedings of the Fourth ACM Workshop on Embedded Sensing Systems for Energy-Efficiency in Buildings. ACM, pp 99–106
32. Agbi C, Krogh B (2014) Decentralized identification of building models. In: American Control Conference (ACC), pp 1070–1075
33. Maasoumy M, Moridian B, Razmara M, Shahbakhti M, Sangiovanni-Vincentelli A (2013) Online simultaneous state estimation and parameter adaptation for building predictive control. In: ASME 2013 Dynamic Systems and Control Conference. American Society of Mechanical Engineers, pp V002T23A006–V002T23A006
34. Koeln JP, Williams MA, Alleyne AG (2015) Hierarchical control of multi-domain power flow in mobile systems: Part I - framework development and demonstration. In: ASME 2015 Dynamic Systems and Control Conference. American Society of Mechanical Engineers, pp V001T08A006–V001T08A006
35. West DB et al. (2001) Introduction to graph theory. Prentice hall Upper Saddle River, vol 2
36. Khalil HK (1996) Noninear Systems. Prentice-Hall, New Jersey
37. Kelman A, Borrelli F (2011) Bilinear model predictive control of a HVAC system using sequential quadratic programming. IFAC Proc Vol 44(1):9869–9874

38. Liao Z, Dexter A (2004) A simplified physical model for estimating the average air temperature in multi-zone heating systems. Build Environ 39(9):1013–1022
39. Sane H, Haugstetter C, Bortoff S (2006) Building HVAC control systems-role of controls and optimization. In: American Control Conference (ACC), pp 6-pp
40. Chandan V, Zak G, Alleyne A (2009) Modeling of complex hydronic systems for energy efficient operation. In: ASME 2009 Dynamic systems and control conference. American Society of Mechanical Engineers, pp 123–130
41. Koeln JP, Williams MA, Pangborn HC, Alleyne AG (2016) Experimental validation of graph-based modeling for thermal fluid power flow systems. In: ASME 2016 Dynamic Systems and Control Conference. American Society of Mechanical Engineers, pp V002T21A008–V002T21A008
42. Goetzler W (2007) Variable refrigerant flow systems. ASHRAE J 49(4):24
43. Shah R, Alleyne AG, Bullard CW (2004) Dynamic modeling and control of multi-evaporator air-conditioning systems. ASHRAE Transactions, vol 110, no 1
44. Lin J-L, Yeh T-J (2007) Identification and control of multi-evaporator air-conditioning systems. Int J Refrig 30(8):1374–1385
45. Lin J-L, Yeh T-J (2009) Control of multi-evaporator air-conditioning systems for flow distribution. Energy Convers Manag 50(6):1529–1541
46. Elliott MS, Rasmussen BP (2008) Model-based predictive control of a multi-evaporator vapor compression cooling cycle. In: American Control Conference (ACC), pp 1463–1468
47. Elliott MS, Rasmussen BP (2009) A model-based predictive supervisory controller for multi-evaporator HVAC systems. In: American Control Conference (ACC), pp 3669–3674
48. Koeln JP, Alleyne AG (2014) Scalable model predictive control for multi-evaporator vapor compression systems. In: American Control Conference (ACC), pp 392–397
49. Pollock DT, Yang Z, Wen JT (2015) Dryout avoidance control for multi-evaporator vapor compression cycle cooling. Appl Energy 160:266–285
50. Zhou J, Burns DJ, Danielson C, Di Cairano S (2016) A reconfigurable plug-and-play model predictive controller for multi-evaporator vapor compression systems. In: American Control Conference (ACC), pp 2358–2364
51. Rasmussen BP (2012) Dynamic modeling for vapor compression systems - Part I: Literature review. HVAC R Res 18(5):934–955
52. Jain N, Koeln JP, Sundaram S, Alleyne AG (2014) Partially decentralized control of large-scale variable-refrigerant-flow systems in buildings. J Process Control 24(6):798–819
53. Koeln J (2013) A decentralized control design approach to a class of large-scale systems. Master's thesis, University of Illinois
54. Corporation ME (2011) Variable refrigerant flow zoning systems: 3–30 tons
55. Rumianowski P, Brau J, Roux J (1989) An adapted model for simulation of the interaction between a wall and the building heating system. In: Proceedings of the Thermal Performance of the Exterior Envelopes of Buildings IV Conference, pp 224–233
56. Wang S (1999) Dynamic simulation of building VAV air-conditioning system and evaluation of EMCS on-line control strategies. Build Environ 34(6):681–705
57. Bengea S, Kelman A, Borrelli F, Taylor R, Narayanan S (2012) Model predictive control for mid-size commercial building HVAC: Implementation, results and energy savings. In: Second International Conference on Building Energy and Environment, pp 979–986
58. Sangi R, Baranski M, Oltmanns J, Streblow R, Müller D (2016) Modeling and simulation of the heating circuit of a multi-functional building. Energy Build 110:13–22
59. Dobbs JR, Hencey BM (2013) Structured building model reduction toward parallel simulation. In: Proceedings of the 13th Conference of the International Building Performance Simulation Association
60. Skogestad S, Postlethwaite I (2007) Multivariable feedback control: analysis and design, vol 2. Wiley, New York
61. Papachristodoulou A, Chang YC, August E, Anderson J (2010) Structured model reduction for dynamical networked systems. In: Conference on Decision and Control (CDC), pp 2670–2675

62. Sandberg H, Murray RM (2008) Model reduction of interconnected linear systems using structured gramians. IFAC Proc Vol 41(2):8725–8730
63. Greenberg D, Pratt K, Hencey B, Jones N, Schumann L, Dobbs J, Dong Z, Bosworth D, Walter B (2013) Sustain: An experimental test bed for building energy simulation. Energy Build 58:44–57
64. Deng K, Goyal S, Barooah P, Mehta PG (2014) Structure-preserving model reduction of nonlinear building thermal models. Automatica 50(4):1188–1195
65. Chandan V, Alleyne AG (2014) Decentralized predictive thermal control for buildings. J Process Control 24(6):820–835
66. Chandan V, Alleyne A (2013) Optimal partitioning for the decentralized thermal control of buildings. IEEE Trans Control Syst Technol 21(5):1756–1770
67. Chandan V (2013) Decentralized thermal control of building systems. Ph.D. dissertation, University of Illinois at Urbana-Champaign
68. Atam E, Helsen L (2016) Control-oriented thermal modeling of multizone buildings: Methods and issues: Intelligent control of a building system. IEEE Control Syst 36(3):86–111
69. Cigler J, Gyalistras D, Široky J, Tiet V, Ferkl L (2013) Beyond theory: the challenge of implementing model predictive control in buildings. In: Proceedings of 11th Rehva World Congress, Clima, vol 250
70. Killian M, Kozek M (2016) Ten questions concerning model predictive control for energy efficient buildings. Build Environ 105:403–412
71. Belic F, Hocenski Z, Sliskovic D (2015) HVAC control methods - a review. In: International Conference on System Theory, Control and Computing (ICSTCC), pp 679–686
72. Jain N, Otten RJ, Alleyne AG (2010) Decoupled feedforward control for an air-conditioning and refrigeration system. In: American Control Conference (ACC), pp 5904–5909
73. Wen JT, Mishra S, Mukherjee S, Tantisujjatham N, Minakais M (2013) Building temperature control with adaptive feedforward. In: Conference on Decision and Control (CDC), pp 4827–4832
74. Chinnakani K, Krishnamurthy A, Moyne J, Gu F (2011) Comparison of energy consumption in HVAC systems using simple on-off, intelligent ON-OFF and optimal controllers. In: IEEE Power and Energy Society General Meeting, pp 1–6
75. Kasahara M, Matsuba T, Kuzuu Y, Yamazaki T et al (1999) Design and tuning of robust PID controller for HVAC systems. ASHRAE Trans 105:154
76. Wang J, An D, Zhang C, Jing Y (2009) Genetic optimization algorithm of PID decoupling control for VAV air-conditioning system. Trans Tianjin Univ 15(4):308–314
77. Jain N, Li B, Keir M, Hencey B, Alleyne A (2010) Decentralized feedback structures of a vapor compression cycle system. IEEE Trans Control Syst Techno 18(1):185–193
78. Rehrl J, Horn M, Temperature control for HVAC systems based on exact linearization and model predictive control. In: Conference on Control Applications (CCA)
79. Privara S, Širokỳ J, Ferkl L, Cigler J (2011) Model predictive control of a building heating system: The first experience. Energy Build 43(2):564–572
80. Ma Y, Borrelli F, Hencey B, Coffey B, Bengea S, Haves P (2012) Model predictive control for the operation of building cooling systems. IEEE Trans Control Syst Technol 20(3):796–803
81. Singh J, Singh N, Sharma J (2006) Fuzzy modeling and control of HVAC systems-a review. J Scientific Ind Res 65:470–476
82. Mirinejad H, Welch KC, Spicer L (2012) A review of intelligent control techniques in HVAC systems. In: Energytech, pp 1–5
83. Alcalá R, Benítez JM, Casillas J, Cordón O, Pérez R (2003) Fuzzy control of HVAC systems optimized by genetic algorithms. Appl Intell 18(2):155–177
84. Villar JR, de la Cal E, Sedano J (2009) A fuzzy logic based efficient energy saving approach for domestic heating systems. Integr Comput Aided Eng 16(2):151–163
85. Mukherjee S, Mishra S, Wen JT Building temperature control: A passivity-based approach. In: Conference on Decision and Control (CDC), pp 6902–6907

# Chapter 7
# Distributed Model Predictive Control for Forced-Air Systems

**Sarah M. Koehler, Frank Chuang, Yudong Ma, Allan Daly and Francesco Borrelli**

## 7.1 Introduction

This chapter focuses on state of the art control design for energy efficient buildings as well as advanced predictive control methodologies that seek to incorporate predictions of weather, occupancy, renewable energy availability, and energy price signals [4, 5]. As was highlighted in Chaps. 4 and 5, model predictive control [6] (MPC) is a control methodology that can systematically use all the aforementioned predictions to improve building thermal comfort, decrease peak demand, and reduce total energy costs.

In buildings, performance improvement using forecasted information is possible through two mechanisms. The first mechanism is referred to as load shifting or active storage. Load shifting consists of shaping the energy profile delivered to a building,

---

Sections 7.4.1, 7.6.1, 7.6. include material reused with permission from [1] (©2013 EUCA), Sect. 7.6.3 includes material reused with permission from [2] (©2015 AACC) and [3] (©2016 John Wiley & Sons, Ltd.).

---

S.M. Koehler (✉) · F. Chuang · F. Borrelli
University of California, Berkeley, CA, USA
e-mail: skoehler@berkeley.edu

F. Chuang
e-mail: fc4321@berkeley.edu

F. Borrelli
e-mail: fborrelli@berkeley.edu

Y. Ma · A. Daly
NEXTracker, Inc., Fremont, CA, USA
e-mail: yma@nextracker.com

A. Daly
e-mail: adaly@nextracker.com



J.T. Wen and S. Mishra (eds.), *Intelligent Building Control Systems*, Advances in Industrial Control, https://doi.org/10.1007/978-3-319-68462-8_7

exploiting the possibility of storing energy for later use. Thermal storage is inherent to building structures with thermal capacity and can be increased by including additional external energy storage devices. The optimal profile of delivered energy depends on various factors including time varying utility prices, availability of renewable energy, ambient temperature variation, and load shedding signals received from the utility grid. The second mechanism is component optimization. Buildings can be large systems with many control variables and degrees of freedom. Predictive models of building thermal dynamics and energy costs of control actuators allow computation of the optimal inputs to each actuator in order to deliver the desired energy profile in the most cost-effective way. The two mechanisms are coupled in an actual MPC control strategy.

In [7–10], the authors have implemented MPC on building heating, ventilation, and air conditioning (HVAC)—systems obtaining reduced peak power consumption and reduced overall energy usage compared to existing production logic. In the aforementioned literature, control computation is performed at a centralized unit. In practice, embedded building control platforms are physically distributed throughout the building. It is standard to have a central processor and router located at the main air supply location (called the Air Handling Unit, AHU) and an embedded control unit at the locations which regulate flow and temperature of each zone (called the Variable Air Volume box, VAV box). In this chapter, we describe the main elements of computation and communication in today's building automation systems. In particular, we highlight the limited computational power and communication speeds that are present in today's buildings. We then discuss a class of distributed control algorithms, including typical control architectures for buildings, as well as distributed MPC control algorithms which fit these existing distributed control platforms. The current state of the art for distributed control in real-world buildings is known as the "Trim and Respond" method. This method has been shown to be more efficient and more intuitive to tune than control logic typically used by equipment manufacturers [7].

Then, we present methodology for design of model predictive controllers for forced-air systems. We analytically show that "Trim and Respond" controllers can be seen as a special class of one-step distributed model predictive control (DMPC) algorithms. A simulation study shows the DMPC control design and its equivalence with "Trim and Respond" heuristics.

The chapter concludes with a summary of an advanced distributed model predictive control method that performs well in the presence of high communication lag and limited computation processors. This chapter complements Chaps. 4 and 5 where a vapor-compression model is used. In this chapter, we focus on more general forced-air systems and present a distributed control design which addresses the presence of substantial communication lags in the control system. Parts of the work in this chapter have been extracted with permission from [1] (©2013 EUCA), [2] (©2015 AACC), and [3] (©2016 John Wiley & Sons, Ltd.).

**Nomenclature**

$x$: State variable
$u$: Input variable
$y$: Output variable
$w$: Exogenous disturbance variable
$T$: Temperature
$\dot{m}$: Mass flow rate
$c_p$: Specific heat capacity
$R$: Thermal resistance
$C$: Thermal capacitance
$\dot{Q}_d$: External heat load disturbance
$\mathbb{R}$: Set of real numbers
$\mathbb{Z}$: Set of integers

## 7.2 Forced Air Systems

There are several approaches to building thermal modeling, some of which can be seen in Chaps. 4, 5 and 6 of this book. These models range in granularity and complexity, from the high-level building envelope down to heating, ventilation, and air conditioning (HVAC) system component modeling. This includes white box modeling such as that seen in Energy Plus [11] or Modelica [12] to black box modeling. We describe the components of forced air systems and review the thermal models that we use for model predictive control. Our thermal models are derived from physics and simplified for control purposes, making them grey box models.

### *7.2.1 HVAC System Components*

This section introduces the components of a typical forced air heating, ventilation, and air conditioning system. Air Handling Units (AHU) and Variable Air Volume (VAV) boxes are the main components used to heat or cool and distribute air in a building (Fig. 7.1). The AHU recirculates return air from building spaces, and mixes it with fresh outside air. The proportion of return air to outside air is controlled by damper positions in the AHU (Fig. 7.1). The mixed air can be cooled by cooling coils that extract the cooling energy from chilled water produced by chillers.

The air temperature after these coils depends on the coil valve positions, the temperature of the chilled or heated water, the temperature of mixed air entering the cooling coil, the mass flow rate of the mixed air, and the physical characteristics as well as thermal effectiveness of the coils. Air is delivered to the building spaces by electrical fans. Before reaching a given space, the air goes through VAV boxes. At each VAV box, air temperature can be increased using reheat coils installed in the

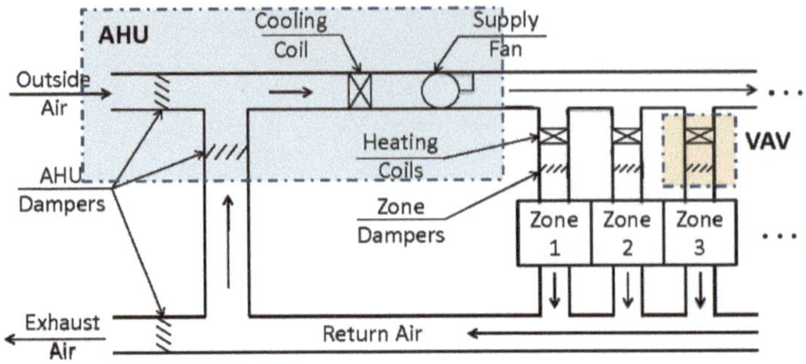


**Fig. 7.1** System Scheme

VAV box when needed. The space served by one VAV box is denoted by a thermal zone. The delivered air enters a zone through diffusers that are designed to mix the incoming air with the air in the thermal zone.

The temperature dynamics of thermal zone $i$ can be modeled by using a resistance-capacitance (RC) circuit analogy (assuming no significant heat exchange with neighboring zones):

$$C_i \dot{T}_i = u_i + \frac{(T_{oa} - T_i)}{R_i} + \dot{Q}_{d,i} \tag{7.1}$$

where $T_i$ is the temperature of zone $i$, $\dot{Q}_{d,i}$ is the external disturbance load on zone $i$ generated for instance by occupants, direct sunlight, and electrical devices, $T_{oa}$ is the temperature of outside air, $C_i$ and $R_i$ are thermal resistive-capacitive parameters of zone $i$, and $u_i$ is the heating and cooling power input to the space. The zone is cooled when $u_i \leq 0$ and heated when $u_i \geq 0$.

In fact, the input to the zone $u_i$ is a combination of the power provided by the AHU and the VAV boxes. The cooling power provided by the AHU is denoted as $u_c$. The local reheating power from the $i$-th VAV box is denoted by $u_{h,i}$. Then, the zone input is as follows:

$$u_i = k_{c,i} u_c + u_{h,i} \tag{7.2}$$

where $k_{c,i}$ is a coefficient to differentiate between the effect the AHU has on each zone. The system dynamics of zone $i \in \{1, ..., n\}$ are modeled using the following differential equation:

$$C_i \dot{T}_i = k_{c,i} u_c + u_{h,i} + \frac{(T_{oa} - T_i)}{R_i} + \dot{Q}_{d,i}. \tag{7.3}$$

The system dynamics of equation (7.1) are discretized using the Euler forward discretization method to obtain

$$x_{i,k+1} = A_i x_{i,k} + B_{c,i} u_{c,k} + B_{h,i} u_{h,i,k} + w_{i,k}, \tag{7.4}$$

where $x_{i,k} = T_{i,k}$ is the state of zone $i$ at time step $k$, $A_i = 1 - \frac{\Delta t}{R_i C_i}$, $B_{c,i} = \frac{\Delta t k_{c,i}}{C_i}$, $B_{h,i} = \frac{\Delta t}{C_i}$, and $w_{i,k} = \frac{\dot{Q}_{d,i}\Delta t}{C_i} + \frac{T_{oa}\Delta t}{R_i C_i}$. A detailed investigation of the effect of discretization methods in MPC can be found in [13].

**Return Air Circulation and Heat Input Calculation**

The zone input $u_i$ is a product of the air mass flow rate and difference between the temperature of the supply air from the VAV box and room temperature. The air enters zone $i$ with a mass flow rate $\dot{m}_i$, and supply air temperature $T_{s,i}$, resulting in a power input of

$$u_i = \dot{m}_i c_p (T_{s,i} - T_i), \tag{7.5}$$

where $c_p$ is the specific heat capacity of air and the supply air temperature $T_{s,i}$ is the temperature of the conditioned air that is supplied by the VAV box to zone $i$.

It is assumed that in the AHU, the outside air fully mixes with the return air without delay, and the mixing proportion $\delta$ between the return air and outside air is controlled by the damper configurations in the AHU system to obtain the following:

$$T_m = \delta T_r + (1 - \delta) T_{oa}, \tag{7.6}$$

where $T_{oa}$ is the outside air temperature, and $T_m$ is the temperature of the mixed air. $T_r$ is the return air temperature calculated as weighted average temperature of return air from each room

$$T_r = \frac{\sum_{i \in \mathcal{I}} \dot{m}_i T_i}{\sum_{i \in \mathcal{I}} \dot{m}_i}, \tag{7.7}$$

where $\mathcal{I}$ is the set of indices for all zones in the building. The return air is not recirculated when $\delta = 0$, and no outside fresh air is used when $\delta = 1$. $\delta$ can be used to save energy through recirculation but it has to be strictly less than one to guarantee a minimal outdoor fresh air delivered to the rooms.

The supply air temperature then is computed as

$$T_{s,i} = T_m - \Delta T_c + \Delta T_{h,i}, \ \forall i \in \mathcal{I}, \tag{7.8}$$

where $\Delta T_c$ represents the supply air temperature difference across the cooling coil in the AHU, and $\Delta T_{h,i}$ is the air temperature difference across the heating coil in the $i$-th VAV box. The cold water in the cooling coil is typically provided by a chiller system whereas the hot water in the heating coil is provided by a boiler system. The colder the cooling coil supply water is, the larger $\Delta T_c$ can be, and likewise, the hotter the heating coil supply water is, the larger $\Delta T_{h,i}$ can be.

The decomposition (7.2) matches equations (7.5)–(7.8). Substituting (7.8) into (7.5), the power input is expressed as:

$$u_i = \dot{m}_i c_p (T_m - \Delta T_c + \Delta T_{hi} - T_i). \tag{7.9}$$

If we assume that $T_m \approx T_i$, then it is clear that $u_i$ has two parts:

$$u_i = -\dot{m}_i c_p \Delta T_c + \dot{m}_i c_p \Delta T_{h,i}. \tag{7.10}$$

The approximation $T_m \approx T_i$ eliminates coupling between $T_i$ of different zones. The difference between $T_m$ and $T_i$ can be treated as part of the disturbance term $\dot{Q}_{d,i}$. $u_c$ and $u_{h,i}$ are computed accordingly as:

$$k_{c,i} u_c = -\dot{m}_i c_p \Delta T_c \tag{7.11}$$

$$u_{h,i} = \dot{m}_i c_p \Delta T_{h,i}. \tag{7.12}$$

Let $u_c = -\dot{m}_{max} c_p \Delta T_c$ where $\dot{m}_{max} = \max_i(\dot{m}_i)$. Then $k_{c,i} = \frac{\dot{m}_i}{\dot{m}_{max}}$.

For simplicity, we consider the control of $\Delta T_c$ and $\Delta T_{h,i}$ where the mass flow rate $\dot{m}_i$ is controlled separately or is otherwise approximated as constant. This approximation may lead to a linear time varying model if $\dot{m}_i$ is not constant (i.e., $B_{c,i}$ in (7.4) is not time invariant). In many cases, the approximation of a constant $\dot{m}_i$ is sufficient.

If $\dot{m}_i$ is also controllable, the system dynamics (7.3) are bilinear. The Euler forward discretization method can also be applied with equations (7.5)–(7.8) substituted into (7.1), obtaining nonlinear discrete time system dynamics.

## 7.3 Building Automation Systems

As discussed in Chaps. 2 and 3, building automation systems (BASs) are large, complex, and distributed systems. The building automation system is responsible for building management including control of the HVAC system, lighting system, and fire alarm system. In this section, we outline the embedded platforms and communication systems that are used for HVAC control in real buildings today.

### *7.3.1 Control Platforms*

In practice, embedded building control platforms are physically distributed throughout the building. In a commercial building, the HVAC typically includes a chiller plant, a boiler plant, AHU(s), and VAVs. Figure 7.1 shows a single AHU. Figure 7.2 generalizes Fig. 7.1 to represent one or more chiller plants, multiple AHUs, VAVs. In practice, there is an embedded controller at each VAV, each AHU, and the chiller plant. The larger and more complex the component or subsytem, the larger the corresponding controller; a common installation uses a small zone controller at each VAV box, a midsize controller at each AHU, and a more powerful control platform at

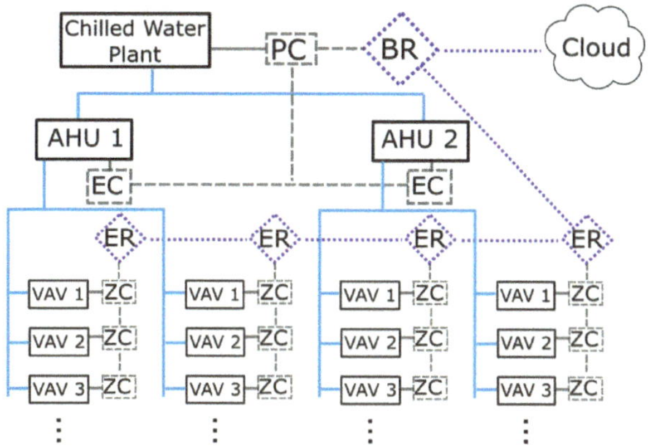


**Fig. 7.2** Typical layout of building embedded platforms and routers. Physical components and connections are denoted by solid lines; control platforms and wired connections are denoted by dashed lines; routers and communication backbone are denoted by purple dotted lines

the chiller plant. For instance, for the Automated Logic Corporation (ALC) vendor [14], The "ZN" controller is often used at the VAV or "zone controller" (ZC) level, the "SE" controller at the AHU or "equipment controller" (EC) level, and the "ME" controller at the chiller plant or "plant controller" (PC) level, as illustrated in Fig. 7.2.

At the VAV box, the zone controller "ZC"has a small processor with limited memory and computational power. A typical "ZC" has 512 kB non-volatile memory, 1 MB Flash memory, and a 16-bit processor [15]. This processor typically handles control of the VAV box (damper position and heating coil valve position $\delta_{h,i}$) as well as reading the sensor information in the VAV box (supply temperature $T_{s,i}$, mass flow rate $\dot{m}_i$) and the zone temperature $T$.

At the AHU level, the equipment controller "EC" reads AHU sensor information such as return air temperature $T_r$, outside air temperature $T_{oa}$, and mixed air temperature $T_m$. The controller also actuates the AHU controllable inputs, such as cooling coil valve position $\delta_c$, the supply fan speed, and the mixing ratio $\delta$ provided by the dampers.

Finally, at the plant level, the plant controller "PC" is used to control multiple independent components that must work together to form a system, such as pumps, cooling towers, chillers, etc. It reads information such as the chilled water supply and return temperatures, and controls the chillers (which run either a vapor-compression or absorption refrigeration cycle) as well as the pump that pushes chilled water to the AHUs and cooling towers or other heat rejection equipment that dump heat to the ambient environment.

Despite trends in computing such as Moore's law which has enabled the rapid advancement of computers and mobile devices, it is unlikely that the existing processors in buildings will change much over time. Hardware elements that are used to implement the controls systems are not changed over the lifetime of a building and even new buildings are installed with these embedded systems. Thus, it is desirable to fit control algorithms on these processors.

## 7.3.2 Communication Systems

Each building comes with its own unique set of embedded platforms and networking. Separate devices in the same building may be provided by different vendors, each of which is packaged with its own embedded platform and software. In order to communicate amongst these devices, a buildings specific communication protocol is used. The standard data communication protocol for building control systems is known as BACnet, or "Building Automation Control Network" [16]. BACnet was developed by ASHRAE (American Society of Heating Refrigerating and Air-conditioning Engineers) and is both an ANSI (American National Standards Institute) and ISO (International Organization for Standardization) standard.

BACnet provides an object-oriented model abstraction for devices to talk to each other over various types of network. The choice of network (both the physical wiring as well as the accompanying lower level communication protocol) has a significant effect on the overall performance of the communications. A simulation analysis of the most popular networking options used in buildings (MS/TP, Ethernet, and ARCNET) is done in [17]. In general, Ethernet is the fastest option, but is also the most expensive. On the other end, MS/TP is the cheapest option but is also the slowest. ARCNET however is only a little more expensive than MS/TP, but is noticeably faster [18]. The authors of this chapter have most commonly encountered ARCNET as the choice of networking.

The communication system of a large-scale forced-air system is pictured in Fig. 7.2. A "BR" building router is connected to the "PC" embedded platform at the chiller plant [14]. This router can communicate with external servers such as cloud servers. The "PC" platform is connected to a daisy chain of the "EC" controllers at the AHUs. The zone controllers are then connected in a daisy chain to one of the "EC" control platforms directly or via a subnetwork router or equipment router "ER" such as the Arcnet to Arcnet Router (AAR) [19]. In very large buildings, multiple subnetwork routers may be used to split up the network; each AAR can handle up to 100 zone controllers [19].

Generally, point-to-point communication between embedded platforms is possible, but in practice, communication is done in a hierarchical fashion. That is, platforms communicate over a star communication network, such as that in Fig. 7.3. The zone controllers "ZC" communicate with an "EC"controller, and the "EC" controllers communicate with the "PC" controller. On the uplink, devices may communicate by using BACnet services such as *ReadProperty* or *ReadPropertyMultiple*. These services are akin to *receive* and *gather* services seen in the Message Passing Interface (MPI) standard, respectively. The operations *receive* or *ReadProperty* allow the central hub to read a single data point from a single leaf node. The *gather* or *ReadPropertyMultiple* operation takes vector pieces from the leaf nodes and aggregates them into one large vector at the central hub. On the downlink, devices may communicate by using BACnet services *WriteProperty* or *WritePropertyMultiple* which are akin to *send* or *scatter* in MPI terminology. The *send* or *WriteProperty* operation sends a single data point from the central hub to a leaf node. A *scatter* or *WriteProp-*

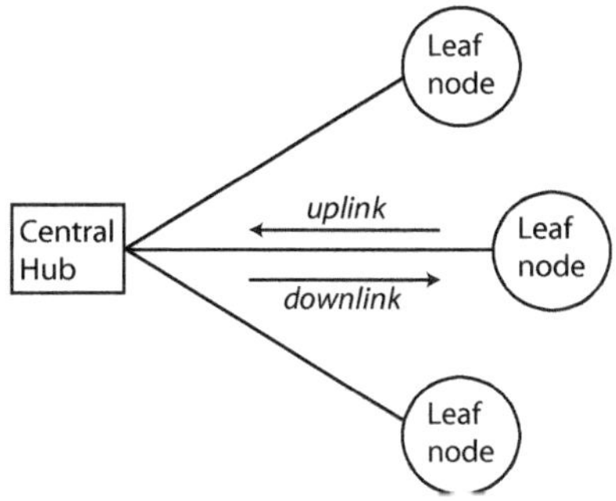


**Fig. 7.3** Star Network Communication Graph

*ertyMultiple* operation takes a vector at the central hub and sends each piece of that vector to the appropriate leaf node.

It is worth highlighting that the communication systems present in buildings are slow. We have found that in practice, *ReadPropertyMultiple* on an ARCNET network results in a communication delay of 0.03 to 0.4 s per floating point value. The simulation results in [17] state similar values for communication delay.

## 7.4 Current Building Control Practices

Due to the large distributed nature of BASs, the most common implementation in practice is to decouple the building controls into separate single-input-single-output proportional-integral-derivative (PID) loops to control variables to track desired fixed setpoints [20, 21]. These setpoints include duct static pressure, AHU supply temperature, VAV supply temperature, etc.

The PID controllers provide setpoint tracking, but do not promote energy efficiency. A more sophisticated building control method which is used in practice is "Trim and Respond" [20, 22]. This method differs from the PID loops because it changes or *resets* setpoints over time. The main idea is to modify setpoints by *responding* to heating or cooling requests and *trimming* energy consumption. The actions of *trimming* and *responding* are enabled by communicating information about demand from the zone level processors to the centralized AHU processor. We next describe this "Trim and Respond" logic; it has been shown to be more efficient and is considered to be more intuitive to tune than control logic used by equipment manufacturers [20].

### *7.4.1 "Trim and Respond" Logic*

In this section, we describe the mechanics of "Trim and Respond" for the linearized model in (7.3). In this logic, each reheating input $u_{hi,k}$ is controlled locally by a

PI loop regulating zone temperature to remain within comfort bounds. Each zone generates a heating request when the reheating coil is approaching its maximum capacity. This maximum capacity is denoted as $\bar{u}_k$ and is changed over time by the "Trim and Respond" logic. Likewise, each zone generates a cooling request when the zone is close to its lower thermal bound. Since there is no cooling generation available at each zone $i$, the zone must request that the central cooling input $u_{c,k}$ add more cool air to the system, and thus $u_{c,k}$ is also modulated over time by the "Trim and Respond" logic as a function of cooling requests.

Next, we describe the "Trim and Respond" logic in detail. A heating request $r_{h,i,k}$ is generated by zone $i$ at time step $k$ if its local reheat control $u_{h,i,k}$ is near its maximum capacity $\bar{u}_k$. Similarly, zone $i$ generates a cooling request $r_{c,i,k}$ if its local reheating input $u_{h,i,k}$ is near zero and the temperature is close to its upper bound $\bar{T}$. Each zone has an importance multiplier $M_i \in \mathbb{Z}_{>0}$ which can be manually increased for a zone to have more influence on central control. The following formulas describe request generation for zone $i$:

$$\text{If } u_{h,i,k} \leq \varepsilon,\ T_{i,k} \geq \bar{T} - 0.1, \qquad r_{c,i,k} = M_i. \tag{7.13a}$$

$$\text{If } u_{h,i,k} \geq (1-\varepsilon)\bar{u}_k, \qquad r_{h,i,k} = M_i. \tag{7.13b}$$

where $\varepsilon$ is a small value such as 0.05. If no cooling request is generated (conditional for (7.13a) is not met), $r_{c,i,k} = 0$, and if no heating request is generated (conditional for (7.13b) is not met), $r_{h,i,k} = 0$.

Requests are sent to centralized platforms associated with the main air handling unit and a central building boiler. Thc value for $u_{c,k}$ is reset at the AHU by adjusting the cooling coil valve position (see AHU cooling coil in Fig. 7.1), and $\bar{u}_k$ is reset at the boiler by updating the supply water temperature that is provided to local reheating coils. In particular, the "Trim and Respond" logic responds to requests by decreasing $u_{c,k}$ by an amount proportional to the total number of cooling requests and by increasing $\bar{u}_k$ by an amount proportional to the number of heating requests.

At every time step $k$, regardless of the number of requests, the values of $u_{c,k}$, $\bar{u}_k$ are also "trimmed", or $u_{c,k}$ and $\bar{u}_k$ are decremented toward 0. Note that $u_{c,k}$ can only take on nonpositive values whereas $\bar{u}_k$ can only take on nonnegative values. This trimming effect ensures reduction in energy usage by decreasing centralized control efforts when no requests are made.

Let the sum of requests be $\text{req}_{c,k}$ and $\text{req}_{h,k}$:

$$\text{req}_{c,k} = \sum_{i=1}^{n} r_{c,i,k}, \qquad \text{req}_{h,k} = \sum_{i=1}^{n} r_{h,i,k}, \tag{7.14}$$

where $n$ is the number of zones. The central controls then "trim" and "respond" to these summed requests as follows:

In cooling mode, $\left(\text{req}_{c,k} > \text{req}_{h,k}\right)$

$$u_{c,k+1} = u_{c,k} + \text{trim}_c + (\text{req}_{c,k} - r_I)\text{res}_c \tag{7.15a}$$

$$\bar{u}_{k+1} = \bar{u}_k + \text{trim}_b, \tag{7.15b}$$

and in heating mode, $\left(\text{req}_{h,k} > \text{req}_{c,k}\right)$

$$u_{c,k+1} = u_{c,k} + \text{trim}_c \tag{7.16a}$$

$$\bar{u}_{k+1} = \bar{u}_k + \text{trim}_b + (\text{req}_{h,k} - r_I)\text{res}_b, \tag{7.16b}$$

where the tuning parameters are: $r_I$, a fixed number of ignored requests; $\text{res}_c$, $\text{res}_b$, the proportional "response" per request for $u_c$ in cooling mode and $\bar{u}$ in heating mode, respectively; and $\text{trim}_c$, $\text{trim}_b$ "trim" amounts. If $\text{req}_{c,k} = \text{req}_{h,k}$, then the control law uses (7.15b) and (7.16a).

The values $u_{c,k}$, $\bar{u}_k$, $\text{req}_{c,k}$ and $\text{req}_{h,k}$ are also limited by the following upper and lower bounds:

$$\text{req}_{c,k} \leq \frac{\text{max}_c}{\text{res}_c} + r_I, \qquad \text{req}_{h,k} \leq \frac{\text{max}_b}{\text{res}_b} + r_I, \tag{7.17a}$$

$$u_c \in \left[u_{c_{min}}, u_{c_{max}}\right], \qquad \bar{u} \in [0, \bar{u}_{max}]. \tag{7.17b}$$

An example set of parameters used in cooling mode on an actual building is $\text{trim}_c = \frac{5}{9}$ °C, $\text{res}_c = -\frac{10}{9}$ °C, $u_{c_{max}} = 0$ °C, $u_{c_{min}} = -12$ °C, $\text{max}_c = -4$ °C, $r_I = 0$.

The logic presented in this section is a simplified version of "Trim and Respond" logic used in buildings today. In particular, we have only considered heating and cooling requests for temperature control. In practice, the logic extends to pressure control as well [20, 22].

## 7.5 Model Predictive Control

MPC has the ability to utilize system models and disturbance forecasts to automatically synthesize an efficient building controller [4, 8, 10, 21]. Sensor feedback is used to inform the system about the current state, capturing unexpected temperature disturbances as well as model errors. Energy price signals and forecasts change from day to day, requiring controller flexibility on at least a day-to-day basis. MPC systematically incorporates models, forecasts, and feedback to flexibly and efficiently handle building thermal control.

In this section, a model predictive controller is designed to minimize an objective (e.g., total energy consumption of the HVAC system), and to satisfy constraints such as thermal comfort constraints and actuation constraints. A model predictive controller solves a constrained finite time optimal control (FTOC) problem at each time step in a receding horizon fashion. Consider the following FTOC problem at time step $t$:

$$\min_{X,U} \quad J(X, U) \tag{7.18a}$$

$$s.t. \quad x_{k+1} = f(x_k, u_k, \hat{w}_k) \tag{7.18b}$$

$$x_k \in \mathcal{X}, \; k = 1, \ldots, N \tag{7.18c}$$

$$u_k \in \mathcal{U}, \; k = 0, \ldots, N-1 \tag{7.18d}$$

$$x_0 = x(t) \tag{7.18e}$$

where the state at time step $k$ is $x_k \in \mathcal{R}^n$ with $X = [x_0^T, \ldots, x_N^T]^T$, the input at time step $k$ is $u_k \in \mathcal{R}^m$ with $U = [u_0^T, \ldots, u_{N-1}^T]^T$, $\hat{w}_k$ is the predicted load at time step $k$, the prediction horizon length is $N$, the dynamics are encoded by $f(\cdot)$, $\mathcal{X}$ is a polyhedron encoding state constraints such as thermal comfort constraints, $\mathcal{U}$ is a polyhedron encoding actuator constraints, and the current state feedback $x(t)$ is used.
Let the optimal solution to problem (7.18) at time step $t$ be

$$\mathbf{U}^* = [u_0^{T^*}, u_1^{T^*}, \ldots, u_n^{T^*}]^T. \tag{7.19}$$

At time step $t$, the solution $u_0^*$ is implemented. The optimization (7.18) is repeated at time $t+1$, with the updated state estimation $x(t+1)$ and estimated load, yielding a receding horizon control strategy.

### 7.5.1 Cost Design

The objective of the controller is directly encoded into the cost function $J(X, U)$ of the MPC problem (7.18). This section describes typical cost functions that are used for building thermal control. We describe the cost functions for power inputs, i.e., $u_k = [u_{1,k}, \ldots, u_{n,k}]$ where $u_{i,k}$ is the input defined in (7.2) at time step $k$.
First, the objective of minimal total energy consumption is encoded by integrating the power inputs over the prediction horizon:

$$J_{energy}(X, U) = \sum_{k=0}^{N-1} |u_k| \Delta_t. \tag{7.20}$$

The total energy may also be minimized by expressing the cost function as a summation of the HVAC component power consumption equations such as the fan power, cooling coil power, heating coil power, etc. Detailed equations of component power consumption are included in [21].
Electrical pricing can be accounted for in a similar fashion as follows:

$$J_{price}(X, U) = \sum_{k=0}^{N-1} c_k(u_k) \tag{7.21}$$

where the cost function $c_k(\cdot)$ in general varies over time. This allows the controller to plan a control trajectory that can take advantage of lower cost periods and lessen power consumption at periods of higher cost. Similarly, the peak power usage of the day can be optimized by minimizing the following cost function:

$$J_{peak}(X, U) = \|U\|_\infty = \max(u_0, \dots, u_{N-1}). \tag{7.22}$$

The infinity norm is used to indicate the largest magnitude power input over the prediction horizon. The usage of $J_{peak}(X, U)$ encourages *precooling*, or shifting the energy consumption to be before a high thermal load period, rather than during such a high thermal load period.

For an illustrative example of the effect of the various cost functions discussed here, see [21]. Note that the prediction horizon $N$ affects the controller's ability to look ahead into the future. For the majority of the cost functions described here, the prediction horizon should be long, covering as much of a day as possible in order to accomplish pre-cooling, load shifting, etc.

In general, a weighted combination of cost functions (7.21) and (7.22) may be used. The inputs may be penalized by a 2-norm cost function rather than a 1 or $\infty$-norm cost function. For example,

$$J(X, U) = p(x_N) + \sum_{k=0}^{N-1} q(x_k, u_k) \tag{7.23}$$

where the stage cost $q(x_k, u_k)$ is defined by the following in the 2-norm case:

$$q(x_k, u_k) = (x_k - x_{ss})^T Q(x_k - x_{ss}) + u_k^T R u_k \tag{7.24}$$

with $Q \succeq 0$, $R \succ 0$, and $x_{ss}$ is the desired steady state for $x_k$. The desired $x_{ss}$ is the desired zone temperature setpoint. The terminal cost $p(x_N)$ is defined by the following in the 2-norm case:

$$p(x_N) = (x_N - x_{ss})^T P(x_N - x_{ss}) \tag{7.25}$$

with $P \succ 0$. The 2-norm cost functions are popularly used in MPC design (e.g., see Chap. 6), especially for the design of an MPC controller with stability guarantees [23].

### 7.5.2 *Constraints*

While the controller optimizes an objective such as energy consumption, it must also obey constraints such as maintaining comfort conditions for occupants. Occupancy comfort is modeled as a band of temperature in which the zone temperature should remain [24]. That is, the zone temperature is constrained by an upper bound $\bar{T}$ and a lower bound $\underline{T}$:

$$x_i \in [\underline{T}, \bar{T}]. \tag{7.26}$$

This constraint (7.26) would define the state constraint $\mathcal{X}$. In some cases, these bounds on zone temperature are implemented as soft constraints since it is acceptable to have some constraint violation. See more details on adding soft constraints into the MPC design in [21].

In addition to state constraints, there are also constraints on the control inputs. The control inputs are physically constrained by saturation limits. For example, the power input $u_i$ is limited by upper and lower bounds:

$$u_i \in [\underline{u}, \bar{u}]. \tag{7.27}$$

The constraints on low level control inputs of a forced air system in equations (7.5)–(7.8) include the following:

- $\dot{m}_i \in [\underline{\dot{m}}_i, \overline{\dot{m}}_i]$: The mass flow rate must be above minimum ventilation requirements and cannot exceed maximum flow capacity
- $\delta \in [0, 1]$: The damper position ranges between completely open and completely closed
- $\Delta T_c \in [0, \Delta \overline{T}_c]$: The cooling coil may cool the air up to a maximum amount
- $\Delta T_h \in [0, \Delta \overline{T}_h]$: The heating coil may heat the air up to a maximum amount

These constraints are encoded in $\mathcal{U}$, the polyhedron describing input constraints. A similar list of constraints is described in Chaps. 4 and 5 for vapor compression systems specifically.

In some cases, the design of a terminal constraint such that $x_N \in \mathcal{X}_f$ is added to the MPC problem (7.18) for persistent feasibility and stability. See [23] for more details on the design of a stable MPC controller.

## 7.6 Distributed Model Predictive Control for Forced Air Systems

In most demonstrations of MPC on building heating, ventilation, and air conditioning (HVAC) systems, control computation is performed at a centralized unit [7–10]. The size of the centralized predictive control problem rapidly grows when a realistic number of rooms together with a meaningful control horizon are considered. Therefore the real-time implementation of an MPC scheme is a challenge for the low-cost embedded platforms and low-speed communication networks currently used for HVAC control algorithms.

In this section, we present algorithms for distributed model-based predictive control (DMPC) for regulating heating and cooling in order to minimize energy consumption while satisfying comfort constraints. These distributed MPC control algorithms fit the aforementioned existing distributed control platforms.

We first propose a distributed solution that is the explicit solution of a one-step (i.e., prediction horizon $N = 1$) MPC problem. We show that this explicit solution matches both the control logic and the distributed form of "Trim and Respond".

This special class of DMPC is nice because it fits the limited computation and communication requirements of real building systems. However, this special class is limited since it does not include a long prediction horizon. We show how to extend this distributed solution for a general prediction horizon by using a primal-dual active-set optimization method. We compare the primal-dual active-set method to other popular distributed optimization methods, dual decomposition and the alternating direction method of multipliers (ADMM). We show that these popular methods for large-scale MPC cannot perform well on real building computing and communication systems whereas the primal-dual active-set method can.

### 7.6.1 One-Step DMPC

Consider the following MPC problem at time step $t$:

$$\min_{u_c, u_i} \; u_c^2 + \sum_{i=1}^{n} u_i^2 \tag{7.28a}$$

$$\text{subject to: } x_{i,1} = A_i x_{i,0} + B_{c,i} u_c + B_{h,i} u_{h,i} + \hat{w}_{i,t} \tag{7.28b}$$

$$x_{i,1} \in [\underline{T}, \bar{T}] \tag{7.28c}$$

$$u_i \geq 0 \tag{7.28d}$$

$$u_c \leq 0 \tag{7.28e}$$

$$x_{i,0} = x_i(t) \tag{7.28f}$$

$$\forall i \in \{1, \ldots, n\},$$

where the prediction horizon length is one, $\hat{w_{i,t}}$ is the estimated load at the $i$-th zone for time $t$, (7.28c) are thermal comfort constraints at the next time step, (7.28d) is the local actuation reheating constraint, and (7.28e) is the central cooling constraint.

Let the optimal solution to problem (7.28) be

$$\mathbf{U}^* = [u_c^*, u_1^*, \ldots, u_n^*]. \tag{7.29}$$

At time step $t$, the solution $u_c^*$ is implemented by the AHU and $u_i^*$ is implemented by the $i$-th VAV box. The optimization (7.28) is repeated at time $t + 1$, with the updated state estimation $x_{i,t+1}$ and estimated load $\hat{w}_{i,t+1}$, yielding a receding horizon control strategy.

In [1], it is shown that the explicit solution of the one step MPC controller in (7.28) and (7.29) is a one-step distributed solution. The centralized MPC controller (7.28) and (7.29) can be reformulated as a distributed algorithm which converges in one step. Moreover, in [1], we show that the online optimizer can be replaced by rule based equations using explicit MPC [23]. This solution is written in Algorithm 7.1 and is described in detail next.

**Algorithm 7.1** Explicit Solution of One-Step DMPC

```
Local zone platform i: Calculate initial guess for dual variable μ_0,i (underbar) and μ̄_0,i from (7.30a) and (7.30b).
Local zone platform i: Send μ_0,i (underbar) and μ̄_0,i to central AHU platform.
Central AHU platform: Compute I_C from (7.31).
if I_C == ∅ then
    Central AHU platform: u_c := 0
else
    Central AHU platform: Compute u_c = −(A_i x_i,0 + d̂_i − T̄)/B_C,i  ∀i ∈ I_C
end if
Central AHU platform: Send u_c to local VAV processors.
Local zone platform i: Compute if i in I_H, as defined by (7.32).
if i ∈ I_H then
    Local zone platform i: Compute u_i = (T_ (underbar) − A_i x_i,0 − ŵ_i − B_C u_c)/B_i  ∀i ∈ I_H
else
    Local zone platform i: u_i := 0.
end if
```

The algorithm is initialized at each local zone controller $i$ by an initial guess for the dual variables $\underline{\mu}_i$ and $\bar{\mu}_i$ as follows:

$$\underline{\mu}_{0,i} = (\underline{T} - A_i x_{i,0} - \hat{d}_i)_+ \tag{7.30a}$$

$$\bar{\mu}_{0,i} = (-\bar{T} + A_i x_{i,0} + \hat{d}_i)_+, \tag{7.30b}$$

where $(\cdot)_+ := max(\cdot, 0)$. Zone $i$ requests heat if $\underline{\mu}_i^0 > 0$, and likewise, $\bar{\mu}_i^0 > 0$ indicates that zone $i$ requests cooling. In fact, given parameter values $x_{i,0}$ and $\hat{d}_i$ for all $i \in \mathcal{I}$, the sets $\mathcal{I}_C$ and $\mathcal{I}_H$ can be determined as follows:

$$\mathcal{I}_C = \left\{ i \in \mathcal{I} \middle| i = \text{argmin}_i \frac{-\bar{\mu}_i^0}{B_{C,i}} \right\} \tag{7.31}$$

where $\mathcal{I}_C = \emptyset$ if the minimizing value is zero, and

$$\mathcal{I}_H = \left\{ i \in \mathcal{I} \middle| \underline{\mu}_i^0 - B_{C,i} u_c > 0 \right\}. \tag{7.32}$$

The zones in $\mathcal{I}_C$ correspond to the zones requesting the maximum cooling required for any zone to maintain the upper temperature bound constraint $\bar{T}$. Similarly, the zones in $\mathcal{I}_H$ request heat in order to maintain the lower temperature bound constraint $\underline{T}$, assuming $u_c$ has been calculated.

The optimal law for $u_c$ is determined from a sorting of $\bar{\mu}_i^0$ and does not rely on $u_i$. Thus, only one local-to-central and one central-to-local communication is required. Therefore, the structure of this problem is a distributed algorithm which converges in one step, as proven in [1]. Throughout the algorithm, it is assumed that each zone only knows its own model parameters ($A_i$, $B_{C,i}$, $B_i$, $\hat{w}_i$), and the central processor knows $B_{C,i}$. Each step is preceded by the location of the computation or the location at which a communication begins. The embedded platform at zone $i$ is denoted as "local zone platform $i$" and the embedded platform at the AHU is denoted as the "central AHU platform".

### 7.6.2 Comparison of "Trim and Respond" and One-Step DMPC

The explicit MPC method and the "Trim and Respond" method lead to controllers with the same structure and similar data communication. A detailed explanation is provided next.

The initial dual variables $\underline{\mu}_{0,i}$, $\bar{\mu}_{0,i}$ and requests $req_{h,k}$, $req_{c,k}$ have the same meaning: initial dual variables are a real-valued measurement of temperature bound violation, whereas requests (7.13) are integer-valued indicators of zone heating or cooling demand. While dual variables are the one-step predicted heating or cooling demand, requests are manually tuned to demand an appropriate amount of central control adjustment. The "Trim and Respond" algorithm handles this coarse measurement of demand by ignoring some requests (using $\text{req}_I$) and setting a maximum change in centralized control, e.g., $\text{max}_C$ to prevent overreaction. If the model is accurate, the dual variables provide more valuable information than the boolean requests values.

The "Trim and Respond" importance factor $M_i$ attempts to manually find the zone that has the highest cooling demand. Typically, one zone tends to request the most cooling (e.g., a computer server room or other zone with regularly high levels of thermal load), and so it is given the highest importance factor. The MPC algorithm automatically finds this zone by calculating $\mathcal{I}_C$.

Even if requests are only a function of the current state $x_i(t)$, predictive capability is introduced in practice. Requests are generated when local controls are close to their limits instead of at saturation. Thus, requests are generated when temperature bounds are likely to be saturated in the near future.

The explicit MPC algorithm and the "Trim and Respond" logics communicate between processors in a similar fashion. "Trim and Respond" sends information from local processors to centralized processors. Communication from central to local processors is not present because each local PI controller will implicitly respond to changes in $u_c$. Explicit MPC requires one communication each way between local and central processors (see Algorithm 7.1).

Both methods address energy efficiency. MPC directly addresses this by minimizing an effective cost of energy use. "Trim and Respond" addresses energy consumption by using a "trim" factor to decrease the control setpoints.

Finally, "Trim and Respond" does not require load estimation or a system dynamics model. However, there are many parameters to manually tune in the "Trim and Respond" algorithm which are in practice tuned as a function of nominal load.

#### 7.6.2.1 Simulation Results

Simulation results are presented in this section for the one-step DMPC algorithm and the "Trim and Respond" algorithm. The simulation is run with $n = 10$ zones. The MPC has a sampling time of 5 min; the "Trim and Respond" logic updates the centralized controls $u_c(k)$, $\bar{u}(k)$ every 5 min with zone-level PI loops running at 1 min intervals. Model parameters from equation (7.1) are: $C_i = 9.2e3$ kJ/°C, $R_i = 50$ °C/kW, $\underline{T} = 21$ °C, and $\bar{T} = 26$ °C.

The disturbance $w_i$ is shown at the top of Fig. 7.4, estimated from data from the UC Berkeley Bancroft Library. Simulation results are provided in Fig. 7.4. The MPC implementation uses a Luenberger observer to estimate the load, allowing a slight violation of temperature bounds. The "Trim and Respond" profile shows many small variations in temperature, while MPC provides a smoother result. The small variations are an artifact of the integer requests and the prefixed resolution by which central controls can change in the "Trim and Respond" logic. Similar comparisons exist between the control inputs of the two methods as well as between the MPC initial dual variables and the "Trim and Respond" requests. A calculation of total power used illustrates that the explicit MPC outperforms the "Trim and Respond" method.

### *7.6.3 Advanced Distributed Model Predictive Control*

The explicit distributed solution of the one-step DMPC solution is not easily extensible to $N$-step DMPC. Thus, we explore general distributed optimization algorithms that can handle solving the $N$-step DMPC via distributed computational splitting and communication. Existing methods include matrix splitting [25], dual decomposition [26], the Alternating Direction Method of Multipliers (ADMM) [27], an alternating minimization algorithm [28], a distributed inexact Newton method [29], and more. The disadvantage of these solutions is high communication costs. These algorithms require the communication of a large number of variables during each iteration. Additionally, it may require many iterations to converge, involving substantial communication to find the solution. Moreover, it is well known in the parallel com-

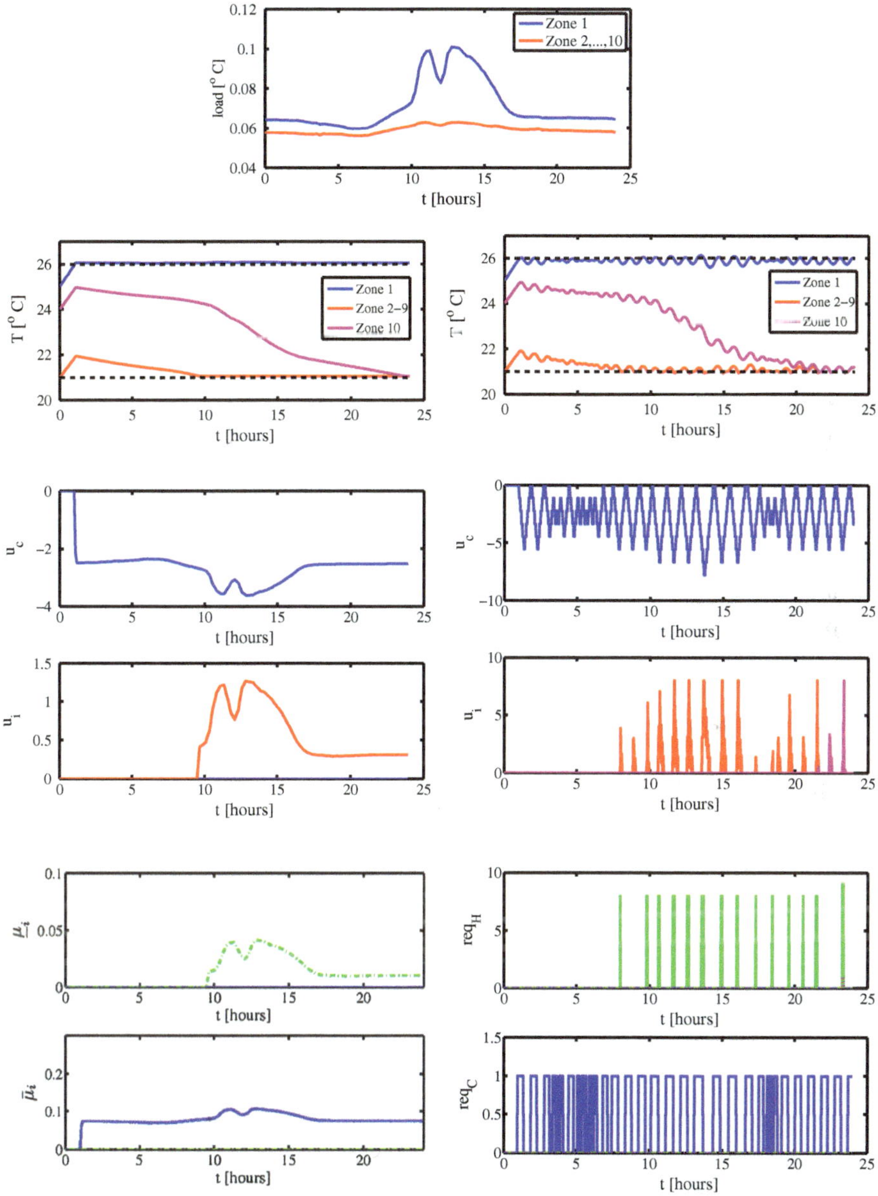


**Fig. 7.4** Top: Disturbance Profile $w$ over one day Left column: MPC simulation results Right column: "Trim and Respond" simulation results

puting community that communication between processors is dramatically slower than computation on a single chip [30, 31].

This section reviews a recently proposed primal-dual active-set algorithm that is particularly well suited for distributed implementation in systems with high communication costs. The algorithm is naturally decomposable for optimization on a star communication network for large-scale systems having dynamics coupled through the control inputs. Additionally, the decomposition fits a typical hardware architecture where the leaf nodes of the network have limited computational resources and the central hub of the star network has substantially more computing power. A detailed presentation and analysis is presented in [2, 3], but we review the main idea and the main results here.

Active-set methods are a class of constrained optimization algorithms. Active-set algorithms work by iteratively finding the subset of inequality constraints for which equality holds at the optimum. This set is called the optimal active set. In traditional active-set methods, only one constraint is added or removed from the active set during each iteration. As a result, traditional active-set methods can require an exponential number of iterations to find the optimal active set. In this chapter, we examine a primal-dual active-set method for strictly convex quadratic programs that was initially proposed by Hintermüller [32] and expanded upon by Curtis [33]. In this algorithm, multiple constraints are added to or removed from the active set during each iteration of the algorithm. As a result, the algorithm often converges in very few iterations and it exhibits local superlinear convergence [34].

The primal-dual active-set method is able to solve the following MPC problem:

$$\min_{X,U} \; (x_N - x_{ss})^T P(x_N - x_{ss}) + \sum_{k=0}^{N-1} \left((x_k - x_{ss})^T Q(x_k - x_{ss}) + u_k^T R u_k\right) \tag{7.33a}$$

$$s.t. \;\; x_{k+1} = Ax_k + Bu_k + \hat{w}_k \tag{7.33b}$$

$$\underline{x} \le x_k \le \bar{x}, \; k = 1, \ldots, N \tag{7.33c}$$

$$\underline{u} \le u_k \le \bar{u}, \; k = 0, \ldots, N-1 \tag{7.33d}$$

$$x_0 = x(t) \tag{7.33e}$$

with symmetric $Q, R, P \succ 0$. This matches the cost function defined by (7.23)–(7.25), and any linearized discrete time model is acceptable. The constraints are bound constraints and can encode the comfort constraints and input saturation constraints described in Sect. 7.5.2.

The distributed algorithm takes advantage of the separability of the constraints. Each leaf node keeps track of a relevant subset of the state variables and input variables. Recall that each leaf node represents a VAV box, so if using model (7.3), leaf node $i$ keeps track of its temperature $T_i$ and reheating input $u_{h,i}$. The algorithm begins by initializing the active sets at the leaf nodes. Each leaf node sends its active sets up to the central hub, where the full active sets are gathered. The central hub computes an update to the state variables and input variables that are not active and scatters the elements of $x_k$ and $u_k$ to the relevant leaf nodes. The $i$-th leaf node then updates its active-set based on these state and input updates. The algorithm terminates when the active sets no longer change.

We present results to compare the total execution time $T$ for the primal-dual active-set method, ADMM, and dual decomposition where $N = 5$ and the number of zones varies. The total problem size is equal to $N(2n + 1)$. The total execution time includes computation time and communication time. We use the best-case communication delay time of $\Delta t_{delay} = 0.03$ seconds per floating point to estimate communication delay. All algorithms converge to the same optimal solution, given enough iterations. Each result is averaged from a set of 10 trials, with the difference between trials being different initial conditions and disturbance profiles. The max number of iterations allowed is 10000.Note that the dual decomposition algorithm reaches the maximum number of iterations for the larger problems; we do not present the numerical results for these larger problems for dual decomposition since the convergence was not reached within the maximum number of iterations.

In practice, the number of iterations $n_{iters}$ is many orders of magnitude less for the primal dual active-set than for ADMM (Fig. 7.5). From these results, it is quite clear that the impact of communication on solver time is not negligible – even in the best case scenario for communication delay $\Delta t_{delay}$. The worst case result for communication delay for ADMM is on the order of days for the HVAC problem and the order of hours for the random problem. This result would be dramatically worse for the worst case communication latency of $\Delta t_{delay} = 0.4$ seconds per point.

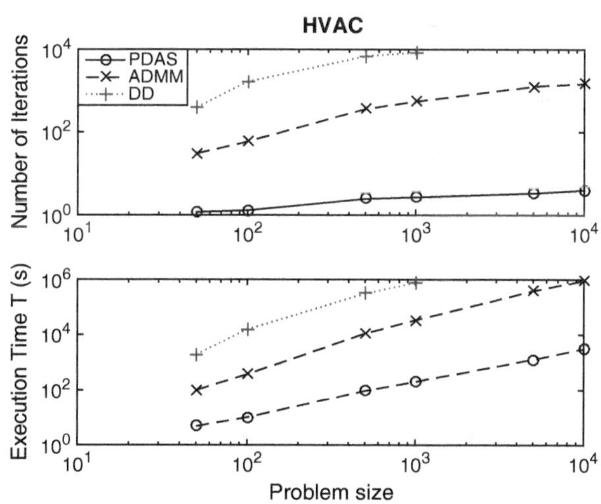


**Fig. 7.5** Communication and Execution Times versus Problem Size

## 7.7 Conclusion

In this chapter, we describe the control of forced-air systems, from currently implemented methods to advanced distributed model predictive control methods. We describe the system dynamics of the heating, ventilation, and air conditioning system. The current embedded platforms and communication systems are detailed. Current building practices, including "Trim and Respond" are described. We describe how model predictive control is designed for forced-air systems. Then, we show that "Trim and Respond" is equivalent to a special class of DMPC. Finally, an advanced distributed model predictive controller that fits the existing embedded platforms and communication systems is proposed.

**Acknowledgements** This material is based upon work supported by the National Science Foundation under Grant No. 1239552. The U.S. Department of Energy (DOE) and the Department of Science and Technology (DST), Government of India (GOI) provided joint funding for work under the U.S.–India Partnership to Advance Clean Energy Research (PACE–R) program's "U.S.–India Joint Center for Building Energy Research and Development" (CBERD) project. The Assistant Secretary for Energy Efficiency and Renewable Energy, Office of Building Technology, State and Community Programs, of the U.S. DOE under Contract No. DE-AC02- 05CH11231 supports the U.S. CBERD activity. The DST, GOI, administered by Indo-U.S. Science and Technology Forum, supports the Indian CBERD activity. The authors acknowledge Dr Philip Haves of Lawrence Berkeley National Laboratory for his advice and continued support.

## References

1. Koehler S, Borrelli F (2013) Building temperature distributed control via explicit MPC and "trim and respond" methods. In: European control conference, July 2013
2. Koehler S, Danielson C, Borrelli F (2015) A primal-dual active-set method for distributed model predictive control. In: American control conference, July 2015
3. Koehler S, Danielson C, Borrelli F (2016) A primal-dual active-set method for distributed model predictive control. Optim Control Appl Methods
4. Henze GP, Felsmann C, Knabe G (2004) Evaluation of optimal control for active and passive building thermal storage. Int J Thermal Sci 43(2):173–183
5. Liu S, Henze GP (2006) Experimental analysis of simulated reinforcement learning control for active and passive building thermal storage inventory: part 1. theoretical foundation. Energy Build 38(2):142–147
6. Mayne D, Rawlings J, Rao C, Scokaert P (2000) Constrained model predictive control: stability and optimality. Automatica 36(6):789–814
7. Ma Y, Borrelli F, Hencey B, Coffey B, Bengea S, Haves P (2012) Model predictive control for the operation of building cooling systems. IEEE Trans Control Syst Technol 20(3):796–803
8. Oldewurtel F, Parisio A, Jones C, Morari M, Gyalistras D, Gwerder M, Stauch V, Lehmann B, Wirth K (2010) Energy efficient building climate control using stochastic model predictive control and weather predictions. In: American control conference, July 2010, pp 5100–5105
9. Ma Y, Kelman A, Daly A, Borrelli F (2012) Predictive control for energy efficient buildings with thermal storage: modeling, stimulation, and experiments. IEEE Control Syst 32(1):44–64
10. Bengea S, Kelman A, Borrelli F, Taylor R, Narayanan S (2012) Model predictive control for mid-size commercial building HVAC: implementation, results and energy savings. In Second international conference on building energy and environment, pp 979–986

11. EnergyPlus. https://energyplus.net/. Accessed 19 Aug 2016
12. Wetter M, Zuo W, Nouidui TS, Pang X (2014) Modelica buildings library. J Build Perform Simul 7(4):253–270
13. Kelman A, Ma Y, Borrelli F (2012) Analysis of local optima in predictive control for energy efficient buildings. J Build Perform Simul
14. ALC control products system architecture. http://www.automatedlogic.com/pages/products.aspx
15. ZN253 zone controller. http://www.automatedlogic.com/specsheets/zn253_cs_r10_hires.pdf. Accessed 19 Aug 2016
16. BACnet Website. http://www.bacnet.org/index.html. Accessed 22 Aug 2016
17. Wong S, Hong S, Bushby S (2003) NISTIR 7038A simulation analysis of BACnet local area networks, National Institute of Standards and Technology
18. Introduction to BACnet for building owners and engineers (2014). https://www.ccontrols.com/pdf/BACnetIntroduction.pdf
19. AAR: high speed ARCNET to ARCNET router. http://www.automatedlogic.com/SpecSheets/csaarrcv4.pdf
20. Taylor S (2007) Increasing efficiency with VAV system static pressure setpoint reset. In: American Society of Heating, Refrigerating and Air-conditioning Engineers
21. Ma Y, Kelman A, Daly A, Borrelli F (2012) Predictive control for energy efficient buildings with thermal storage: modeling, simulation, and experiments. IEEE Control Syst Mag 32(1):44–64, 2
22. Taylor S (2015) Resetting setpoints using trim and respond logic. In: American Society of Heating, Refrigerating and Air-conditioning Engineers
23. Borrelli F, Bemporad A, Morari M (2014) Predictive control. http://www.mpc.berkeley.edu/mpc-course-material
24. R. American Society of Heating and A.-C. Engineers (2013) Standard 55–2013 - thermal environmental conditions for human occupancy. https://www.ashrae.org/resources-publications/bookstore/standard-55
25. Bertsekas DP, Tsitsiklis JN (1989) Parallel and distributed computation, vol 290. Springer Englewood Cliffs
26. Boyd S, Xiao L, Mutapcic A (2003) Notes on decomposition methods. http://web.stanford.edu/class/ee392o/decomposition.pdf
27. Boyd S, Parikh N, Chu E, Peleato B, Eckstein J (2011) Distributed optimization and statistical learning via the alternating direction method of multipliers. Found Trends Mach Learn 3(1):1–122
28. Pu Y, Zeilinger M, Jones C (2014) Inexact fast alternating minimization algorithm for distributed model predictive control. In: 53rd conference on decision and control, pp 5915–5921
29. Wie E, Ozdaglar A, Jadbabaie A (2013) A distributed newton method for network utility maximization. IEEE Trans Autom Control 58
30. Asanovic K, Catanzaro B, Gebis J, Husbands P, Patterson D, Plishker W, Shalf J, Williams S W, Yelick K (2006) The landscape of parallel computing research: a view from Berkeley. Electrical Engineering and Computer Sciences, University of California at Berkeley, UCB/EECS-2006-183, December 2006. http://www.eecs.berkeley.edu/Pubs/TechRpts/2006/EECS-2006-183.html
31. Hager G, Wellein G (2010) Introduction to high performance computing for scientists and engineers. CRC Press, Boca Raton
32. Hintermüller M, Ito K, Kunisch K (2002) The primal-dual active set strategy as a semismooth Newton method. SIAM J Optim 13
33. Curtis F, Han Z, Robinson D (2015) A globally convergent primal-dual active-set framework for large-scale convex quadratic optimization. Comput Optim Appl 60(2):311–341
34. Hintermüller M (2010) Semismooth Newton methods and applications. [Online]. http://www.math.uni-hamburg.de/home/hinze/Psfiles/Hintermueller_OWNotes.pdf

# Chapter 8
# Human-in-the-Loop Thermal Management for Smart Buildings

**Santosh K. Gupta and Koushik Kar**

## 8.1 Introduction

In spite of the tremendous advances in HVAC system modeling and control, the occupant dissatisfaction associated with the prevailing indoor thermal conditions has been highlighted by several studies [2, 3]. A more recent news article [4] titled, "This cold war, too, has roots in the past," highlights the problem with thermal comfort model that is being followed by commercial buildings. The occupants (employees), in turn, have to face discomfort with regard to indoor thermal environment leading to direct loss of productivity and reduced overall satisfaction.

Occupant dissatisfaction associated with the indoor building environment (both residential and commercial) can be mainly attributed to the existing methods for indoor environment management employed commonly by the Building Management Systems (BMS). Most of the multi-occupant space environment control systems today are based on setpoint scheme of thermal management. Either there is a centralized control by operation managers of the building, or the occupants set a specific temperature set point for the indoor environment agnostic of the changes in ambient conditions and occupancy patterns. However, a static setpoint does not seem to fit the dynamic lifestyle that we normally follow. This calls for the design of smart systems that can achieve desired human comfort levels without putting additional pressure on energy resources, which the existing set point scheme fails to achieve.

---

Parts of the work in this chapter have been extracted from [1] (©[2015] IEEE. Reprinted, with permission, from S.K. Gupta, K. Kar, S. Mishra, J.T. Wen, "Collaborative energy and thermal comfort management through distributed consensus algorithms," IEEE Trans. Automation Science and Eng., vol. 12, no. 4, pp. 1285–1296, Oct. 2015.).

---

S.K. Gupta (✉) · K. Kar
ECSE Department, Rensselaer Polytechnic Institute, Troy, NY 12180, USA
e-mail: santoshgpt@gmail.com

K. Kar
e-mail: koushik@ecse.rpi.edu



J.T. Wen and S. Mishra (eds.), *Intelligent Building Control Systems*,
Advances in Industrial Control, https://doi.org/10.1007/978-3-319-68462-8_8

There have been studies based on weather prediction models that use the concept of preheating and cooling [5] to reduce energy consumption. However, this does not necessarily take into account occupant behavior and/or preferences. Some of the more recent studies based on human physiology [6] do propose an aggregate or average model for occupants, but fails to provide a mechanism to incorporate individual differences in body metabolism and preferences. Smart and learning thermostats like Google Nest® [7] that claim to learn occupant preference dynamically, may however fail to perform in multi-occupant spaces.

The framework proposed in this chapter has the following novel components:

- Simultaneous minimization of energy cost and maximization of occupant comfort, with the flexibility to modify weights associated with each (say for instance to incorporate real-time energy pricing).
- Multi-zoning solution that takes into account the thermal correlation among different zones in a residential home or a commercial building.
- Consensus algorithm to enable deployment to multi-occupant spaces (corporate office buildings, university buildings, etc.). This algorithm drives co-located occupants to a logical consensus on thermal setting based on individual preferences.

Personalized comfort level expectations pose a conflicting situation in multi-occupant spaces such as residential homes, research laboratories, corporate office buildings, student dorms, etc., where occupants have their own range of thermal comfort and other environmental settings. This range generally depends on individual occupant body type, external factors such as attire, physical and mental condition, and level of tolerance; and can also vary depending on other environmental factors such as time of the day, lighting conditions etc. [8–10]. This personalized thermal comfort range can be best captured by individual occupant feedback in real time. Further, in shared multi-occupant spaces personal comfort levels are affected both by the presence of co-occupants and the correlation between temperatures in different zones and rooms occupied [6, 11, 12]. Arriving at temperature set points to minimize the aggregate discomfort among all occupants of different rooms or zones in a building is an important yet challenging problem. With rising energy cost and emphasis on energy conservation, the total energy cost also needs to be accounted for when trying to determine the optimal temperature set points in different zones of a building.

We present a solution to optimally trade off the overall energy cost with the aggregate occupant discomfort. Our solution is occupant centric and focuses on designing an algorithm for efficient human interaction and utilizing the feedback. We start by depicting a high level work flow of the proposed solution for energy efficient operation of buildings in Fig. 8.1.

*Occupant comfort and sensor measurement:* User experience is of utmost importance, and our solution also starts with the perception of thermal comfort level by the occupants to the current temperature settings. Simultaneously static sensors if present in the zones can feed real-time zonal temperature (sensors can also be used to measure humidity and other related comfort parameters) to the central processor.

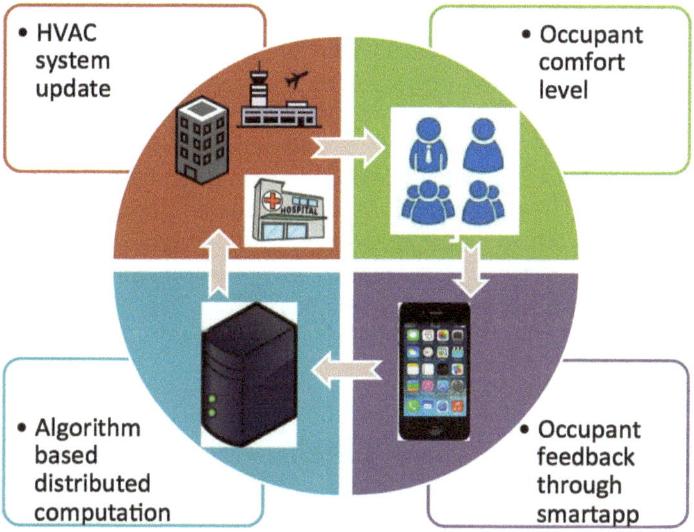


**Fig. 8.1** High-level process flow depiction of the proposed algorithm. The process runs in a loop incorporating real-time changes to occupancy pattern, occupant preferences, and/or environmental changes

*Occupant feedback through smart phone application:* The occupants can use their smart phones to access our mobile application and provide their temperature-preference feedback. It also shows the percentage distribution of the occupants based on their preferences for the given zone of the building. The displayed information gets auto updated at specific intervals. An update request can also be generated by the user manually.

*Algorithm for Optimal temperature determination:*

The minimization objective is an aggregate of all the occupant discomfort functions and the total energy cost, subject to the constraint of a common temperature set point for each zone. Pricing per unit temperature change serves as the feedback signal to the occupants, to drive them to declare their truthful temperature preference that optimizes the overall discomfort plus energy cost objective. The occupant feedback obtained through smart phone application are used by the central server under building management running the algorithm to determine the optimal zonal temperatures.

*HVAC system update:* The HVAC system is then set to the predicted optimal zonal temperature settings. For an integrated system, the thermostat(s) can be automatically set to the suggested temperature settings. Our solution is designed to work for any multi-occupant space such as corporate building, airports, aircrafts, homes, apartments, etc.

For the rest of the chapter we start by presenting the overall system model in Sect. 8.2 that includes the occupant discomfort function and building heat transfer model. In Sect. 8.3, we present our novel algorithm that is developed and analyzed using *gradient optimization* and *singular perturbation theory*. We use simple feedback from users in the form of "heat up" or "cool down", which are further consolidated to estimate their comfort ranges or the discomfort functions. In Sect. 8.4, we discuss the consensus algorithm that we developed through the use of *alternating direction method of multipliers*. In Sect. 8.5, we share results from simulation and experimental study based on our Watervliet-based test facility. Finally, we present some concluding remarks in Sect. 8.6. Parts of the work in this chapter have been extracted from [1] (©[2015] IEEE. Reprinted, with permission, from S. K. Gupta, K. Kar, S. Mishra, J. T. Wen,"Collaborative energy and thermal comfort management through distributed consensus algorithms," *IEEE Trans. Automation Science and Eng.*, vol. 12, no. 4, pp. 1285–1296, Oct. 2015.).

### 8.1.1 Existing Solutions Incorporating Occupant Feedback

We will first briefly review the existing building thermal control studies that incorporate occupant feedback in some form. The predicted mean vote (PMV) index has been used as the metric for user comfort integration in multiple studies [13–15]. Some studies proposed sensor network solutions to increase the accuracy of PMV calculation [16, 17]. Owing to the complexity of sensor network deployment, a number of studies have proposed utilizing occupant feedback for thermal comfort integration into the control logic of building systems. Through custom keyboards in each room, Guillemin and Morel made use of occupant preferences in the form of temperature set points [18]. Murakami et al. used binary preference of warmer and cooler along with a logic to build consensus for controlling the air-conditioning set point [19]. Daum et al. utilized too hot/too cold occupant complaint along with a probabilistic approach for determining user comfort profiles [20]. Thermovote [6] utilized a seven- level occupant comfort voting to integrate with the building control logic. Purdon et al. developed a smart phone interface to receive 3-point scale comfort feedback from occupants and determine the direction for temperature drift with a system defined step-size [21]. Some more recent works have conducted experimental study for a group of occupants [22], and have presented thermal comfort model for a single person [23]. However, these are based on thermal complaint behavior using one-class classifier. Jazizadeh et al. used a fuzzy predictive model to learn occupant comfort profiles and a complementary control strategy for the HVAC control [24]. Zhao et al. conducted a simulation study tying occupant subjective thermal comfort feedback with MPC control algorithm for the active HVAC system against a baseline rule-based control algorithm [25]. Experimental facilities, such as Living Lab [26] at Purdue University are actively utilized by researchers to understand human behavior and productivity relationship to indoor environment and personalized control. Choi et al. have presented HVAC thermal models based on sensing of human vitals such as heart rate [27].

## 8.2 Human-in-the-Loop System Model

### 8.2.1 Occupant Discomfort Modeling

Occupant thermal comfort modeling has been extensively researched and can be summarized into the following three major approaches: (1) the chamber study model, based on mapping thermal comfort from environmental and personal factors to a 7-level comfort value scale, viz., the Predicted Mean Vote - Predicted Percent Dissatisfied (PMV-PPD) [8, 12]; (2) human body physiology based models such as Gagge's core to skin model [28], Stolwijk's comfort model for multi-human segments [29], and Zhang et al.'s sensation on human body segments [30]; and (3) adaptive comfort models developed in field study, viz., Humphreys [31, 32]. Recent work based on thermal complaint behavior using one-class classifier [22, 23] have also been presented. However, existing work mainly focus on average thermal comfort models instead of individual comfort modeling. Such group comfort models only capture average behavior and are not particularly useful in maximizing aggregate comfort for multi-occupant spaces, with individual thermal preferences differing from each other.

In the current work, therefore, we take into consideration discomfort functions of the occupants individually modeled as convex quadratic functions of temperature variation based on the PMV-PPD model. Our model captures the difference across occupants in their comfortable temperature range. For simulation and experimental study, we adopt occupant discomfort function of the form:

$$D_i(y_j) = \begin{cases} (y_j - y_s^U)^2 & \text{if } y_j > y_s^U, \\ 0 & \text{if } y_s^L \leq y_j \leq y_s^U, \\ (y_j - y_s^L)^2 & \text{if } y_j < y_s^L, \end{cases} \tag{8.1}$$

where $y_s^U$ and $y_s^L$ are the upper and lower limit temperatures, respectively, of the user $i$ located in zone $j$.

Note that the boundary values can be either explicitly indicated by the user, or can be conveyed implicitly to the system through user feedbacks in a simple binary form of whether he/she is feeling hot or cold in the current setting.

### 8.2.2 Optimization Objective

Consider a building with $m$ zones, and let $S_j$ represent the set of occupants located in zone $j$ of the building. Let $D_i$ represent the discomfort function of occupant $i$, and function $E$ the overall energy cost. Then a reasonable objective is to attain (in steady state) the zonal temperature vector $y$ that achieves the following objective:

$$\text{minimize} \sum_{j=1}^{m} \sum_{i \in S_j} D_i(y_j) + E(u). \quad (8.2)$$

where $y_j$ is the temperature of zone $j$, and $u$ is the heat input vector that is required to attain those zonal temperatures. Note that an occupant $i$ located in zone $j$ (i.e., $i \in S_j$) experiences temperature $y_j$, and therefore its discomfort can be represented as $D_i(y_j)$. We assume the discomfort function $D_i(y_j)$ as convex in its argument $y_j$. It is worth noting that the discomfort function *need not be* "strictly" convex. This allows for the occupants to be insensitive to temperature fluctuations over a certain range; or in other words, the discomfort function could be flat over the occupant's "comfort range".

In the above, $E(u)$ is assumed to be a convex function of the control input vector $u$. For the sake of definiteness, we use $E(u)$ to be of the following quadratic form (although other convex forms of the function $E(u)$ are also allowed by our framework):

$$E(u) = u^T \Gamma u, \quad (8.3)$$

where $\Gamma$ is a positive definite matrix. The $\Gamma$ matrix captures the weight of the energy cost relative to the total discomfort cost. In practice, it could be determined by the actual cost of energy, as well as additional input from the building operator to determine how much relative weight to associate with the energy cost as compared to the occupant discomfort cost.

Finally, since the optimization variable in the objective function (8.2) is only the zonal temperature vector $y$, the relationship between the heat input vector $u$ and the zonal temperature vector $y$ needs to be stated to make the formula meaningful. We can express $u = g(y)$ and using that write the energy cost $E(u)$ as $G(y) = E(g(y))$, where the function $G(y)$ is convex in $y$. Taking the case of an RC model, we would express function $g(.)$ in terms of model parameters for the purpose of simulation later in the chapter.

### *8.2.3 Building Heat Transfer Model*

Multiple building modeling strategies have been proposed in the literature, which include the finite element method-based model [33], lumped mass and energy transfer model [34], and graph theoretic model based on electrical circuit analogy [35–39]. The system model selection entails a trade-off between computational efficiency and accuracy of representation of the temperature dynamics. For the purpose of this study, we take the electrical circuit analogy approach, and combine it with the distributed consensus algorithm to achieve collaborative temperature control of buildings. A building is modeled as a collection of interconnected zones, with energy/temperature dynamics evolving according to a lumped heat transfer model.

In the lumped heat transfer model, a single zone is modeled as a thermal capacitor and a wall is modeled as an RC network. This results in the standard lumped 3R2C wall model [40]. The heat flow modeling is based on temperature difference and thermal resistance: $Q = \Delta T/R$, where $\Delta T$ is the temperature difference, $R$ is the thermal resistance and $Q$ is the heat transferred across the resistance. This is analogous to the current due to voltage difference across a resistor. Also, note that the thermal capacitance denotes the ability of a space to store heat: $C\frac{\mathrm{d}\Delta T}{\mathrm{d}t} = Q$.

The heat flow and thermal capacitance model can be written for all the thermal capacitors in the system, with $T_i$ as the temperature of the $i$th capacitor. Consider the system to have $n$ thermal capacitors and $l$ thermal resistors. Taking the ambient temperature ($T_\infty$) into account, and neglecting any "thermal noise" in the system, we can write the overall heat transfer model of the system with $m$ zones as [41]:

$$C\dot{T} = -DR^{-1}D^T T + B_0 T_\infty + Bu \tag{8.4}$$

where $T \in \mathbb{R}^n$ is the temperature vector (representing the temperature of the thermal capacitors in the model), $u \in R^m$ is the vector of heat inputs into the different zones of the building, and $B \in \mathbb{R}^{n \times m}$ is the corresponding input matrix. Also, note that $(T, u)$ are functions of time $(T(t), u(t))$ and accordingly $\dot{T} = \frac{dT}{dt}$. Note that positive values of $u$ correspond to heating the system while negative values of $u$ correspond to cooling. In the above equation, $C \in \mathbb{R}^{n \times n}$ consists of the wall capacitances and is a diagonal positive definite matrix; $R \in \mathbb{R}^{l \times l}$ consists of the thermal resistors in the system and is a diagonal positive definite matrix as well. Also, $D \in \mathbb{R}^{n \times l}$ is the incidence matrix, mapping the system capacitances to the resistors, and is of full row rank [37], and $B_0 = -DR^{-1}d_0^T \in \mathbb{R}^n$ is a column vector with nonzero elements denoting the thermal conductances of nodes connected to the ambient.

In our model, the zones are picked such that each of them has a heating/cooling unit, which in turn implies that $B$ is of full row rank. Also, since matrix $D$ is of full row rank the product $DR^{-1}D^T$ is a positive definite matrix. The vector of zone temperatures, denoted by $y$ (which is a function of $T$) can be expressed as

$$y = B^T T. \tag{8.5}$$

## 8.3 Singular Perturbation Method

In this section, we present a novel algorithm using gradient optimization and analyzed with *singular perturbation theory*. The user input when combined with the resulting energy cost, determines the direction in which the energy control input is adjusted. We consider a multi-zone building, and use a lumped heat transfer model based on thermal resistance and capacitance for system analysis. Collecting user feedback before the system has settled down sufficiently close to the equilibrium value (for a given control input) is problematic however, as such user feedback do not correctly reflect the effect of the current control input on user discomfort. This implies that for convergence of our control algorithm to the optimal, sufficient separation between the user feedback frequency and the dynamics of the system is necessary. We use singular perturbation theory to analyze the system, with temperature evolution on a faster time scale and user input on a relatively slower time scale. With such time scale separation, we establish the stability condition under which the proposed control algorithm achieves convergence to a desired temperature that minimizes the sum of total energy cost and the aggregate occupant discomfort. We run simulations (using parameters of our smart building testbed) as well as conduct experimental study in the testbed to establish viability and evaluate the performance of our proposed algorithm.

### *8.3.1 Solution Approach*

Using equilibrium condition (setting $\dot{T} = 0$ in (8.4)) we obtain:

$$T = h(u) = (DR^{-1}D^T)^{-1}(B_0 T_\infty + Bu). \tag{8.6}$$

$$\text{Define, } J(u) = U(u, h(u)), \tag{8.7}$$

i.e., $J(u)$ is obtained by plugging in $T = h(u)$ from (8.6) into (8.2). Note that energy cost term in (8.2) is strictly convex in $u$; and the aggregate occupant term is convex in $T$, and therefore convex in $u$ when $T$ is set to $h(u)$, since $h(u)$ is affine in $u$. This implies that $J(u)$ is strictly convex in $u$. Therefore, $J(u)$ has a unique optimal solution $u^*$. Define

$$T^* = h(u^*), \tag{8.8}$$

which is also unique by definition.

With the goal of driving the system to $(u^*, T^*)$, we propose the control input $u$ be updated once every $\Delta$ time units as

$$u_{k+1} = u_k - \eta\big(\Gamma u + Y \Lambda F(y)\big), \tag{8.9}$$

where $\eta$ is a scalar that can be loosely interpreted as the "feedback gain" of the system.

Furthermore, $Y \in \mathbb{R}^{m\times m}$ in the above is the Jacobian obtained using (8.5) and the equilibrium condition (8.6), expressed as

$$Y = (\frac{\partial y}{\partial u}) = B^T(DR^{-1}D^T)^{-1}B. \tag{8.10}$$

Also, $\Lambda \in \mathbb{R}^{m\times \rho}$ is the zone-occupant matrix that indicates which occupants are present in a zone ($\Lambda_{js} = 1$ if $s \in S_j$, and 0 otherwise), and $F(y) \in \mathbb{R}^{\rho\times 1}$ is the "marginal discomfort" vector of the occupants, obtained by taking partial derivative of the occupant discomfort functions with respect to $y$. In other words, the $s^{\text{th}}$ element of $F(y)$, where $s \in S_j$, is obtained as

$$F_s(y_j) = \frac{\mathrm{d}D_i(y_j)}{\mathrm{d}y_j}, \quad s \in S_j. \tag{8.11}$$

Comparing (8.9) with (8.2) provides the motivation of our control algorithm: roughly speaking, (8.9) updates $u$ in the gradient direction of $U(u, T)$, while taking in account the relationship between $T$ and $u$ at equilibrium, as given by (8.6). In other words, it attempts to update $u$ is the direction of $-\nabla J(u)$, where $J(u)$ is defined by (8.7). In this interpretation, $\eta$ represents the constant "step size" associated with the gradient descent.

Note however that using (8.2)–(8.7), $\nabla J(u)$ is expressed as

$$\nabla J(u) = \Gamma u + Y \Lambda F(B^T h(u)). \tag{8.12}$$

From (8.12) we note that update of $u$ in the gradient direction of $J(u)$ requires user discomfort feedback at $y = B^T h(u)$, the equilibrated zone temperatures corresponding to $u$. In practice, however, a user $s \in S_j$ will provide a comfort feedback at the current temperature it experiences, $y_j = [B^T T]_j$ (different in general from the equilibrated temperature $[B^T h(u)]_j$), which is what we incorporate into our control algorithm as stated in (8.9). This implies that our control algorithm as described in (8.9) does not exactly move $u$ in the gradient direction ($-\nabla J(u)$).

The effect of this difference (error) can be analyzed using *singular perturbation theory* [42, 43], which in our case requires (for convergence to optimality) that the occupant feedback be collected after long intervals (i.e., $\Delta$ is large), allowing the temperature $T$ to settle down close to $h(u)$ before the next occupant feedback collection.

Toward developing a singular perturbation model of our system, we first consider a continuous approximation to the evolution of the control input $u$:

$$\dot{u} \approx \frac{u_{k+1} - u_k}{\Delta} = -\frac{\eta}{\Delta}\Big(\Gamma u + Y \Lambda F(y)\Big). \tag{8.13}$$

Note that time step $\Delta$ is the interval at which user feedback is solicited and the control input $u$ is updated. A larger $\Delta$ implies a slower evolution of $u$.

We next express the system evolution in the time scale of the evolution of $u$ (slower time scale as compared to the time scale at which $T$ evolves). Define $\varepsilon = \frac{1}{\Delta}$ as the perturbation parameter; then $\tau = \frac{t}{\Delta} = \varepsilon t$ is the slower time scale. Then

$$\frac{\mathrm{d}\tau}{\mathrm{d}t} = \varepsilon \implies \dot{u} = \frac{\mathrm{d}u}{\mathrm{d}t} = \varepsilon\Big(\frac{\mathrm{d}u}{\mathrm{d}\tau}\Big); \; \dot{T} = \frac{\mathrm{d}T}{\mathrm{d}t} = \varepsilon\Big(\frac{\mathrm{d}T}{\mathrm{d}\tau}\Big). \tag{8.14}$$

Control input Eq. (8.13) can now be expressed in terms of $\tau$ as follows, using the fact that $y = B^T T$:

$$\frac{du}{d\tau} = -\eta\Big(\Gamma u + Y\Lambda F(B^T T)\Big). \tag{8.15}$$

Similarly, Eq. (8.4) modeling the temperature evolution of the building can now be expressed as:

$$C\varepsilon\frac{dT}{d\tau} = -DR^{-1}D^T T + B_0 T_\infty + Bu. \tag{8.16}$$

Equations (8.15) and (8.16) represent a singularly perturbed system. Note, $\Delta \uparrow \Longrightarrow \varepsilon \downarrow$ leading to steady state condition for temperature evolution. In the next section, as we establish the global asymptotic stability of our system as given by Eqs. (8.15) and (8.16).

Finally, note that implementation of our control algorithm would require that $F_s(y_j)$, the "marginal discomfort" value of user $s$ in zone $j$ at the current zonal temperature $y_j = [B^T T]_j$, be reasonably estimated from the discomfort feedback of $s$ at any time. In practice, the occupants may provide the feedback in some simple form describing their actual level of discomfort ("I am feeling hot," "I am feeling very cold," etc.). This feedback must be processed to estimate the marginal discomfort (derivative of the actual discomfort function), as we do in our simulation study described in Sect. 8.5.

### *8.3.2 System Analysis*

The system evolution is governed by the set of Eqs. (8.15) and (8.16). In Eq. (8.16) the coefficient $DR^{-1}D^T$ is positive definite, which makes the unforced system (with $u = 0$) exponentially stable. We use singular perturbation analysis [43] to establish the condition for stability of the system.

**Theorem 1** *There exists an $\varepsilon^* > 0$ such that $(u^*, T^*)$ is a globally asymptotically stable equilibrium of the system given by (8.15) and (8.16) for all $\varepsilon < \varepsilon^*$.*

We just introduce the Lyapunov functions $V(u)$ and $W(u, T)$ that is used in the stability analysis:

$$V(u) = J(u) - J(u^*), \text{ and} \tag{8.17}$$

$$W(u, T) = (T - h(u))^T P(T - h(u)), \tag{8.18}$$

where $P$ in the above equation is a symmetric positive definite matrix (the exact choice of matrix $P$ will be determined at a later stage). We now define a combined Lyapunov function $L(u, T)$:

$$L(u, T) = (1 - \alpha)V(u) + \alpha W(u, T), \tag{8.19}$$

where $\alpha$ satisfies $0 < \alpha < 1$.

Detailed analysis involves evaluating the conditions to establish stability using Theorem 2.1 and Corollary 2.1 from Chap. 7 of [43], and can be referred to in our work [44]. Finally, note that $L(u, T)$ is minimized uniquely at $(T^*, u^*)$. It follows therefore that $(T^*, u^*)$ is a globally asymptotically stable equilibrium point for all $\varepsilon < \varepsilon^*$, or all $\Delta > \Delta^*$.

## 8.4 Consensus Algorithms

In the singular perturbation method, the occupants have no incentive to provide truthful feedback. A malicious user can gain the system by providing irrational feedback. Moreover, users or their agents have to provide constant feedback to maintain the temperature at their desired levels. We address these drawbacks through our consensus algorithm developed in this section.

Achieving a common temperature set point in a distributed framework, where the exact discomfort functions are held privately by each occupant, remains an open question which we address in this section. Pricing per unit temperature change serves as the feedback signal to the occupants, to drive them to a consensus on zonal temperatures that optimize the overall discomfort plus energy cost. The consensus algorithm that we develop, through the use of the alternating direction method of multipliers (ADMM), is amenable to distributed implementation and has the following appealing properties. First, occupants (or their agents) are only assumed to be rational, in that they choose their preferred temperature set points so as to minimize their personal discomfort plus energy cost, given the pricing signals. In other words, the occupants are not required to explicitly declare their discomfort functions (which can be held privately), but only react rationally to the pricing signals by choosing their preferred temperature set point. On the other hand, the building thermal management system (BTMS) chooses the zonal temperature set points to maximize the overall profit of the building operator (for the current prices); the price signals are then updated so as to attain consensus among the occupants, and with the building operator, on the zonal temperatures. Finally, as we formally show, the algorithm converges to the optimal zonal temperatures, from which rational occupants would not have any incentive to deviate. In terms of practical implementation, occupant feedback could be obtained through a smart phone application, and the zonal temperature set points could be calculated by the building operator on a central server.

### 8.4.1 Solution Approach

If the individual occupant discomfort functions are assumed to be known to the building operator, the optimal zonal temperature vector $y^*$ could be computed directly. Such a centralized approach suffers from several practical limitations, however. Firstly, reporting the entire discomfort function to the building operator is complex, and the occupant may not even be able to correctly estimate its discomfort function. Second, even if we assume that the occupant knows its discomfort function exactly, there is no incentive for it to report the same truthfully. In practice, therefore, it may be more desirable to have a mechanism through which the building operator indirectly learns about the true discomfort functions of the occupants, who are providing their temperature preference feedback in a simple and convenient format, acting in best response to some price signals provided by the building operator. Furthermore, the price signals should be such that it guides the occupants toward a *consensus*, i.e., rational users (acting in self-interest) in a zone will end up agreeing on their temperature choice for each zone. The distributed consensus algorithm that we describe in the next subsection works according to the above principles.

To provide an overview of our approach, we first introduce new notations to denote the choice of zonal temperatures by the occupants and the building thermal management system (BTMS); these temperature choices will in general be different from the actual (current) zonal temperatures. Let $x_{ij}$ denote the desired temperature of occupant $i \in S_j$ located in zone $j$. Let $z_j$ denote the target temperature of zone $j$ as set by the BTMS. Then, vector $z$ represents the target temperature of the entire building consisting of $m$ zones. In general, $x_{ij}$ for any occupant $i \in S_j$ can differ from $z_j$; the actual zonal temperature $y_j$ could also differ from these temperatures. On convergence however, the consensus algorithm ensures that $x_{ij}$ for all occupants $i \in S_j$ equals $z_j$, which optimizes the objective function in (8.2). The zonal temperatures obtained through consensus is then attained in the building by utilizing some temperature set point-based HVAC control system.

In the distributed consensus algorithm, the BTMS provides pricing signal $p_{ij}$ and target zonal temperature $z_j$ to each occupant $i$ located in zone $j$. Based on that, occupant $i$ chooses its desired temperature $x_{ij}$ and sends it back to the BTMS, which is then used to recompute the new target zonal temperatures and pricing signals. Figure 8.2 provides an overview of how the distributed consensus algorithm works. Upon convergence of the distributed consensus algorithm, let $x^*$, $z^*$, and $p^*$, respectively, denote the (vectors of) occupant temperature choices, zonal target temperatures, and pricing signals. Then, we desire that $x^*, z^*, p^*$ satisfy the following properties:

- *(Individual Rationality)* Each occupant chooses its desired temperature so as to minimize its total cost function, represented as the sum of its discomfort plus the energy price paid to the building operator:

$$x_{ij}^* = \arg\min_{x_{ij}} \left\{ D_i(x_{ij}) + p_{ij}(x_{ij} - z_j) \right\}.$$

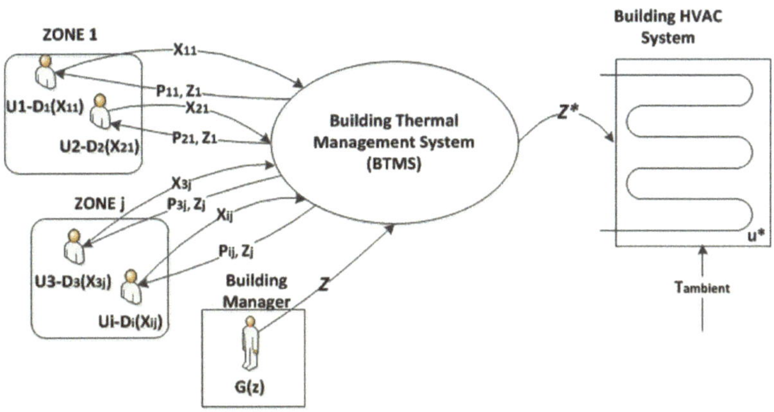


**Fig. 8.2** Distributed consensus algorithm flow between the building occupants and the central building management system

- *(Consensus)* For each zone, the temperature choices of the occupants of the zone agree with each other, and with the target zone temperature set by the BTMS:

$$x_{ij}^* = z_j^*, \ \forall i \in S_j.$$

- *(Optimality)* The target zone temperatures minimize the aggregate occupant discomfort plus the building energy cost, given by (8.2):

$$z^* = \arg\min_z \sum_{j=1}^{m} \sum_{i \in S_j} D_i(z_j) + E(g(z)).$$

#### 8.4.1.1 Distributed Consensus Algorithm

To develop the consensus algorithm, we rewrite the minimization objective in (8.2) in terms of the zonal temperature choices of the occupants and the BTMS, as:

$$\begin{aligned} &\text{minimize} \sum_{j=1}^{m} \sum_{i \in S_j} D_i(x_{ij}) + G(z) \\ &\text{subject to } x_{ij} = z_j, i \in S_j, \end{aligned} \tag{8.20}$$

where function $G(z) = E\big(g(z)\big)$ represents the total energy cost in terms of the target zonal temperature vector $z$.

We can now solve (8.20) through the ADMM approach as described in [45]. The ADMM approach blends the decomposability of dual ascent with the superior convergence properties of the method of multipliers, to develop an algorithm that is amenable to distributed implementation, and also has good convergence properties.

To motivate the ADMM-based consensus algorithm, let us consider the augmented Lagrangian:

$$L_\rho(x, z, p, \rho) = \sum_{j=1}^{m} \sum_{i \in S_j} \Big( D_i(x_{ij}) + p_{ij}(x_{ij} - z_j) + (\rho/2)|x_{ij} - z_j|^2 \Big) + G(z) \tag{8.21}$$

where $p_{ij}$ is the dual variable, $\rho > 0$ is a constant.

The ADMM-based consensus algorithm can then be derived as iterations of coordinate-wise optimization of this augmented Lagrangian along each $x_{ij}$ and $z$ directions, followed by update of the dual variable in a gradient direction. More precisely, in our consensus algorithm, in iteration $k = 1, 2, \ldots$, the variable vector $z$, and the variables $x_{ij}$, $p_{ij}$ for all $i \in S_j$, $j = 1, \ldots m$, are updated as follows:

$$x_{ij}^{k+1} := \operatorname*{argmin}_{x_{ij}} \Big( D_i(x_{ij}) + p_{ij}^k x_{ij} + (\rho/2)|x_{ij} - z_j^k|^2 \Big), \tag{8.22}$$

$$z^{k+1} : = \operatorname*{argmin}_{z} \Big( G(z) + \sum_{j=1}^{m} \big( - \sum_{i \in S_j} p_{ij}^k z_j + \sum_{i \in S_j} (\rho/2)|x_{ij}^{k+1} - z_j|^2 \big) \Big), \tag{8.23}$$

$$p_{ij}^{k+1} := p_{ij}^k + \rho(x_{ij}^{k+1} - z_j^{k+1}). \tag{8.24}$$

The above set of update equations has a nice game theoretic (price-driven rational-response) interpretation, as follows. The BTMS iteratively communicates to each occupant $i$ in any zone $j$ two parameters, $p_{ij}$ and $z_j$, based on which the occupant's cost (price paid) for a chosen temperature set point $x_{ij}$ would be computed as $p_{ij}x_{ij} + (\rho/2)|x_{ij} - z_j|^2$. A rational occupant then chooses its personal temperature preference $x_{ij}$ to minimize their individual cost function:

$$\text{minimize } D_i(x_{ij}) + p_{ij}x_{ij} + (\rho/2)|x_{ij} - z_j|^2. \tag{8.25}$$

The BTMS, acting on behalf of the building operator, would choose the target building temperature vector $z$ so as to minimize

$$\text{minimize } G(z) - \langle p, z \rangle + (\rho/2)|x - z|^2, \tag{8.26}$$

which on convergence (when consensus is attained) would equate to the total energy cost incurred by the building operator, when the payments made by the occupants are taken into account. Finally, the per-unit prices ($p_{ij}$) are updated in a way that helps in the consensus, i.e., in bringing $x_{ij}$ and $z_j$ close to each other in each zone $j$, for each occupant $i \in S_j$.

### 8.4.2 System Analysis

The convergence proof presented in this section assumes that the functions $D(.)$ and $G(.)$ are closed, proper, and convex, and the un-augmented Lagrangian $L_o$ in (8.27) below has a saddle point.

$$L_o(x, z, p) = \sum_{j=1}^{m} \sum_{i \in S_j} \Big( D_i(x_{ij}) + p_{ij}(x_{ij} - z_j) \Big) + G(z). \tag{8.27}$$

Based on these assumptions we establish the objective convergence, the residual convergence, and the convergence of the dual variables, for our consensus algorithm as described in Sect. 8.4.1.1. In doing so, we utilize the convergence analysis of the ADMM approach as described in [39], suitably adapted to our model. Consider the objective,

$$\begin{aligned} O^* &= \text{minimum} \sum_{j=1}^{m} \sum_{i \in S_j} D_i(x_{ij}) + G(z) \\ &= \sum_{j=1}^{m} \sum_{i \subset S_j} D_i(x_{ij}^*) + G(z^*), \end{aligned} \tag{8.28}$$

where $x_{ij}^*$ and $z^*$ denote the corresponding optimal values of temperature choices. Note that for any zone $j$, $x_{ij}^* = z_j^*$ for all $i \in S_j$. Also, define residual for zone $j$ as

$$r_{ij} = x_{ij} - z_j \tag{8.29}$$

We prove our result through a sequence of lemmas, each involving an inequality (refer to [1] for complete proof of the lemmas).

**Lemma 8.1**

$$O^* - O^{k+1} \leq \sum_{j=1}^{m} \sum_{i \in S_j} p_{ij}^* r_{ij}^{k+1}. \tag{8.30}$$

**Lemma 8.2**

$$O^{k+1} - O^* \le -\sum_{j=1}^{m} \sum_{i \in S_j} \Big( p_{ij}^{k+1} r_{ij}^{k+1} + \rho (z_j^{k+1} - z_j^k)(-r_{ij}^{k+1} - (z_j^{k+1} - z_j^*)) \Big). \tag{8.31}$$

Next, define Lyapunov function $V$ for the ADMM algorithm as:

$$V^k = (1/\rho) \sum_{j=1}^{m} \sum_{i \in S_j} |p_{ij}^k - p_{ij}^*|^2 + \rho \sum_{j=1}^{m} |z_j^k - z_j^*|^2 \tag{8.32}$$

This Lyapunov function satisfies the inequality as stated in the lemma below.

**Lemma 8.3**

$$V^{k+1} \le V^k - \rho \sum_{j=1}^{m} \sum_{i \in S_j} |r_{ij}^{k+1}|^2 - \rho \sum_{j=1}^{m} |z_j^{k+1} - z_j^k|^2 \tag{8.33}$$

Now, since $V^k \le V^0$, $p_{ij}^k$ and $z_j^k$ are bounded. Iterating (8.33) gives

$$\rho \sum_{k=0}^{\infty} \big( (r_{ij}^{k+1})^2 + |z_j^{k+1} - z_j^k|^2 \big) \le V^0, \tag{8.34}$$

which implies $r_{ij}^k \to 0$ and $|z_j^{k+1} - z_j^k| \to 0$ as $k \to \infty$. Further, from inequalities (8.30) and (8.31) we have $\lim_{k \to \infty} O^k = O^*$ or the objective convergence.

Hence, the inequalities (8.30), (8.31) and (8.33) implies the convergence of our algorithm.

#### 8.4.2.1 Profit Analysis for the Building Operator

Let the vectors $y^*$ and $p^*$ respectively denote the optimal zonal temperatures and the optimal prices. It is easy to establish that the optimal pricing feedback signal $p_{ij}^*$ satisfies:

$$p_{ij}^* = -D_i'(y_j^*). \tag{8.35}$$

We can further show that,

$$\sum_{i \in S_j} p_{ij}^* = \frac{\partial G(y^*)}{\partial y_j}. \tag{8.36}$$

Now, we can obtain an expression for the difference of the net revenue from occupants and the operational cost of maintaining the building at the consensus temperature set point. Note, that this expression represents the operational profit of the building operator and is given by:

$$\sum_j \sum_{i \in S_j} p_{ij}^*(y_j^* - T_\infty) - (G(y^*) - G(T_\infty)). \tag{8.37}$$

Using Taylor series:

$$\begin{aligned} G(T_\infty) &= G(y^*) + (T_\infty - y^*)^T \nabla G(y^*) \\ &\quad + \frac{1}{2}(T_\infty - y^*)^T \nabla^2 G(y^*)(T_\infty - y^*), \end{aligned} \tag{8.38}$$

the building operator's profit can be further expressed as

$$\frac{1}{2}(T_\infty - y^*)^T \nabla^2 G(y^*)(T_\infty - y^*). \tag{8.39}$$

Since $G(y)$ is convex in $y$, $\nabla^2 G(y^*) \geq 0$, and (8.39) is upper bounded by $\leq \frac{\lambda_{max}}{2}||y^* - T_\infty||^2$, where $\lambda_{max}$ is the maximum eigenvalue of $\nabla^2 G(y^*)$. From (8.39) we can assert that if the pricing signals were to be translated to real money (or equivalent credit) transaction between the building operator and the occupants, the building operator does not lose money, and instead may end up making a small profit that is bounded by the convexity of the energy cost function $G(y)$. The expression in (8.39) equals zero ($= 0$) when $G(y)$ is affine in $y$. Therefore, when $G(y)$ is affine in $y$, perfect budget balance is attained, i.e., payments (credits) of the users are just redistributed between themselves, and the building operator does not make any profit or loss.

## 8.5 Simulation Study

### *8.5.1 Testbed Layout*

We consider our six zone physical testbed of an intelligent building located in Watervliet, NY for simulation and experimental study of our proposed algorithm and validating performance of the same. Figure 8.3 represents the dimensions of the facility as generated using the BRCM toolbox [46]. BRCM toolbox is used to generate the RC model of the six zone test facility, mapping it to 31 building elements resulting in a total of 93 capacitive elements.

Each zone of the testbed (except for zone 2, which is the hallway) is actuated with thermoelectric coolers. Real-time temperature sensing is enabled through

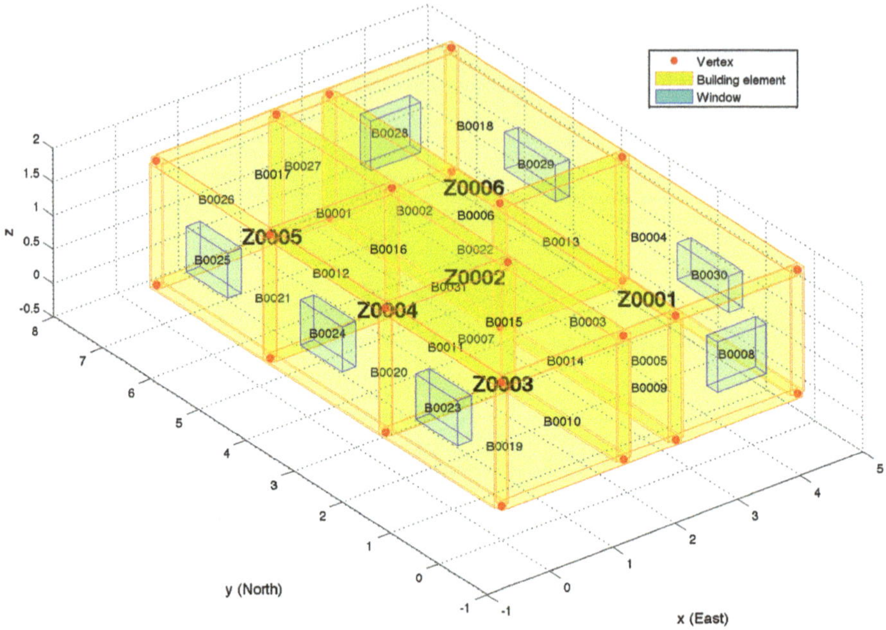


**Fig. 8.3** Watervliet-based smart building test bed with the building elements as generated by the BRCM Toolbox. This setup is used for the simulation and experimental study of the solution framework

J-type thermocouples spread across the test facility. Sensor data is acquired through wireless communication in real time to a central server, which also runs the control loops to operate the coolers and achieve the desired ambient condition. Further details of the test bed layout, instrumentation, and software architecture can be referred to in [47].

The simulated occupancy of the building is represented in Fig. 8.4. Zones 1 and 6 are occupied by two occupants each and the other zones 3, 4 and 5 have one occupant each. Occupants U1 and U2 are in Zone 1, U3 in Zone 3, U4 in Zone 4, U5 in Zone 5, and finally U6 and U7 in Zone 6.

All the occupants have their own specific temperature preference as depicted in Fig. 8.5. Note that the occupants U1 and U2 co-located in Zone 1 have no common range of comfort preferences, whereas the occupants U6 and U7 co-located in Zone 6 have an overlapping region of comfort preference. This distribution enables us to capture all possible scenarios in terms of conflicting and common preferences among co-located occupants.

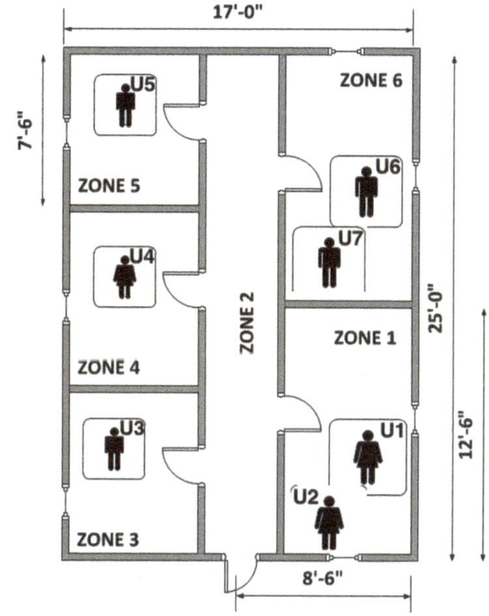


**Fig. 8.4** Layout of the Watervliet test facility with the simulated occupancy pattern of each zone used for simulation and experimental study

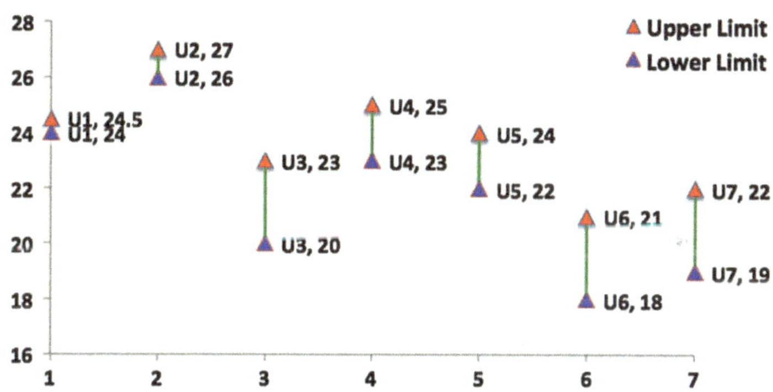


**Fig. 8.5** Preferred temperature comfort range of each occupant in °C. U6, U7 in Zone 6 have an overlapping comfort preference, whereas U1, U2 in Zone 1 have no common range

## *8.5.2 Simulation Results*

For simulation purpose we consider both heating and cooling actuation in each zone with a typical power rating of 1000 W. The system dynamics and the control algorithm are simulated using MATLAB and Simulink. Using the occupancy distribution as per Fig. 8.4, occupant preference from Fig. 8.5 and the model parameters for the Jacobian generated using BRCM toolbox, we first simulate temperature dynamics for a 48 h

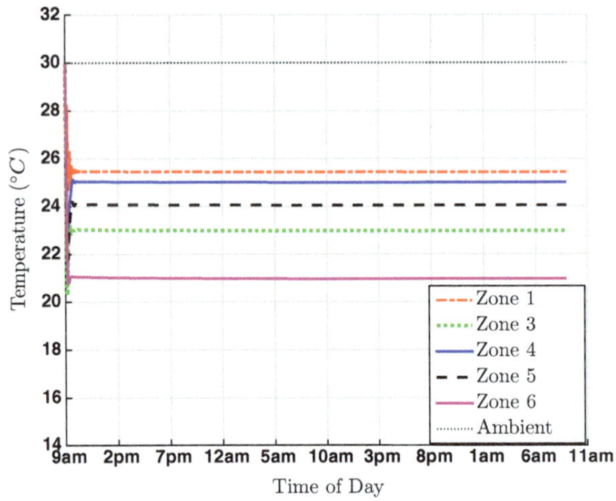


**Fig. 8.6** Temperature dynamics for a 48 h period simulation with ambient condition higher than the occupant preferences with occupancy pattern as per Fig. 8.4

period with a fixed ambient condition of 30 °C. Figure 8.6 represents the temperature dynamics for the 48 h period simulation run.

The temperature of zone 6 settles at 21 °C, which is acceptable to both the occupants U6 and U7, and simultaneously energy optimal being closer to the ambient temperature. Note that anything between 19 and 21 °C would have been comfortable for both U1 and U2 based on their comfort preferences, with 21 °C being optimal for the given ambient condition. Zone 1 settles around 25.5 °C, which tends to minimize the aggregate discomfort of both users U1 and U2 and simultaneously minimizes energy consumption considering the thermal correlation among all the zones. Similarly, the temperature of other zones also settle at a point to minimize aggregate occupant discomfort and the energy cost. In this simulation we also set the initial condition of each zone to the ambient temperature as that represents the extreme scenario, thus providing a good performance evaluation of the proposed algorithm. In most practical condition the initial zonal temperatures would be much closer to the corresponding desired temperatures.

To demonstrate the energy saving in this scenario, we compare it to the prevalent set point-based method of temperature control in buildings. With the given occupant preferences, set point method would consider the mid point of each occupants comfort range as the zonal set point. In case of multiple occupants an average of the occupant set point can be used. Using this approach the corresponding zonal set points are presented in Table 8.1. Compared to the set point based approach, our algorithm achieves energy optimal temperature for the zones and in this particular case results in an energy saving of 12.1%.

Next we simulate another case with fixed ambient temperature at 15 °C. The corresponding temperature dynamics are represented in Fig. 8.7. Note that in this case the temperature of Zone 6 settles at 19 °C, compared to 21 °C in Fig. 8.6 as 19 °C is more energy optimal for the ambient condition of 15 °C.

**Table 8.1** Temperature set point for each zone when using the set point based method of building temperature control

| Location | Occupant set points (°C) | Zonal set points (°C) |
|---|---|---|
| Zone 1 | 24.25, 26.5 | 25.4 |
| Zone 3 | 21.5 | 21.5 |
| Zone 4 | 24 | 24 |
| Zone 5 | 23 | 23 |
| Zone 6 | 20.5, 19.5 | 20 |

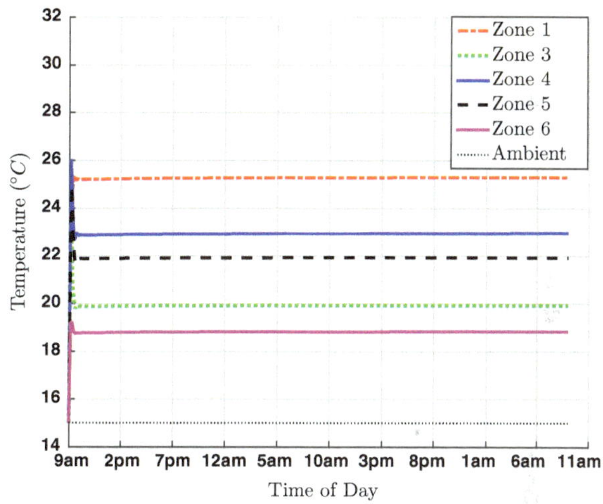


**Fig. 8.7** Temperature dynamics for a 48 h period simulation with ambient condition lower than the occupant preferences with occupancy pattern as per Fig. 8.4

Fixed ambient temperature is an over simplification and does not represent true variations in ambient conditions. Next, we consider a sinusoidal variation of the ambient with $T_\infty$ taking the form: $T_\infty = 20\,°C + 5\,°C \sin(2\pi t/t')$, with $t' = 24$ h. The corresponding temperature dynamics are presented in Fig. 8.8. The control algorithm lets the zonal temperature vary with ambient till it hits the comfort limit, at which point appropriate heating/cooling input is applied. This approach is more optimal than maintaining a fixed set point as it harnesses the ambient variation without the need to constantly readjust the fixed set point. For our particular setup we observed a relative energy saving of 5.3%.

Another simplification that has been applied to our study so far has been the uninterrupted occupancy of the occupants throughout the period of 48 h. We next consider a typical work environment occupancy schedule of the occupants entering their respective zones at 9 am in the morning and departing at 5 pm in the evening. Further, we also consider lunch time and simulate non-occupancy during 12 to 1 pm. In the evening after 5 pm we increase the heat cost factor $\Gamma$. For the lunch break heat cost factor remains unchanged but due to non-occupancy no occupant feedback is generated. The results are presented in Fig. 8.9.

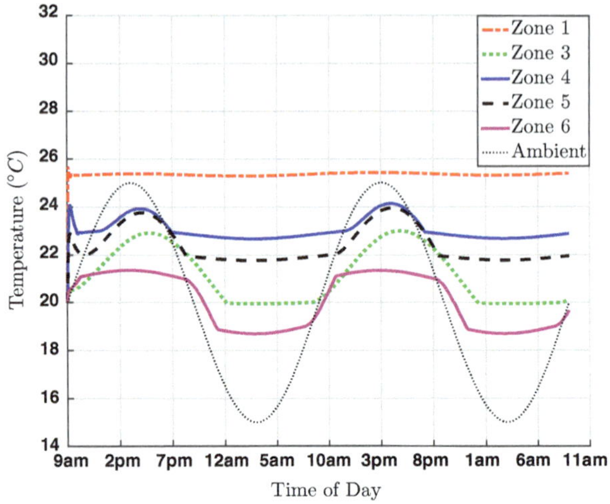


**Fig. 8.8** Temperature dynamics for a 48 h period simulation where ambient condition follows a sinusoidal variation with period of 24 h with occupancy pattern as per Fig. 8.4

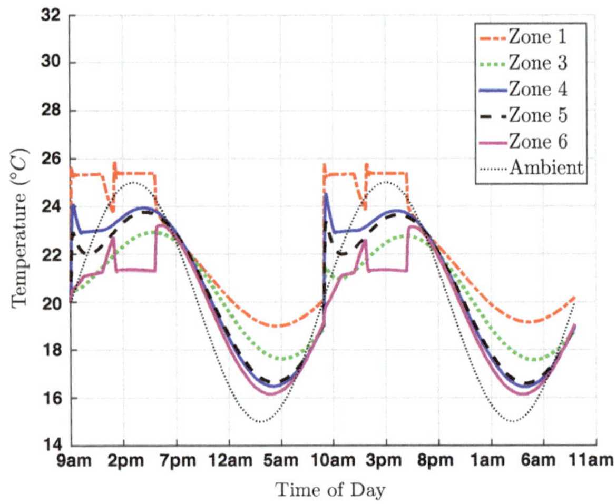


**Fig. 8.9** Temperature dynamics corresponding to 48 hour simulation of occupancy schedule along with sinusoidally varying ambient with occupancy pattern as per Fig. 8.4. Energy cost is increased between 5 pm and 9 am with no occupant feedback during lunch time

Due to increase in energy cost factor after 5 pm, the zonal temperatures tend to follow ambient variation resulting in immense energy savings. The kink in Zone 1 and Zone 6 temperatures during lunch time is attributed to thermal correlation as no occupant feedback is generated resulting in temperature variation as per prevailing thermal conditions. Compared to a set point-based approach (with predetermined energy saving set point after 5pm) this approach can result in energy savings of 6.1%. Note that if the set point based approach does not implement energy saving mode after 5pm then the relative saving through our approach would be even higher. Compared to scenario in Fig. 8.8 additional energy saving can be attributed to efficiency during lunch break.

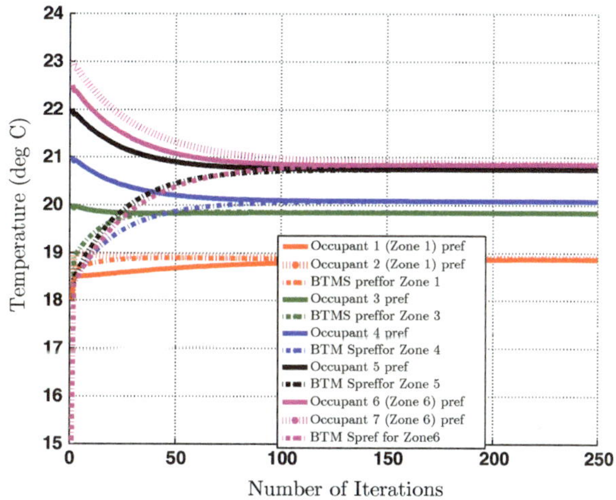


**Fig. 8.10** Convergence of temperature set point preferences in each zone, for the occupants and the BTMS. The solid lines depict the occupant temperature preferences, and the dashed ones the BTMS' corresponding preferences

**Table 8.2** Ideal temperature setting in °$C$ of each zone as per its occupant and the building operator

| Zone | Occupant(s) preference(°C) | Building Operator preference(°C) |
|---|---|---|
| Zone 1 | 18.5 and 19 | 15 |
| Zone 3 | 20 | 15 |
| Zone 4 | 21 | 15 |
| Zone 5 | 22 | 15 |
| Zone 6 | 22.5 and 23 | 15 |

In Fig. 8.10 we present the result of the distributed consensus algorithm using ADMM approach. Each zone (room) occupant agent starts with the ideally preferred temperature set point of the corresponding occupant as per Table 8.2 and the BTMS with the preferred set point of the building operator for the corresponding zones.

Each iteration in Fig. 8.10 represents one round of communication (computation) between (at) the smart applications and the BTMS. So 100 iterations would represent 100 round-trip communication (100 computation) rounds, and typically evaluate to a few seconds. The occupant at their end would only see the final (converged) temperature and penalty factor signal, the intermediate values being internal to the algorithm would not be visible to the occupants.

With each iteration of the algorithm, the difference between the corresponding zonal temperature preference of the occupant and that of the BTMS narrows and finally compromise is attained in all the zones. Note that in Fig. 8.10 the consensus temperature of zone 6 comes down to that of zone 5, irrespective of the zone 6 occupants preferring much higher temperature than the occupant of zone 5. This is in accordance with the energy cost attributed to maintaining zone 6 relative to zone 5. The trend can also be reasoned from the penalty factor curve in Fig. 8.11, as the penalty factor feedback to zone 6 occupants is much higher compared to the occupant

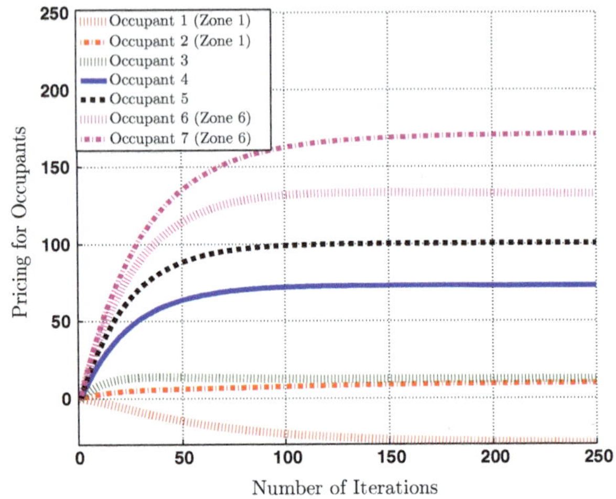


**Fig. 8.11** Variation in penalty factor for the zone occupants for desired change in the zonal temperatures. A negative penalty factor indicates the corresponding occupant receiving reward from the building operator

of zone 5. Further, the consensus temperature for both the occupants of zone 1 and zone 6 converge to the respective consensus zonal temperature.

The penalty factor for unit change in temperature varies with each iteration, as shown in Fig. 8.11. The penalty factor increases for the zone occupant if the temperature choice is away from the BTMS' preference and the ambient temperature. In Fig. 8.11 the per-unit penalty factor for occupant 1 (located in zone 1) turns negative. This can be attributed to the fact that on consensus, the temperature for that zone moves away from the ambient and building operator's preferred temperature for the zone, even beyond the occupant's preferred value.

The penalty factor signals in the form of notifications/information can serve as a means for the building operator to communicate with rational occupants into selecting a lower/higher preferred temperature. An occupant of a particular zone might not be aware of their zone's thermal correlation and the temperature preference of the occupants of their neighboring zones. However, the building operator based on the energy cost data can share that picture with the occupants, through penalty factor signal, without disclosing any private information. Using this information, rational occupants and the building operator can work together to modify their preference and accommodate users with different and at times extreme (which in general incurs greater overall operating energy cost to the building) thermal preferences.

## 8.6 Concluding Remarks

In this chapter, we have demonstrated that the building temperature and energy usage can be controlled successfully and efficiently through dynamic feedback from the occupants based on their comfort levels. The collaborative temperature control in multi-occupant spaces based on the consensus algorithm attains temperature set points that minimizes the sum of the aggregate discomfort of the occupants and the total energy cost in the building. Through simulations we demonstrated the energy efficiency achievable through our algorithms.

Key to real world implementation would be an end-to-end framework designed for enabling occupant feedback collection and incorporating the feedback data towards energy efficient operation of a building. A mobile application that occupants can use on their smart phones and other wearables to provide their thermal preference feedback would be needed, that can also use indoor location techniques to tie the occupant preference to their current thermal zone when relaying the feedback to central server. Using sensors for real-time zonal temperature readings, and based on the information from mobile application the central server can use real-time learning algorithm to learn the environment and use occupant feedback to calculate the optimal temperature set point. The entire process needs to be triggered upon change of occupancy, environmental conditions, and/or occupant preference. The learning algorithm can be scheduled to run at regular intervals to respond dynamically to environmental and occupancy changes.

**Acknowledgements** This work was supported in part by the National Science Foundation under Award - 1230687.

## References

1. Gupta SK, Kar K, Mishra S, Wen JT (2015) Collaborative energy and thermal comfort management through distributed consensus algorithms. IEEE Trans Autom Sci Eng 12(4):1285–1296
2. Guo W, Zhou M (2009) Technologies toward thermal comfort-based and energy efficient HVAC systems: A review. In: Proceedings of the IEEE international conference on systems, man and cybernetics, Oct. 2009, San Antonio, TX
3. Karjalainen S, Koistinen O (2007) User problems with individual temperature control in offices. Build Environ 42(8):2880–2887
4. Times union (2016). http://www.timesunion.com/news/article/This-cold-war-too-has-roots-in-the-past-6422917.php/. Accessed Mar 21 2016
5. Ma Y, Anderson G, Borrelli E (2011) A distributed predictive control approach to building temperature regulation. In: Proceedings of the american control conference, San Francisco, CA, Jun–Jul 2011
6. Erickson VL, Cerpa AE (2012) Thermovote: Participatory sensing for efficient building hvac conditioning. In: BuildSys 2012 Proc. Fourth ACM Workshop on embedded sensing systems for energy efficiency in buildings, Toronto, ON, Canada, 2012
7. Home nest (2016). http://www.nest.com. Accessed 21 Mar 2016
8. Hoff JV (2008) Forty years of fanger's model of thermal comfort: Comfort for all? Indoor Air 18(3):182–201
9. Humphreys MA, Hancock M (2007) Do people like to feel neutral?: Exploring the variation of the desired thermal sensation on the ASHRAE scale. Energy Build 39(7):867–874
10. McIntyre DA (1982) Chamber studies reductio ad absurdum? Energy Build 5(2):89–96
11. ASHRAE (2004) Thermal environmental conditions for human occupancy, ASHRAE Standard 55-2004, 2004
12. Fanger PO (1972) Thermal comfort. Analysis and applications in environmental engineering. McGraw-Hill, New York

13. Kolokotsa D, Niachou K, Geros V, Kalaitzakis K, Stavrakakis GS, Santamouris M (2005) Implementation of an integrated indoor environment and energy management system. Energy Build 37(1):93–99
14. Dalamagkidis K, Kolokotsa D, Kalaitzakis K, Stavrakakis GS (2007) Reinforcement learning for energy conservation and comfort in buildings. Building Environ 42(7):2686–2698
15. Bermejo P, Redondo L, Ossa L, Puerta JM (2012) Design and simulation of a thermal comfort adaptive system based on fuzzy logic and online learning. Energy Build 49(6):367–379
16. Tse WL, Chan WL (2008) A distributed sensor network for measurement of human thermal comfort feelings. Sens Actuators A: Phys 144(2):394–402
17. Kang J, Kim Y, Kim H, Jeong J, Park S (1997) Comfort sensing system for indoor environment. In: International conference on solid state sensors and actuators, Chicago, IL, Jun 1997
18. Guillemin A, Morel N (2002) Experimental results of a self-adaptive integrated control system in buildings: a pilot study. Sol Energy 72(5):397–403
19. Murakamia Y, Terano M, Mizutani K, Harada M, Kuno S (2007) Field experiments on energy consumption and thermal comfort in the office environment controlled by occupants requirements from PC terminal. Build Environ 42(12):4022–4027
20. Daum D, Haldi F, Morel N (2011) A personalized measure of thermal comfort for building controls. Build Environ 46(1):3–11
21. Purdon S, Kusy B, Jurdak R, Challen G (2013) Model-free HVAC control using occupant feedback. In: IEEE 38th Conference on LCN workshops, Sydney, NSW, Oct 2013
22. Zhao QC, Cheng Z, Wang F, Jiang Y, Ding J (2014) Experimental study of group thermal comfort model. In: IEEE international conference on automation science and engineering, Taipei, Aug 2014
23. Zhao QC, Zhao Y, Wang F, Jiang Y, Jiang Y, Zhang F (2014) Preliminary study of learning individual thermal complaint behavior using one-class classifier for indoor environment control. Build Environ 72(2):201–211
24. Jazizadeh F, Ghahramani A, Becerik-Gerber B, Kichkaylo BT, Orosz M (2014) User-led decentralized thermal comfort driven HVAC operations for improved efficiency in office buildings. Energy Build 70(2):398–410
25. Zhao J, Lam KP, Ydstie BE, Loftness V (2016) Occupant-oriented mixed-mode energy - plus predictive control simulation. Energy Build 117(4):362–371
26. Sustainable buildings technology laboratories (Living Labs): Center for high performance buildings. https://engineering.purdue.edu/CHPB/Facilities/LivingLab. Last Accessed 21 Mar 2016
27. Choi J, Loftness V, Lee D (2012) Investigation of the possibility of the use of heart rate as a human factor for thermal sensation models. Build Environ 50(4):165–175
28. Gagge AP, Nishi Y (1971) An effective temperature scale based on a simple model of human physiological regulatory response. ASHRAE Trans 77:247–262
29. Stolwijk JAJ (1980) Mathematical model of thermo regulation. Ann NY Acad Sci 355:98–106
30. Zhang H, Arens E, Huizeng C, Han T (2010) Thermal sensation and comfort models for non-uniform and transient environments, part III: Whole body sensation and comfort. Build Environ 45(2):399–410
31. Humphreys MA, Nicol JF, Raja IA (2007) Field studies of thermal comfort and the progress of the adaptive model. Adv Build Energy Res 1(1):55–88
32. Dear RJd, Brager GS (1998) Towards an adaptive model of thermal comfort and preference. ASHRAE Trans 104(1):145–167
33. Mebee B (2011) Computational approaches to improving room heating and cooling for energy efficiency in buildings, Ph.D. Dissertation, Dept Math, Virginia State and Polytech Inst
34. Riederer P, Marchiob D, Visier JC, Husaunndee A, Lahrech R (2002) Room thermal modeling adapted to the test of hvac control systems. Build Environ 37(8):777–790
35. Boyer H, Chabriat JP, Grondin-Perez B, Tourrand C, Brau J (1996) Thermal building simulation and computer generation of nodal models. Build Environ 31(3):207–214
36. Fraisse G, Viardot C, Lafabrie O, Achard G (2002) Development of simplified and accurate building model based on electrical analogy. Energy Build 34(10):1017–1031

37. Xu B, Fu L, Di H (2008) Dynamic simulation of space heating systems with radiators controlled by TRVs in buildings. Energy Build 40(9):1755–1764
38. Athienitis AK (1985) Modeling and analysis of thermal networks through subnetworks for multi zone passive solar buildings. Appl Math Model 9(2):109–116
39. Ghiaus C (2013) Causality issue in the heat balance method for calculating the design heating and cooling load. Energy 50(1):292–301
40. Chandan V (2010) Modeling and control of hydronic building HVAC systems, Master's thesis, Dept Mech Eng, Univ Illinois Urbana-Champaign
41. Privara S, Siroky J, Ferkl L, Cigler J (2011) Model predictive control of a building heating system: The first experience. Energy Build 43(2):564–572
42. Khalil H (2002) Nonlinear Systems. Prentice Hall, New Jersey
43. Kokotovic P, Khalil HK, O'Reilly J (1986) Singular perturbation methods in control. Analysis and design. Academic Press, London
44. Gupta SK, Kar K, Mishra S, Wen J (2014) Building temperature control with active occupant feedback. In: Proceedings of the 19th IFAC world congress, Cape Town, South Africa, Aug. 2014
45. Bertsekas D, Tsitsiklis J (1989) Parallel and distributed computation. Prentice Hall, New Jersey
46. Sturzenegger D, Gyalistras D, Semeraro V, Morari M, Smith RS (2014) BRCM MATLAB toolbox: Model generation for model predictive cuilding control. In: Proceedings of the american control conference, Portland, OR, Jun 2014
47. Minakais M, Mishra S, Wen JT, Lagace L, Castiglia T (2015) Design and instrumentation of an intelligent building testbed. In: IEEE international conference on automation science and engineering, Gothenberg, Sweden, Aug 2015

# Part III
# Beyond HVAC: Lighting, Grid, and Distributed Intelligence

# Chapter 9
# Smart Lighting Control Systems

**M.H. Toufiq Imam, Sina Afshari and Sandipan Mishra**

Illumination is a key factor for creating a comfortable building interior. It affects health, productivity, mood, comfort, security, and safety of the occupants. Moreover, about 404 billion kilowatthours (kWh) of electricity were used for artificial lighting by the residential sector and the commercial sector in the United States in 2015; which was about 15% of the total electricity consumed by both of these sectors and about 10% of total U.S. electricity consumption [1]. Traditional lighting systems were mainly designed to provide sufficient illumination for the occupants of commercial and residential spaces. However, the new generation of lighting systems leverage advances in solid state lighting and spectral and occupancy sensing technology to deliver functionality beyond just illumination. These lighting fixtures can mix light from different light-emitting diode (LED) sources to create efficient, high-quality, and healthy illumination. Moreover, such systems also aim to use the data acquired by various sensors in the illuminated space to estimate the number and location of the occupants, user preferences, available daylight, etc. The lighting control algorithm then uses this information to intelligently adjust the input signals to individual LEDs and achieve the desired control objective in an energy efficient manner.

For more than a century, traditional lighting systems have been installed and utilized ubiquitously in commercial and residential buildings. While illumination is the crucial functionality in a lighting system, the new generation of advanced smart lighting systems aim to accomplish this in an energy efficient way. Although LEDs are

M.H.T. Imam (✉) · S. Mishra
Rensselaer Polytechnic Institute, Troy, NY, USA
e-mail: imamm@rpi.edu

S. Mishra
e-mail: mishrs2@rpi.edu

S. Afshari
University of Michigan, Ann Arbor, MI, USA
e-mail: sinaafsh@umich.edu



J.T. Wen and S. Mishra (eds.), *Intelligent Building Control Systems*,
Advances in Industrial Control, https://doi.org/10.1007/978-3-319-68462-8_9

inherently more efficient than other lighting technologies (such as incandescent and fluorescent bulbs), using sensor measurements for feedback control of lights allows for even further savings. Two typical mechanisms that contribute to such power savings are (1) occupancy-based lighting and (2) daylight harvesting. The former refers to the dimming of the light fixtures in unoccupied areas in the space, while the latter refers to the harvesting of the incoming daylight from windows or skylights for illumination, so that the fixtures can be dimmed where and when daylight is available. Due to the varying nature of the space occupancy, light transport within the illuminated space, and changing daylight conditions, use of feedback control is an essential requirement for realizing these capabilities. Besides energy efficiency, modern buildings have a greater focus on human comfort, mood, productivity, and health from a lighting perspective. Although lighting comfort depends on several parameters such as illumination, correlated color temperature (CCT), perception of color, glare etc., two major lighting quality metrics that have been studied in literature to ensure comfortable lighting are uniform (and bright) illumination and color matching. Thus, most of the feedback control algorithms for lighting focus on accomplishing these objectives.

Figure 9.1 shows the major components and the architecture of an advanced smart lighting system. The illuminated space is equipped with lighting (typically LED) fixtures, color sensors, and occupancy sensors. The intensity of the light generated by individual LEDs in the fixtures can be controlled by the input signal to the fixture drivers. The occupancy sensors estimate the number and/or location of the occupants, and the color sensors measure the intensity and spectral content of the incident light. The occupancy information is used to determine the desired lighting condition for the ongoing use scenario in the space, while the measured color sensor data is used to efficiently generate and maintain the desired condition subject to varying daylight levels. The controller takes in the desired sensor setpoint values and the sensor measurements, and based on a set of parameters determined by the building manager (e.g., energy cost, uniformity requirement, etc.), generates the required input for the fixtures to achieve the control objective.

Feedback control algorithms for lighting have been implemented in testbeds with various size and configurations, with different control objectives, and various solution approaches. An overview of these studies is presented in Sect. 9.1. Several studies on their performance have also been reported, in [3–5], among others.

This chapter will present a comparative study of several state-of-the-art feedback control algorithms for advanced lighting systems through experimental evaluation on a standard testbed (described in Sect. 9.2). We investigate four candidate algorithms that capture the diversity of the existing lighting control approaches, implement them on a full-scale testbed and experimentally evaluate their performance based on metrics such as power consumption, color quality, etc. in Sects. 9.5–9.8. The algorithms were implemented under identical natural lighting conditions in a controlled environment for multiple days and the results from these experiments are discussed in Sect. 9.9. Finally, conclusions are drawn in Sect. 9.10.

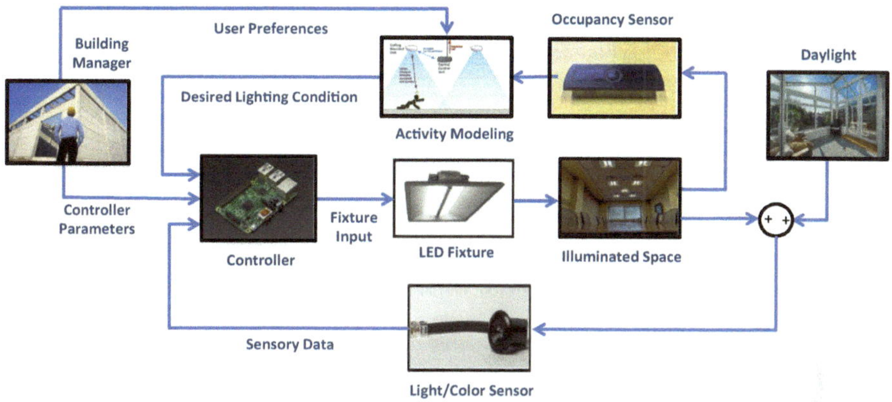


**Fig. 9.1** A typical advanced or *smart* lighting system. [Reprinted from Energy and Buildings, 130 (2016), M.H. Toufiq Imam, Sina Afshari and Sandipan Mishra, An experimental survey of feedback control methodologies for advanced lighting systems, 600–612, Copyright (2016) [2], with permission from Elsevier.]

## 9.1 Smart Lighting Systems – A Brief Survey

Several studies in the past decades have focused on the plant modeling and feedback controller design for lighting systems, aiming to improve power efficiency, human comfort, and control implementation. The modeling methods, control algorithms, and system architectures in these studies are diverse, and aim to address a variety of control objectives under different constraints and assumptions. This chapter presents a review of the existing literature on feedback control algorithms for lighting systems. We will focus on the control objectives, control methodologies, actuation technologies, and sensor technologies that are used in these studies. In the following subsections, we will introduce each of these and the typical alternative approaches that have been proposed in existing literature on lighting control.

### *9.1.1 Control Objectives for Lighting*

Prior to design of the control algorithm, we first outline the typical types of *control objectives* for lighting systems, i.e., the predefined set of goals that the controller aims to achieve. A simple example is maintaining a specified illuminance level in the space subject to varying levels of incoming daylight from windows and/or skylight [6–22]. In these studies, it is assumed that the comfort level of the user solely depends on the illuminance level of the work surface. In other studies, more complex representations of the human factors are used to either guarantee or maximize the occupant comfort level [16, 23–29]. Uniformity of the illuminance pattern on the work surface is another control objective that has previously been studied [16, 22]. Another common control objective is minimization of the electrical power consumed by the lights [15–17, 24, 25, 28–32]. Occupancy-based lighting has also been studied as a method to improve power efficiency [10, 14, 20, 21, 23, 33–36].

### 9.1.2 Control Architectures for Lighting

Once an objective has been specified, the*control architecture* to be used for feedback control of the lights must be established. The proposed algorithms can be classified into centralized, decentralized, and distributed control. In a centralized closed-loop lighting system, the data from all the sensors are sent to a single main controller, where the input for all the light fixtures is calculated and transmitted to the fixtures [6–9, 16–19, 21, 27, 30–32, 34, 37, 38]. In a decentralized closed-loop lighting system, each light fixture has its own light source, sensing module, and controller, and the input to the light source is determined based solely on the fixture's own sensor measurement [14, 15, 22]. A distributed lighting system is similar to a decentralized system, with the difference that the fixtures can use local communication to transmit data to and receive data from the nearby fixtures [14, 22, 29, 36].

A second classification is based on the model dependence of the control law. While many algorithms are based on a mathematical model that describes the generation and distribution of the light in the space, in some studies the control rule is independent of the plant model [10, 11, 22, 23, 30]. Some approaches exploit various techniques for estimation of the occupancy state or the light field in the illuminated space and generate the control input based on the estimated values [17, 33].

### 9.1.3 Control Methodologies for Lighting

From a mathematical perspective, the proposed control methodologies can be classified into three categories: logic-based controllers, regulation-based controllers, and optimization-based controllers. In logic-based methods, various decision-making techniques are used to infer the correct action for the lighting system in different situations based on the measured values from the sensors [10, 11, 23–25, 30]. Regulation-based control algorithms are those that aim to achieve closed-loop stability while guaranteeing that the generated illumination tracks a predetermined reference or setpoint value [6–9, 14, 15, 22]. Optimization-based algorithms are an alternative approach to controller design for closed-loop lighting systems where the control problem is posed as an optimization problem and solved using optimization techniques [16–19, 21, 27, 29, 31, 32, 34, 36]. A classification of these three categories and some example studies that belong to each category are illustrated in Fig. 9.2.

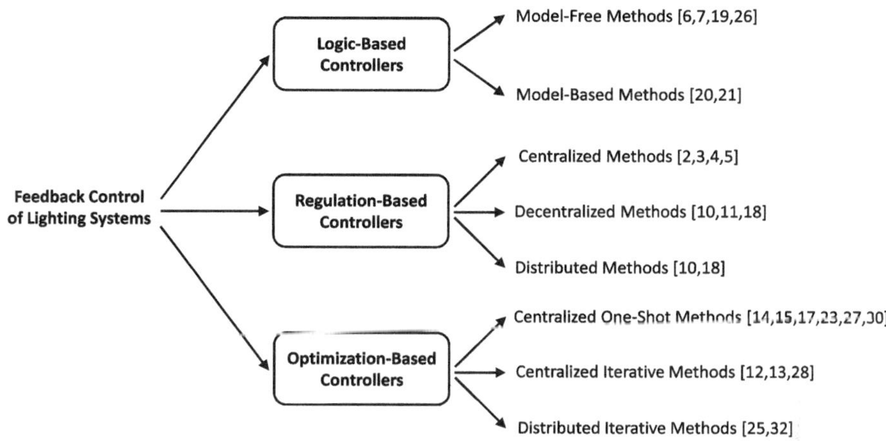


**Fig. 9.2** A classification of the mathematical approaches to controller design for advanced smart lighting systems and example studies that belong to each category

Optimization-based algorithms can be further classified into two subcategories. In *one-shot optimization* algorithms, the control law is obtained by choosing the optimal solution at every time-step of the control loop implementation. This process is repeated in every time-step to ensure maintenance of optimal solution throughout the closed-loop operation [18, 19, 21, 27, 28, 31, 34, 35]. In *iterative optimization* algorithms, a gradient-based technique is used to obtain the optimal direction for the evolution of the system at each time-step. The input for the next time-step is obtained by making an incremental change in the current input in a gradient-related direction. This process is repeated until the system converges to the global optimum of the cost function [16, 17, 29, 32, 36]. While the one-shot optimization algorithms result in faster convergence, they are typically more sensitive to noise as well as modeling uncertainty and are difficult to analyze from a stability and performance standpoint in the presence of model uncertainty.

### *9.1.4 Actuation and Sensing Techniques for Lighting*

Another important aspect that governs the type of control scheme used is the type of actuation and sensing technology that the algorithms have been designed for and verified with. For example, illumination using incandescent [22, 27, 29], fluorescent [6–10, 18, 19, 28, 30], and LEDs [15–17, 31, 32] has been used to implement and verify the proposed algorithms. While the typical actuation systems are light sources, a few studies have focused on control of other lighting-related actuators such as automatic blinds [11, 24, 25], light pipes [10], and electro-chromic windows [26]. Typical sensors used for lighting control feedback include photosensors [6–9, 11], a wireless network of ambient light sensors [18, 19, 23, 27, 30], network of infrared motion and occupancy sensors [10, 14, 21, 23, 33–36], a network of RGB color sensors [15–17, 31, 32], light- dependent resistors [22], and cameras [20].

### 9.1.5 Recent Surveys on Lighting Control Systems

Given the myriad alternatives for lighting control algorithm design, several surveys and comparative studies have been published on lighting control. In an attempt towards this goal, Williams et al. [3] published a meta-analysis of the existing literature on power savings in lighting by including 240 power savings estimations from 88 papers and case studies. The gathered data of power savings were categorized into 4 strategy classes including occupancy (28% of the data), daylighting (32% of the data), personal tuning (8% of the data), and institutional tuning (7% of the data). The remaining 25% of the studied literature were reported to be on daylighting together with occupancy (8%), occupancy together with personal tuning (7%), and other strategies (10%). Another review was presented by Asif ul Huq et al. [4], in which the control strategies were divided into two broader categories: daylight linked and occupancy based systems. Various methods of occupancy detection and daylight harvesting along with their respective power savings were presented. Recently, Pandharipande et al. [5] presented a comparative study of luminaire-based (where sensors are co-located with the light sources) sensing strategies in office lighting applications. The spectral content of the generated lights was not taken into account in these comparative studies and the addressed control methodologies were implemented and evaluated in testbeds of various sizes and configurations which creates a big challenge to their performance evaluation.

In this chapter, in order to give the reader a comprehensive overview of lighting control, we develop and investigate four lighting control algorithms, chosen as candidates that broadly represent classes of design methodologies.

- The Illumination Balancing Algorithm (IBA) [22] uses a regulation-based solution approach to a tracking problem that deals with single-color lighting.
- The daylight and occupancy adaptive lighting control scheme proposed in [21] is an iterative optimization technique for single-color lighting systems.
- The energy efficient control of polychromatic lighting systems algorithm proposed in [31] uses a one-shot optimization technique for color-tunable lighting systems.
- Finally, the control algorithm design a non-square lighting system developed in [32] deals with an constrained optimization problem with multi-channel sources and sensors and proposes an iterative technique to solve the problem.

These state-of-the-art strategies were implemented in a room-scale standard testbed (described in Sect. 9.2) and their performance evaluations are presented in the subsequent sections. Day-long experiments were also carried out for all of these algorithms under identical natural lighting condition. The results from these experiments are presented in Sect. 9.9.

## 9.2 The Smart Conference Room – A Testbed for Lighting Control

Several testbeds with varying sizes and configurations have been used in existing literature to implement and validate the proposed feedback control algorithms. A standard testbed that may be used to experimentally validate the lighting control algorithms should be a full-scale space that is used by occupants on a consistent basis. Ideally, the illumination system for such a space should be dimmable and spectrally tunable to generate a range of lighting conditions. It should also have windows or

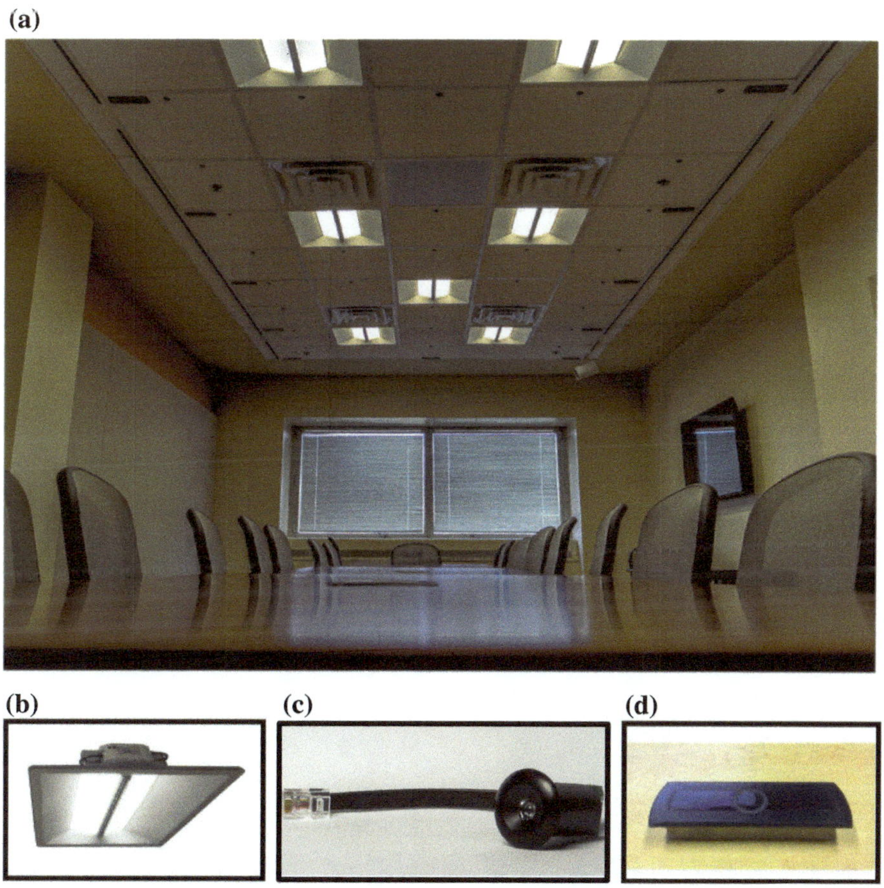


**Fig. 9.3** **a** The ceiling of the Smart Conference Room, **b** the 5-channel LED light fixture, **c** the RGB color sensor, **d** the multi-pixel time-of-flight sensor. [Reprinted from Energy and Buildings, 130 (2016), M.H. Toufiq Imam, Sina Afshari, and Sandipan Mishra, An experimental survey of feedback control methodologies for advanced lighting systems, 600–612, Copyright (2016) [2], with permission from Elsevier.]

skylights for natural light in order to evaluate the effect of ambient lighting on the performance of the algorithms. The actuators (lighting fixtures), sensors, and the controller should be connected through a fast reliable network to ensure a robust controlled environment.

An example of such an advanced smart lighting system is the Smart Conference Room (SCR),a full-size conference room [39], which is equipped with color-tunable LED fixtures, RGB (Red-Green-Blue channel) color sensors and Time-Of-Flight sensors, located in Rensselaer Polytechnic Institute campus in Troy, New York. This room is used for meetings and conferences, and also serves as one of the testbeds within the Center for Lighting Enabled Systems and Applications (LESA) Engineering Research Center, where experiments on feedback control of lighting systems, human factors, and occupancy-based lighting are conducted. The room is equipped with ten LED light fixtures retrofitted in $2' \times 2'$ cases, each containing five different LED colors (i.e., input channels). The five LED channels are red, green, blue, amber, and phosphor-converted white. Also, 53 RGB color sensors are installed on the ceiling to measure the light field in the room, one per each ceiling tile. These sensors consist of a TCS34725 RGB sensor, an optical lens, and a ceiling mount. Additionally, 18 multi-pixel Time-Of-Flight sensors are installed on the ceiling, from which occupancy information is extracted. The room has windows on the eastern and northern walls, allowing for natural daylight to serve as ambient external light. Figures 9.3 and 9.4 demonstrate the SCR, the different components used in it, and the layout of the different components.

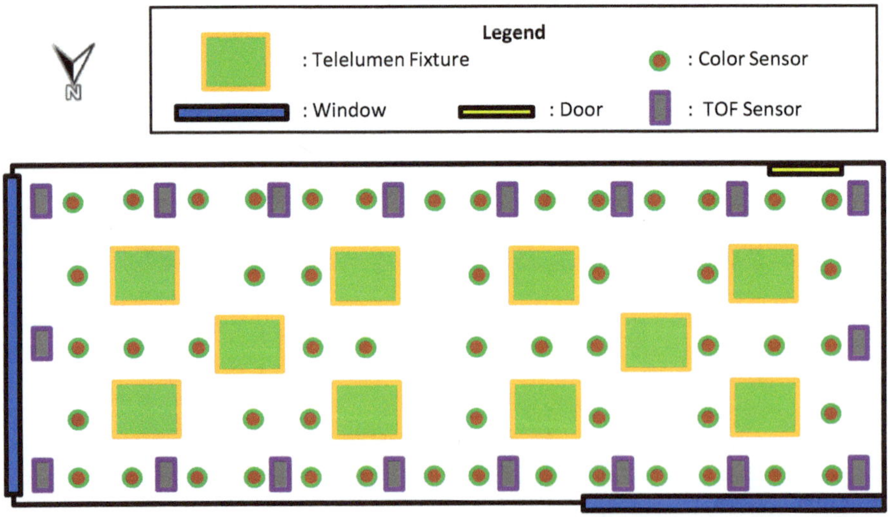


**Fig. 9.4** A top-view layout schematic of the different components in the Smart Conference Room. [Reprinted from Energy and Buildings, 130 (2016), M.H. Toufiq Imam, Sina Afshari and Sandipan Mishra, An experimental survey of feedback control methodologies for advanced lighting systems, 600–612, Copyright (2016) [2], with permission from Elsevier.]

The LED fixtures are connected to a main computer via a local Ethernet connection. Further, all the sensors transmit their measured data to a single main Raspberry Pi processor via $I^2C$ bus. The main Raspberry Pi is programmed to act as a multiplexer, acquiring the data from all the sensors and sending it to the main server using local Ethernet connection.

In view of the above definition given for a standard testbed, the SCR possesses all the qualifications. Hence, the selected four algorithms were implemented in this testbed to experimentally evaluate their performances on a standard ground.

## 9.3 Light Transport Modeling

The first step in performance analysis and controller design for a system is modeling of the input–output relationship. The model is a mathematical representation of the underlying physical phenomenon in the system, and can be linear or nonlinear, time-variant or time-invariant, static or dynamic, deterministic or stochastic, etc. Depending on these properties and the complexity of the model, different analysis and design methods may or may not be suitable for an application. In the case of a lighting system, ray tracing software such as Zemax may be used to directly simulate the underlying physical phenomenon, i.e., generation and flow of the light in the space. However, these models are not suitable for real-time feedback control of the lights due to their computationally heavy nature. Additionally, most of the controller design methodologies require a mathematical description of the system. In this section, one such description based on the principles of light transport is introduced.

The light field at a point in space is characterized by the plenoptic function, $\phi(r, \theta, \lambda)$, which is the spectral irradiance (in $\frac{W}{m^2 \cdot nm}$) along the ray given by the location of the point, $r \in \mathbb{R}^3$, and solid angle of the incoming light direction, $\theta \in S^2$, for the wavelength, $\lambda \in \Lambda$, where $\Lambda$ is the visible light range $[390, 750]$ nm [40]. Consider a space with $n$ light fixtures, each of which contains $p$ adjustable intensity channels (e.g., different color LEDs). Assume that the input to each fixture is represented as a vector, $u^i \in \mathbb{R}^p, i = 1, \dots, n$, and each entry in $u^i$ is normalized to the range $[0, 1]$. Each light fixture generates a light field throughout the space. Let the unit plenoptic light field generated by fixture $i$ be $S^i(r, \theta, \lambda) \in \mathbb{R}^p$, with the spectral dependence for each channel given by the spectral characteristics of the corresponding light source. Denote the ambient light field as $\psi(r, \theta, \lambda)$. Then, the total light field is the linear combination of the generated and ambient light,

$$\phi(r, \theta, \lambda) = \sum_{i=1}^{n} S^i(r, \theta, \lambda)^T u^i + \psi(r, \theta, \lambda). \tag{9.1}$$

Assume that there are $m$ locations of interest for the overall light field in the space, and that there is a sensor with $q$ channels of spectral sensing at each location. The overall output vector is thus a vector with $qm$ elements, consisting of the different $y^j \in \mathbb{R}^q$, $j = 1, \ldots, m$. Let the responsivity function for each sensor be $C^j(r, \theta, \lambda) \in \mathbb{R}^q$. The output at each location is thus given by

$$y^j = \langle C^j, \phi \rangle + v^j \tag{9.2}$$

where $\langle \cdot, \cdot \rangle$ denotes integration over the spatial, angular, and spectral ranges of the color sensor, and $v^j$ is the measurement noise. For a point color sensor at $r^j$, the $r$ dependence in $C^j$ is simply $\delta(r - r^j)$. The $\theta$ and $\lambda$ dependence correspond respectively to the angular sensitivity, which is typically governed by the sensor optics, and the spectral characteristics of the sensor channels. Without loss of generality, and because human color perception is based on three color-sensitive photoreceptors (cones), we will assume $q = 3$. Therefore, $y^j \in \mathbb{R}^3$ represents the RGB sensor measurements at the different locations corresponding to $j = 1, \ldots, m$.

Substituting the light field in (9.1) into the sensor equation in (9.2), we obtain

$$y = Gu + w + v \tag{9.3}$$

where $y = [y^1, y^2, \ldots, y^m]^T \in \mathbb{R}^{3m}$ is the output light measurement vector, $u = [u^1, u^2, \ldots, u^n]^T \in \mathbb{R}^{pn}$ is the input light intensity control vector, $G \in \mathbb{R}^{3m \times pn}$ is the Light Transport Matrix [41] (LTM), $w \in \mathbb{R}^{3m}$ is the effect of ambient light on the sensors with $w^j = \langle C^j, \psi \rangle$, and $v \in \mathbb{R}^{3m}$ is the measurement noise. The LTM is used to characterize the input–output relationship for the lighting system. A schematic diagram illustrating the closed-loop system and the different variables is shown in Fig. 9.5.

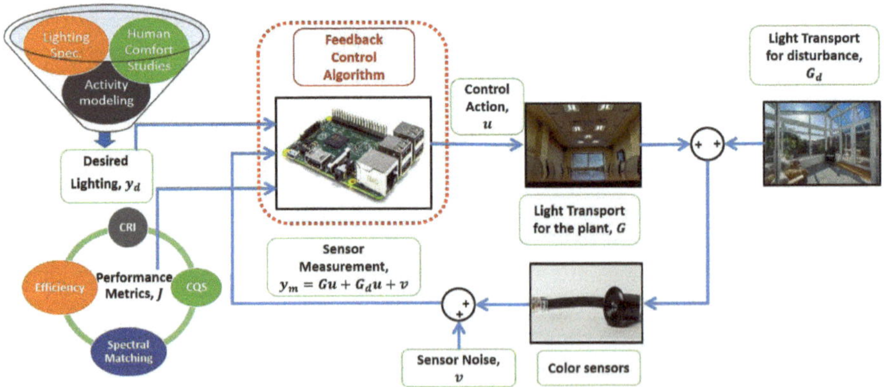


**Fig. 9.5** Lighting Control architecture, illustrating the Light Transport Matrix (LTM) based system model; sensing and actuation; ambient disturbances; performance metrics; and key factors governing the choice of desired illumination. [Reprinted from Energy and Buildings, 130 (2016), M.H. Toufiq Imam, Sina Afshari and Sandipan Mishra, An experimental survey of feedback control methodologies for advanced lighting systems, 600–612, Copyright (2016) [2], with permission from Elsevier.]

The LTM for an illuminated space can be estimated experimentally by posing the model identification problem as a least squares problem. Let $U = [u_1, u_2, \ldots, u_N]$ be a set of $N$ input vectors to the lighting system, and $Y = [y_1, y_2, \ldots, y_N]$ be the corresponding sensor measurements from the space. Then the least square estimate of the LTM is simply $\hat{G} = YU^+$, where $U^+$ is the Moore-Penrose pseudo-inverse of $U$. Note that $U$ has to be full row rank and well-conditioned in order for this estimate to be accurate. This is usually ensured by using randomly generated lighting inputs, $u_i$.

For the remainder of this chapter, we will first start with a brief description of the proportional-integral control of lighting systems, which is one of the most basic controller design methodologies for a lighting system. Then, the subsequent four sections will present the problem formulation, controller design methodology, and performance evaluation for the four selected algorithms (mentioned in the previous subsection).

## 9.4 Decentralized Integral Control

The simplest control methodology for dimmable lights is the proportional (P) controller. At each time-step of the closed-loop operation, the output of this controller (i.e., the input to the lighting system) is proportional to the error signal at that particular time-step only. The error signal is the difference between the desired and the measured light levels for the sensors. This controller always results in a nonzero steady-state error between the desired and measured values, and hence is not suitable for any practical lighting control system.

The most basic control methodology with guarantee of zero steady state error is pure integral control, described as

$$\begin{aligned} u(k) &= \alpha \sum_{j=0}^{k-1} e(j) \\ \textit{where} \quad e(j) &= y_d(j) - y(j). \end{aligned} \tag{9.4}$$

Here $\alpha$ is the controller gain, and $e(j)$ represents the error between the desired ($y_d$) and measured ($y$) sensor measurements at $j$-th time-step. Note that, the integral controller considers the history of the error, i.e., how far the measured light level $y$ has been from the desired value $y_d$ over time. By considering this error history, the integral controller eliminates the steady-state error.

For decentralized implementation of the integral controller, putting the value of $u(k-1)$ into $u(k)$, the control law for $i$-th location can be stated as

$$\begin{aligned} u_i(k) &= u_i(k-1) + \alpha e_i(k-1) \\ \textit{where} \quad e_i(k-1) &= y_d(k-1) - y_i(k-1). \end{aligned} \tag{9.5}$$

here $\alpha$ is a parameter to adjust the speed of convergence, and $y_d(k-1)$ and $y_i(k-1)$ are respectively the desired and measured light levels at $i$-th location at the $(k-1)$-th time-step. This control law will be termed as Decentralized Integral Controller (DIC) for the rest of this chapter. We implemented this algorithm in the SCR to compare its performance with the Illumination Balancing Algorithm from [22]. The results of this comparison are discussed in the next section.

## 9.5 Illumination Balancing Algorithm

For lighting systems with collocated light sources and sensors, the cross-illumination effect between the different fixtures is a major obstacle in achievement of uniform illumination. In [22], Koroglu and Passino addressed this issue by proposing a distributed control algorithm which uses the illumination levels of the neighboring fixtures to generate the control input for each fixture. While the basic DIC fails to achieve the uniformity objective under cross-illumination effects, the proposed algorithm balances the illumination level by taking the local sensor measurements of the neighbor fixtures into account. This algorithm is termed as the Illumination Balancing Algorithm (IBA).

### 9.5.1 Problem Statement

The control objective of IBA was defined [22] as tracking a desired light level while ensuring uniform illumination across all the zones of the work surface, with communication among the light fixtures being restricted within a close neighborhood. The work surface was divided into predefined zones, where each zone consisted of a monochromatic (white channel only) light source and a Light-Dependent Resistor (LDR) was used as light sensor.

### 9.5.2 Solution Approach

The proposed solution begins with picking an arbitrary zone $l$ as "leader", only which contains the information of desired illumination level. Illumination balancing along with integral control are applied in this zone, which is described in the first equation of (9.6). The rest of the zones can only balance their illumination levels within their pre-defined neighborhood (described in the second equation of (9.6)). Without explicitly knowing the value, these zones eventually track the desired illumination level by following the "leader" zone $l$. The proposed solution is

$$u_l(k) = u_l(k-1) + \alpha e_l(k-1) - \sum_{j \in N(l)} \gamma (y_l(k-1) - y_j(k-1))$$
$$u_i(k) = u_i(k-1) - \sum_{j \in N(i)} \gamma (y_i(k-1) - y_j(k-1)) \qquad i \neq l. \tag{9.6}$$

where $N(i)$ represents the set of neighboring zones adjacent to the $i$-th zone, and $\gamma$ is a scaling factor that adjusts the stability margin and convergence speed for IBA. The proposed algorithm in IBA is similar to the DIC, described in (9.5), with an additional term that takes into account the illumination difference between the neighboring zones. If zone $i$ has higher illumination than its neighbors at a particular time-step, this additional term compensates for the existing difference by decreasing the input intensity of the light fixture in the $i$-th zone proportionally. Note that only one zone, denoted by $l$ in (9.6), is required to compensate for the error between the desired and the measured levels. As long as $\gamma \gg \alpha$, the rest of the zones corresponding to $i \neq l$ follow zone $l$, which in steady state leads to uniform illumination across the work surface [22]. The closed- loop stability condition for $\gamma$ is proved to be $\gamma \leq \frac{1}{N_l+1}$, where $N_l$ denotes the maximum number of neighbors for an individual zone. In SCR, $N_l$ is equal to 3 as shown in Fig. 9.6.

Similar to the setup described in [22], we chose a subset of eight light fixtures and eight sensors in the SCR to implement this algorithm ($n = m = 8$). The zone configuration and the neighborhood topology that we used in our implementation is shown in Fig. 9.6. Note that in the SCR, the light sensors are co-located with the fixtures. However, in [22], the LDRs used as illumination sensors were placed on the work surface directly under the lights. Therefore, an additional calibration step was taken to map the illumination values on the work surface to the illumination values measured by the sensors on the ceiling. This calibration procedure (adapted from [22]) normalized both the input intensities $u_i$ and the sensor measurements $y_i$ in the absence of cross-illumination effects.

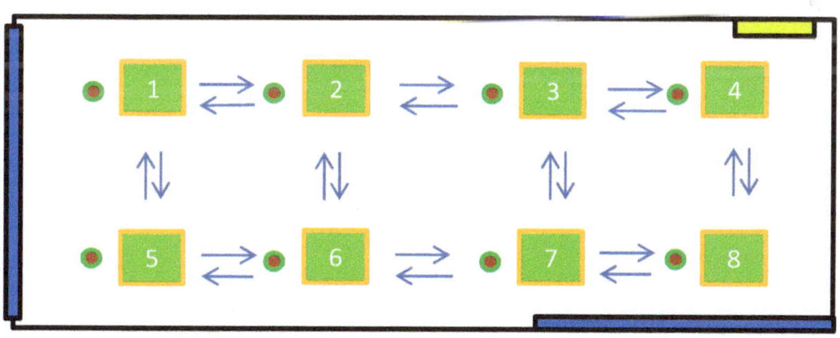


**Fig. 9.6** The zone configuration and neighborhood topology for the implementation of IBA. Zones 1, 4, 5, and 8 have two neighbors each, while each of the other zones have three neighbors. [Reprinted from Energy and Buildings, 130 (2016), M.H. Toufiq Imam, Sina Afshari, and Sandipan Mishra, An experimental survey of feedback control methodologies for advanced lighting systems, 600–612, Copyright (2016) [2], with permission from Elsevier.]

### 9.5.3 Results

First, as a benchmark, the DIC algorithm described in (9.5) was implemented in each of the individual zones in SCR depicted in Fig. 9.6. Note that in this implementation, it was assumed that there was no communication between the light fixtures in different zones. The results with $\alpha = 0.5$ are shown in Figs. 9.7 and 9.8. The cross-illumination effect between the zones prevented achievement of uniform illumination across the work surface, as shown in Fig. 9.7. Zone 4 remained over-illuminated although the input intensity of light 4 reached its minimum level. Similarly, zone 3 stayed under-illuminated although the input intensity of light 3 reached its maximum saturation level.

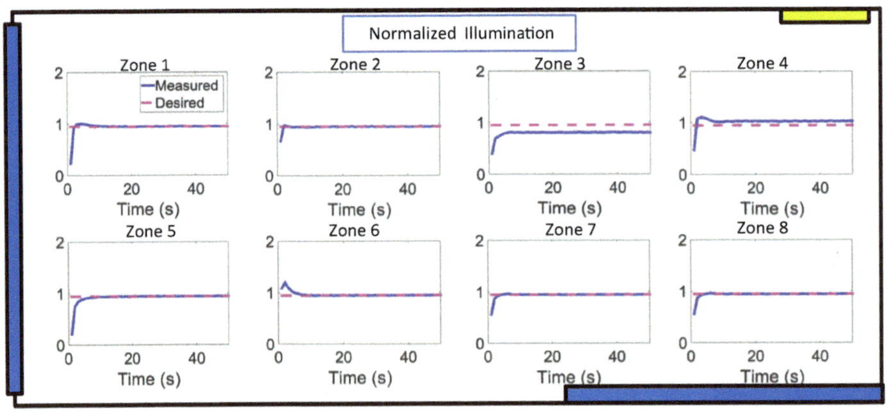


**Fig. 9.7** Sensor measurements during the implementation of the DIC in the SCR. Uniform illumination could not be achieved for zones 3 and 4

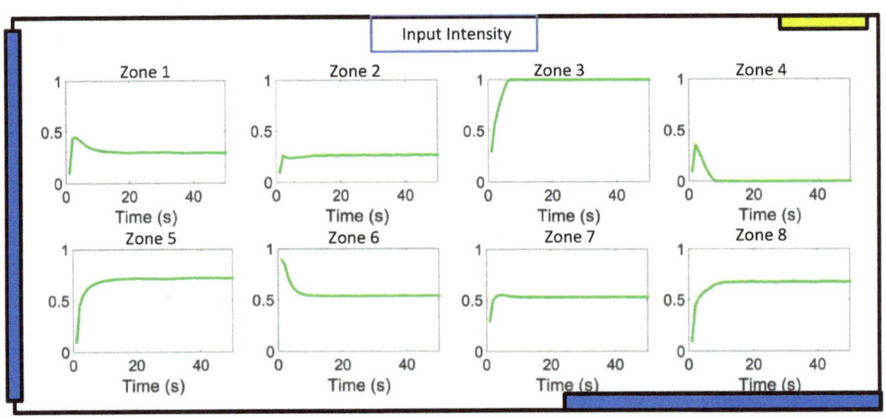


**Fig. 9.8** Input intensities during the implementation of the DIC in the SCR. Zones 3 and 4 reached their maximum and minimum values

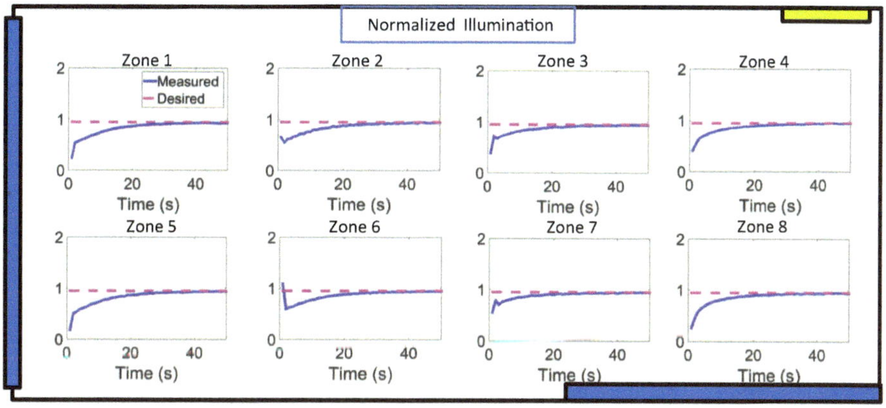


**Fig. 9.9** Sensor measurements during the implementation of the IBA in the SCR. All the zones achieved desired illumination level

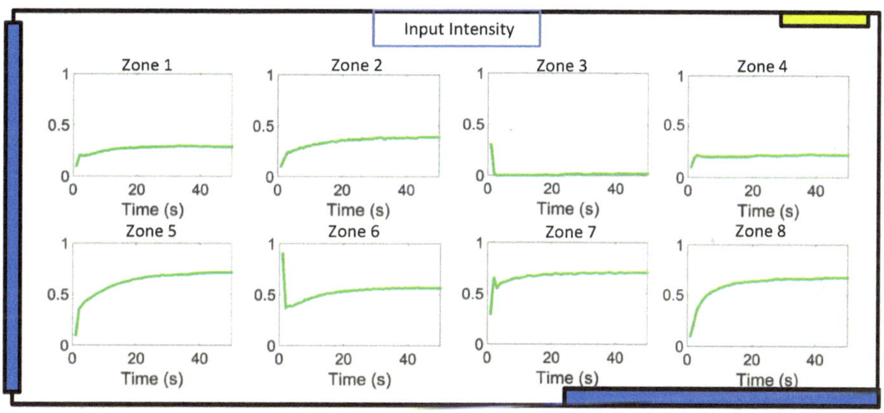


**Fig. 9.10** Input intensities during the implementation of the IBA in the SCR. The input intensities are not uniform for all the light fixtures although they produce uniform illumination

Next, the IBA was implemented in the SCR with $\alpha = 0.1$, $\gamma = 0.2$, and $l$ arbitrarily chosen as 7. The results of this implementation are shown in Figs. 9.9 and 9.10. Figure 9.9 shows that in the presence of cross-illumination effects, IBA not only resulted in tracking of the desired light level for all the zones, but also achieved uniform lighting across the work surface. For all the zones except zone 7, the tracking was achieved without the knowledge of the reference light level, and simply by balancing their light input with the input intensity of their respective neighboring fixtures.

As shown in Fig. 9.8, for the implementation of DIC fixture 3 could not provide enough light to achieve the desired illumination as it already reached its maximum level. However, fixture 7 maintained a fixed input intensity of 0.5 to provide sufficient illumination for its own zone since it was not aware of the requirements of other zones.

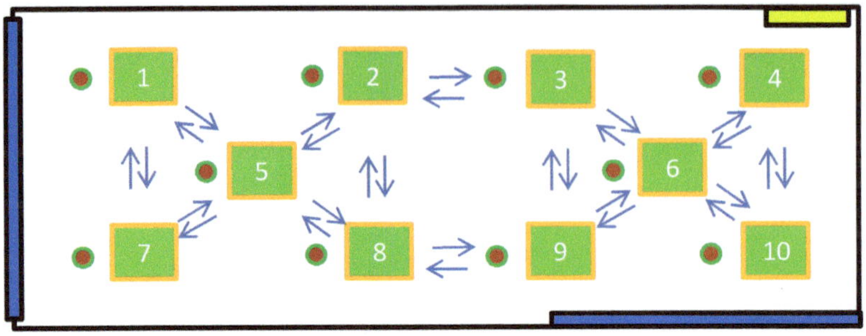


**Fig. 9.11** The zone configuration and neighborhood topology used to define uniformity among the light fixtures. [Reprinted from Energy and Buildings, 130 (2016), M.H. Toufiq Imam, Sina Afshari, and Sandipan Mishra, An experimental survey of feedback control methodologies for advanced lighting systems, 600–612, Copyright (2016) [2], with permission from Elsevier.]

In case of IBA, as shown in Fig. 9.10, the input intensity of fixture 7 went high enough to provide the required light for both zones 3 and 7 so that fixture 3 could almost be turned off during the whole period of closed-loop control.

To examine how much cross- illumination it could compensate, we implemented IBA with a set of ten lights and ten sensors. The topology and sensor configuration is shown in Fig. 9.11. In the new configuration, IBA could not achieve uniform illumination across the work surface due to the excessive cross-illumination among the lights.

## 9.6 Daylight and Occupancy Adaptive Lighting Control

Unlike the distributed approach taken in IBA, Caicedo et al. [35] adopted a centralized approach to optimize power efficiency subject to daylight harvesting and occupancy adaptation. The objective was to minimize the total power consumption of the system while ensuring a variable level of minimum illuminance at each sensor and keeping the input intensities of the neighboring light fixtures close together. The desired minimum level of illuminance depended on the occupancy state of the zones.

### *9.6.1 Problem Formulation*

The light fixtures in SCR use pulse width modulation (PWM) dimming, for which the power consumption by each light channel is proportional to its input intensity. Based on this fact, we chose to represent the total power consumption of the system as the summation of the input intensities of the light fixtures $\sum_{i=1}^{n} u_i$, where $n = 10$. The overall optimization problem can be formulated as follows:

$$
\begin{aligned}
\Delta\mathbf{u}(\mathbf{k}) = \operatorname*{argmin}_{\Delta u} & \sum_{i=1}^{n} \Delta u_i(k) \\
\text{subject to:} & \\
\mathbf{G}\Delta\mathbf{u}(k) &\geq \mathbf{y}_d^x(k) - \mathbf{y}(k-1), \\
-\Delta\mathbf{u}(k-1) &\leq \Delta\mathbf{u}(k) \leq \mathbf{1} - \Delta\mathbf{u}(k-1), \\
\left|\Delta u_i(k) - \Delta u_j(k)\right| &\leq \delta u - \Delta u_i(k-1) + \Delta u_j(k-1), \\
\text{where} \quad & i = 1, \ldots, n \quad and \quad \forall j \in N(i)
\end{aligned}
\tag{9.7}
$$

Where $\Delta\mathbf{u}(k) = [\Delta u_1(k), \ldots, \Delta u_n(k)]^T$ is the incremental step vector for the input intensity, **y**(k–1) is the sensor measurement, $\mathbf{y}_d^x(k)$ is the desired illuminance based on the occupancy state, $x$ represents the occupancy state (0 and 1 mean unoccupied and occupied respectively), $\delta u$ is the threshold for the input intensity differences within the neighborhood fixtures, $N(i)$ defines the set of neighboring fixtures to the $i$-th fixture, and $\mathbf{G} \in \mathbb{R}^{qm \times pn}$ is the LTM described in Sect. 9.3.

### 9.6.2 Solution Approach

The optimization problem with the linear objective function and linear constraints shown in (9.7) was solved using the interior-point method in an iterative approach [35]. The update equation for the input intensity level at iteration $k$ is shown in (9.8).

$$
\mathbf{u}(k) = \mathbf{u}(k-1) + \alpha\Delta\mathbf{u}(k) \tag{9.8}
$$

At each iteration $k$, the centralized controller solves the linear optimization problem in (9.7) to obtain the incremental step vector for the input intensity $\Delta\mathbf{u}(k)$, and uses the update Eq. 9.8 to obtain the complete input intensity vector **u**(k) for the next time-step.

This algorithm assumed co-located light fixtures and sensors and an occupancy sensor for each zone. For the implementation of this algorithm in SCR, we selected ten zones of interest directly under the ten light fixtures ($n = m = 10$). The zone configuration and the neighborhood topology used in this implementation is shown in Fig. 9.11. Since this algorithm was proposed for single-color lighting systems, the white channels ($p = 1$) of the fixtures were selected as the source and the clear channels ($q = 1$) of the color sensors were assumed to be the illuminance sensors. Since SCR was not equipped with occupancy sensors that co-locate with the fixtures, the occupancy state was assumed to be known.

The design parameter $\alpha$ determines the speed of adaptation, and can be adjusted to achieve smoother transition of the intensity levels. To solve this iterative optimization problem, the central controller does not require the knowledge of daylight contribution at each time-step as long as the variation in daylight contribution is negligible between two consecutive iterations.

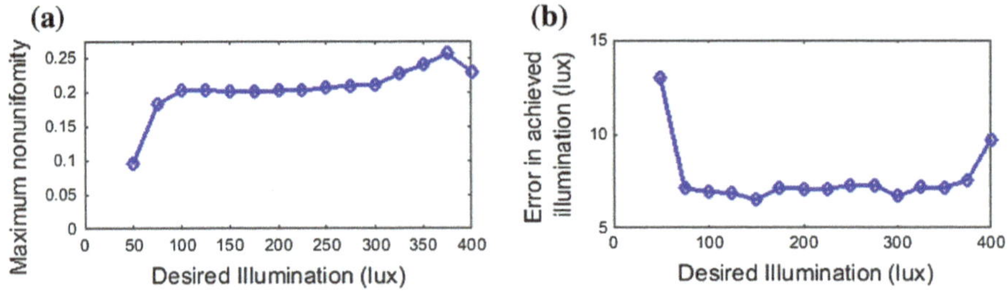


**Fig. 9.12** Evaluation in several parameters for the daylight and occupancy adaptive lighting algorithm as desired illumination varies from 50 to 400 lux with uniformity threshold, $\delta u$, kept constant at 0.2. **a** Maximum nonuniformity among the input intensities, **b** Average error between desired and achieved illumination

### *9.6.3 Results*

To demonstrate the effect of the desired illumination level, $y_d$, the experiment was repeated with the value of $y_d$ varied from 50 to 400 lux with a 25 lux increment. The uniformity threshold, $\delta u$, was fixed at 0.2 throughout the experiments. The results are shown in Fig. 9.12. Figure 9.12a shows the maximum nonuniformity among the steady state input intensities. It demonstrates that, except for very high desired illumination, the constraint on uniformity threshold holds for all the experiments. Note that the maximum illumination error is 14 lux for $y_d = 50lux$ (Fig. 9.12b). The illumination error is less than 10 lux for the rest of the set points, which is quite satisfactory for general lighting applications.

## 9.7 Spectral Optimization for Polychromatic Lighting

Both of the control strategies discussed in the previous sections were designed for single-color lighting systems. Aldrich et al. [31] studied the problem of controlling color-tunable LED fixtures, which can generate a gamut of colors. The authors posed the control problem as a constrained optimization problem, similar to [35]. The objective was to either minimize the energy consumption or maximize the color rendering index (CRI) for a given color temperature ($T_d$), determined by the user.

### *9.7.1 Problem Formulation*

As shown in Eq. 9.9, the piece-wise cost function for the optimization problem includes three different terms. These terms are the 2-norm of the difference between the desired and measured color temperatures ($T_d$ and $T$), a nonlinear term ($\Delta uv(T)$)

that quantifies the distance between the generated color and the black body curve, and an additional term that may be chosen as either the normalized power consumption or the normalized CRI of the generated light depending on the optimization goal. The optimization goal is determined by the parameter $z$. Note that the black body curve characterizes the set of all color points of the light radiated from a black body in different temperatures. The overall controller design problem was formulated as the following optimization problem:

$$\min_{u} \quad \mathbf{J}(u) = \begin{cases} \left(\frac{T_d - T(u)}{T_d}\right)^2 + \alpha_1 \Delta uv(T(u))^2 + \alpha_2 \left(\frac{\Gamma_E u}{P_t}\right)^2, z = 0 \\ \left(\frac{T_d - T(u)}{T_d}\right)^2 + \alpha_1 \Delta uv(T(u))^2 - \alpha_2 \left(\frac{R_a(u)}{100}\right)^2, z \neq 0 \end{cases} \tag{9.9}$$

subject to

$$y_d - y_{dark} \leq \mathbf{Gu} \leq (1 + \beta)(y_d - y_{dark}),$$
$$and \qquad \mathbf{0} \leq \mathbf{u} \leq \mathbf{1}.$$

where $\mathbf{u} = [u_1, \ldots, u_p]^T$ is the input intensity vector for the p LED channels, $\Gamma_E$ is a row vector consisting of the power consumption of each LED channel in the full-on state, $P_t = \sum_{i=1}^{p} \Gamma_E$ is the maximum power consumption for an individual light fixture, $\alpha_1$ and $\alpha_2$ are two weighting parameters, $y_d$ and $y_{dark}$ are, respectively, the desired minimum illumination and ambient illumination measured at the sensor node position and $\mathbf{G} \in \mathbb{R}^{qm \times pn}$ is the LTM described in Sect. 9.3. The linear constraints ensure that the achieved illumination level on the work surface falls within a certain boundary (defined by the parameter $\beta$) around the desired minimum illumination level $y_d$. It also ensures that the achieved input intensities falls within the fixture saturation range.

### 9.7.2 Solution Approach

The standard nonlinear solver in MATLAB (interior-point algorithm) was used to achieve the optimum input intensity

$$\mathbf{u} = \underset{u}{\operatorname{argmin}} \quad \mathbf{J}(u) \tag{9.10}$$

satisfying all the constraints specified in (9.9).

This algorithm does not take the spatial variation of the light field into account and is designed for a system with a single light source and a single sensor placed at a predefined position. Therefore, for implementation of this algorithm in SCR, all ten light fixtures were assumed to be a single color-tunable light source with 5 channels (red, green, blue, amber, and phosphor-converted white). Also, the algorithm only requires measurement of the illumination level from a single sensor. In our experiments, we used the clear channel measurement of the color sensor located in zone 5 of Fig. 9.11. The choice of the particular sensor on the ceiling was arbitrary. Unlike the testbed used in [31], all the sensors in SCR are located on the ceiling. To estimate the illumination level on the work surface from the measurement of the sensor on the ceiling, a mapping between these two values was obtained during the calibration stage. It was also assumed that the user had the option to change the desired minimum illumination level, $y_d$, and the desired CCT, $T_d$, at any time.

### 9.7.3 Results

In order to study the effect of different desired CCTs ($T_d$) on this algorithm, a number of experiments were conducted in the SCR. The algorithm was implemented for 17 different desired CCTs from 2800 to 6000 K with steps of 200 K. The desired minimum illumination level was chosen as 400 lux on the work surface. For each desired CCT, two experiments were carried out, one minimizing power consumption and one maximizing CRI. Next, different performance parameters were measured for the achieved light field in the illuminated space. Figure 9.13a and b show that for lower desired CCT values, the CRI optimization achieved higher illumination level with a higher level of power consumption compared to the higher desired CCTs. This effect of CCT on power consumption and illumination was more subtle for the power optimization case. The largest achieved power saving for the two optimization cases was 70 W for the desired CCT of 5000 K using the power optimization algorithm. The

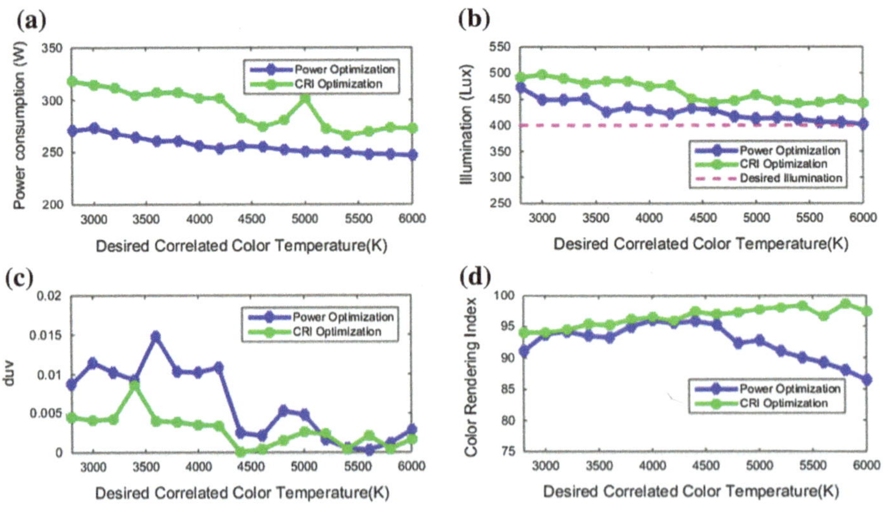


**Fig. 9.13** Variation in **a** Power Consumption at steady- state set points, **b** Desired and achieved illumination, **c** Matching error between the desired and achieved CCTs, and **d** Achieved CRI values for for power and CRI optimization techniques, as desired CCT varies from 2800 to 6000 K with desired illumination kept constant at 400 lux

controller satisfied the minimum illumination level constraint for the whole range of the desired CCTs in both cases, as shown in Fig. 9.13b. The controller had a penalty on the error between the desired ($T_d$) and the achieved ($T$) CCTs as well as the distance from the black body curve ($duv$). This resulted in a maximum $duv$ of 0.015, which was achieved for CCT of 3600 K using the power optimization algorithm, as shown in Fig. 9.13c. In the case of CRI optimization, the maximum achieved $duv$ was 0.008 for 3400K.

Figure 9.13d shows that for the CCT range [3600 K,4600 K], both power and CRI optimization achieved CRI of more than 95. There was a large difference in achieved CRIs between the two algorithms for other desired CCTs. In case of power optimization, the achieved CRI went down to 85 for the maximum desired CCT of 6000 K, while maximum value of 96 was achieved at 4000 K. On the other hand, the CRI optimization guaranteed high CRI throughout the entire desired CCT range with the maximum of 98 for 5800 K. Figure 9.14a and b demonstrate the individual LED channel intensities for the optimal input vectors in the case of CRI and power optimization respectively. The reason for the difference in achieved CRI levels for the two optimization algorithms can be observed in these two plots. For power optimization, the controller turned all the channels except blue and amber off for the high desired CCTs in order to achieve power savings, which resulted in low CRI level. However, the CRI optimization kept a higher number of LED channels on, which resulted in high CRI throughout the range of the desired CCTs with the cost of higher power consumption.

Next, we examined the effect of the number of LED channels in the performance of the controllers. In this experiment, spectral information for only four of the input LED channels was provided to the controller. These channels were red, green, blue, and white. The achieved CRI for the same range of desired CCTs are shown in Fig. 9.15a and b for power and CRI optimization respectively. The achieved CRI in

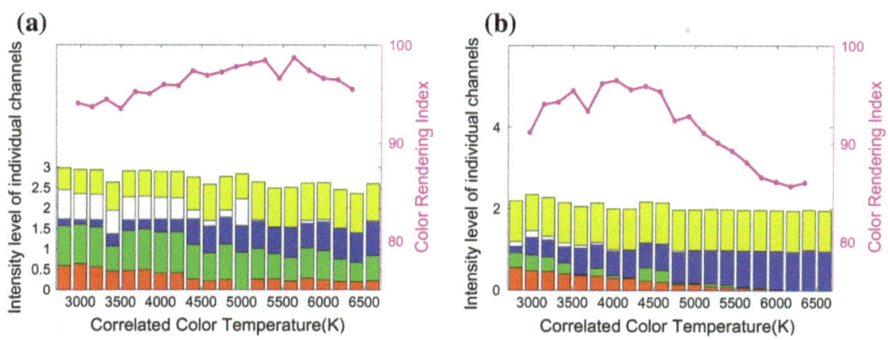


**Fig. 9.14** Achieved CRI and individual LED channel intensities for the case of **a** CRI optimization and **b** Power optimization, as desired CCT varies from 2800 to 6000 K at fixed minimum desired illumination level of 400 lux with all 5 available input channels

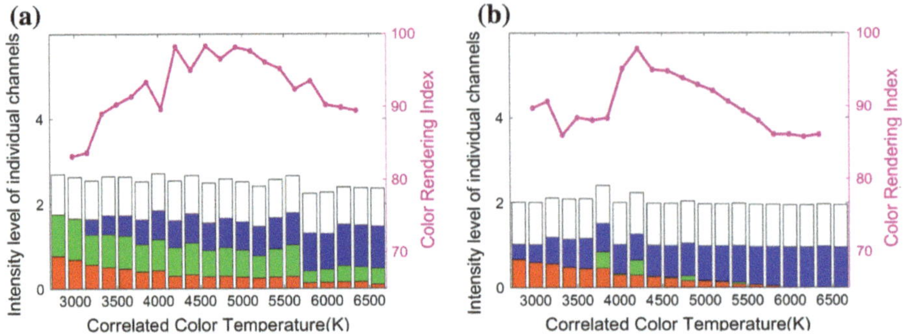


**Fig. 9.15** Achieved CRI and individual LED channel intensities for the case of **a** CRI optimization and **b** Power optimization, as desired CCT varies from 2800 to 6000 K at fixed minimum desired illumination level of 400 lux with only RGBW input channels

the case of power optimization went down significantly compared to the case with all the 5 channels. The absence of amber channel resulted in a significant loss of CRI in the lower CCT range. The results for CRI optimization follow a similar but more subtle pattern.

## 9.8 Hierarchical Optimization for Multi-channel LEDs

In [32], Afshari and Mishra proposed an optimization-based framework for control of LED lighting systems with higher number of source channels than sensor channels. The authors presented several control algorithms for performance optimization in these systems, exploiting the inherent redundancy due to the color metamerism that results from the high number of LED channels. The overall optimization problem was posed as choosing the appropriate light input to minimize the weighted sum of two metrics $\mu_Q$ and $\mu_F$, representing the light quality and a user-defined performance metric respectively.

### *9.8.1 Problem Formulation*

Given the LTM, $\mathbf{G} \in \mathbb{R}^{qm \times pn}$, as described in Sect. 9.3, and a vector $y_d \in \mathbb{R}^{qm}$ consisting of the desired RGB values for each of the sensors, the light quality metric was formulated as the Euclidean norm of the difference between the desired and measured sensor values ($y_d$ and $y$). On the other hand, the performance metric was formulated as a function of the input intensities ($u$) that might optimize different performance parameters, such as power efficiency, CRI or CCT, depending on different applications.

Thus the overall controller design problem was formulated as

$$\begin{aligned} \min_{u} \quad & \|y_d - y_k\|_2^2 + \alpha_f f(u_k) \\ \text{subject to} \quad & 0 \le u \le 1 \end{aligned} \tag{9.11}$$

where $\mathbf{u} = [u_1, u_2, \ldots, u_{pn}]^T$ is the input intensities of the light fixtures, $f(u)$ denotes a user-defined performance metric, and $\alpha_f$ is a weighting coefficient. $u_k$ and $y_k$ denote the input intensity and RGB sensor measurement at $k\text{-}th$ time-step respectively.

## 9.8.2 Solution Approach

To achieve precise color matching between the desired and measured light fields, the optimization problem in (9.11) was solved in a two-step hierarchical approach. In the first step, a candidate solution ($\bar{u}_k$) to drive the output of the system toward the desired light field is calculated using a gradient-based method. This step is demonstrated with the red arrows in Fig. 9.16. The update equation for this step is defined as

$$\bar{u}_k = u_k + \varepsilon(G^T(GG^T)^{-1}(y_d - y_k) \tag{9.12}$$

where $\varepsilon$ is the step size. Note that the system has more number of source channels than sensor channels. This redundancy in the generation of color results in existence of an infinite number of candidate solutions, all resulting in the same sensor measurements. Each of these solutions may successfully achieve the desired color setpoint, as long as achievement of $y_d$ is feasible. This creates the opportunity of further optimization of the system performance within the set of all the candidate solutions.

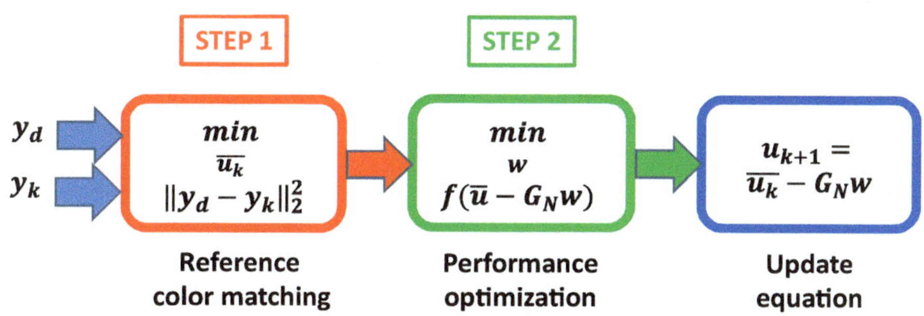


**Fig. 9.16** The schematic demonstration of hierarchical optimization. Red and green arrows represent the first step (as shown in (9.12)) and second step (as shown in (9.14)) of the algorithm respectively. [Reprinted from Energy and Buildings, 130 (2016), M.H. Toufiq Imam, Sina Afshari, and Sandipan Mishra, An experimental survey of feedback control methodologies for advanced lighting systems, 600–612, Copyright (2016) [2], with permission from Elsevier.]

In the next step, another optimization problem is solved to find the input intensity vector from the set of all candidate solutions that optimizes the user-defined performance metric. This optimal solution, $u$, is obtained using the following equation:

$$u_{k+1} = \bar{u}_k - G_N w \tag{9.13}$$

where $G_N$ is an orthogonal matrix the columns of which span the null space of $G$ and the $w$ is obtained from solving the following optimization problem

$$\begin{aligned} &\min_{w} && f(u - G_N w) \\ &\text{subject to} && 0 \leq (u - G_N w)(i) \leq 1, \quad i = 1, \ldots, pn. \end{aligned} \tag{9.14}$$

The second step is demonstrated with the green arrows in Fig. 9.16. As an example of the user-defined performance metric $f(u)$, we chose to optimize the total power consumption of the system. The LED light fixtures in SCR use PWM dimming. Therefore, the power consumption of individual LED channels of the fixtures can be approximated as a linear function of the channel's input intensity. Hence, we can model the performance metric $f(u)$ as $\Gamma_E u$. Here the row vector $\Gamma_E$ represents the power consumption by the individual LED channels when they are in full-on state. Using this formulation, the hierarchical approach results in convergence to the most power efficient solution that obtains $y_d$ at the sensor locations. Note that this approach ensures zero steady state error as well as power minimization, as opposed to trading off light quality with power savings in a single step.

In order to implement this algorithm in SCR we selected ten zones directly under the ten light fixtures ($n = m = 10$). The zone distribution was identical to Fig. 9.11. However, unlike the previous three algorithms, this algorithm is designed for lighting systems with multi-channel light sources and sensors, hence all the five channels ($p = 5$) of the fixtures and three channels ($q = 3$) of the sensors were considered for this implementation. The five LED channels were red, green, blue, amber, and phosphor-converted white, while each sensor had red, green, and blue channels to measure the spectral content of the incident light.

### 9.8.3 Results

A number of experiments were conducted to study the effect of variation in desired sensor measurements, $y_d$, on the performance of this algorithm. In these experiments, a desired input setpoint was generated at first, and the corresponding sensor measurements were recorded ($y_d$). While the desired input intensities for the five LED channels were chosen to be identical for all the fixtures, the recorded values in $y_d$ were different for different zones. Next, closed-loop control operation was initialized from an arbitrary point.

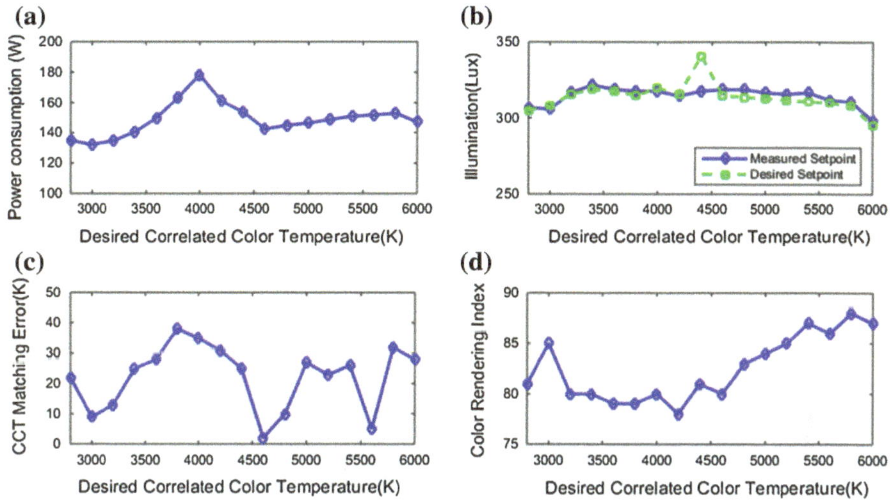


**Fig. 9.17** Variation in **a** Power Consumption at steady- state set points, **b** Desired and achieved illumination, **c** Matching error between the desired and achieved CCTs, and **d** Achieved CRI values for hierarchical optimization technique, as desired CCT varies from 2800 to 6000 K with desired illumination kept constant at 400 lux. [Reprinted from Energy and Buildings, 130 (2016), M.H. Toufiq Imam, Sina Afshari and Sandipan Mishra, An experimental survey of feedback control methodologies for advanced lighting systems, 600–612, Copyright (2016) [2], with permission from Elsevier.]

The algorithm was implemented for 17 different $y_d$ that corresponded to the desired CCT from 2800 to 6000 K with steps of 200 K. Various performance parameters were measured for the desired and achieved steady-state light fields on the work surface in zone 5. Figure 9.17a shows that for lower desired CCT values, the controller achieved smaller power consumption than that for higher desired CCT values. For lower CCTs, the controller had the option of choosing between all the available LED channels including red and amber, as shown in Fig. 9.18. However, at the higher CCT range, the controller had no option but choosing the blue channel to ensure color accuracy. The maximum CCT matching error was 38 K, corresponding to the desired CCT of 3800 K, as shown in Fig. 9.17c. Figure 9.17b shows that for the whole range of desired CCTs, the generated illuminations remained close to the desired value of 400 lux. The maximum discrepancy was 10 lux for the desired CCT of 4400 K. Note that the hierarchical controller does not take into account the spectral information of the fixtures. Therefore, the achieved CRI levels in these experiments, shown in Fig. 9.17d, were not as high as the values achieved by the CRI optimization algorithm described in Sect. 9.7. Figure 9.18 shows the input intensity levels for the individual LED channels of light fixture 5. The low CRI is a result of eliminating the amber channel from the solution by the controller for most of the desired CCT values in an effort to achieve power saving. If CRI had been chosen as the user-defined performance metric instead of power consumption, we would have found a different combination of channel intensities to achieve the same $y_d$ with a higher CRI level.

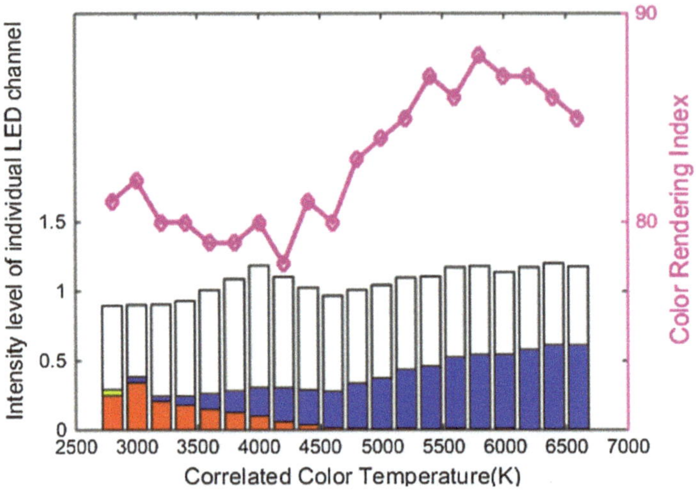


**Fig. 9.18** Achieved CRI and individual LED channel intensities as desired CCT varies from 2800 to 6000 K at fixed desired illumination level of 400 lux for hierarchical optimization

## 9.9 Effect of Ambient Daylight: Day-Long Experiments

Daylight harvesting is one of the major drivers for developing feedback control for lighting systems. By sensing the available ambient daylight, the feedback control systems can dim down the input intensity, while ensuring minimum desired illumination and color quality. To investigate the effect of ambient daylight on the performances and energy savings of the above mentioned algorithms, 12 hour-long daylight harvesting experiments were carried out, with varying ambient lighting conditions. Daylight was allowed to enter the conference room through two windows from 6 am to 6 pm. The desired illumination level was set to five preset values for different periods of the day (as a simulation of varying usage scenarios). Throughout the day, the four algorithms were simultaneously running in a multiplexed way, so that their performances could be validated under the *same* ambient lighting conditions (which was measured every 10 min).

Figure 9.19 shows the ambient, measured and desired illumination and Fig. 9.20 shows the total power consumption for the four algorithms for one particular day. Although the IBA suffered significant overshoot during the transition periods, the other algorithms can track the desired illumination under varying ambient lighting conditions quite successfully. Note that, for this particular day (shown in Fig. 9.20), there was little ambient daylight available due to cloudy weather, so there was significant power consumption for all four cases.

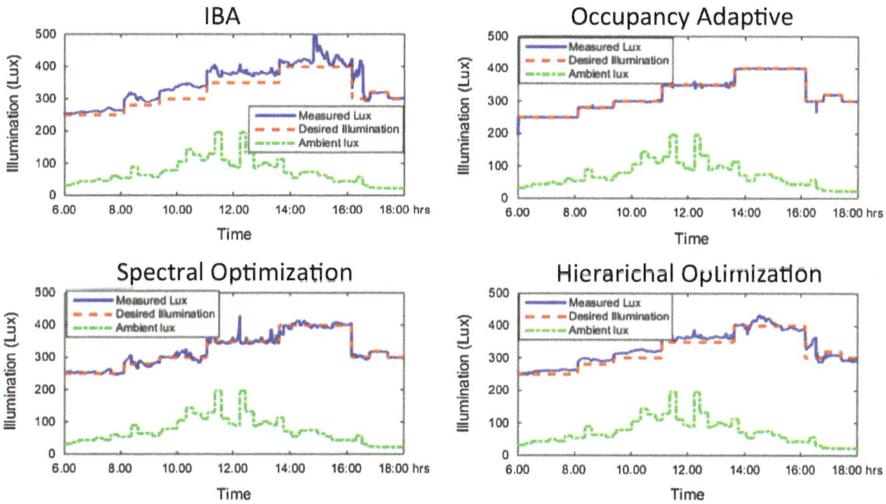


**Fig. 9.19** Results from 12 h long daylight harvesting experiment for the four algorithms during a cloudy day. Measured and desired illuminations are shown along with the ambient daylight measurement. [Reprinted from Energy and Buildings, 130 (2016), M.H. Toufiq Imam, Sina Afshari, and Sandipan Mishra, An experimental survey of feedback control methodologies for advanced lighting systems, 600–612, Copyright (2016) [2], with permission from Elsevier.]

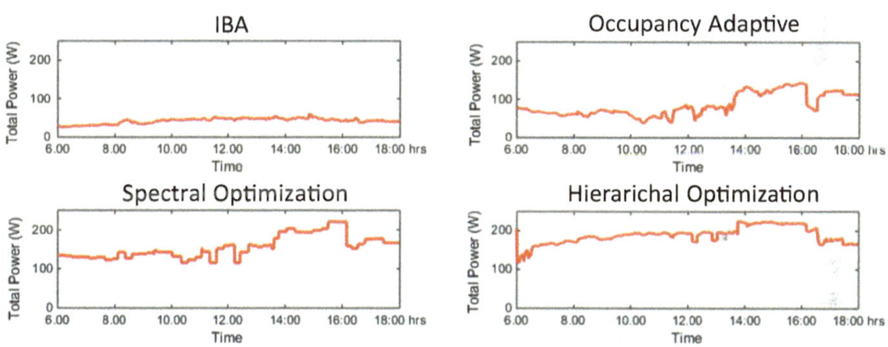


**Fig. 9.20** Total power consumption from 12 h long daylight harvesting experiment for the four algorithms during a cloudy day. [Reprinted from Energy and Buildings, 130 (2016), M.H. Toufiq Imam, Sina Afshari, and Sandipan Mishra, An experimental survey of feedback control methodologies for advanced lighting systems, 600–612, Copyright (2016) [2], with permission from Elsevier.]

## 9.10 Summary and Conclusion

This chapter provides an overview of various existing feedback control algorithms for LED- based lighting systems. We presented a brief review of state-of-the-art control methodologies for lighting systems followed by experimental results of four recently proposed control algorithms.

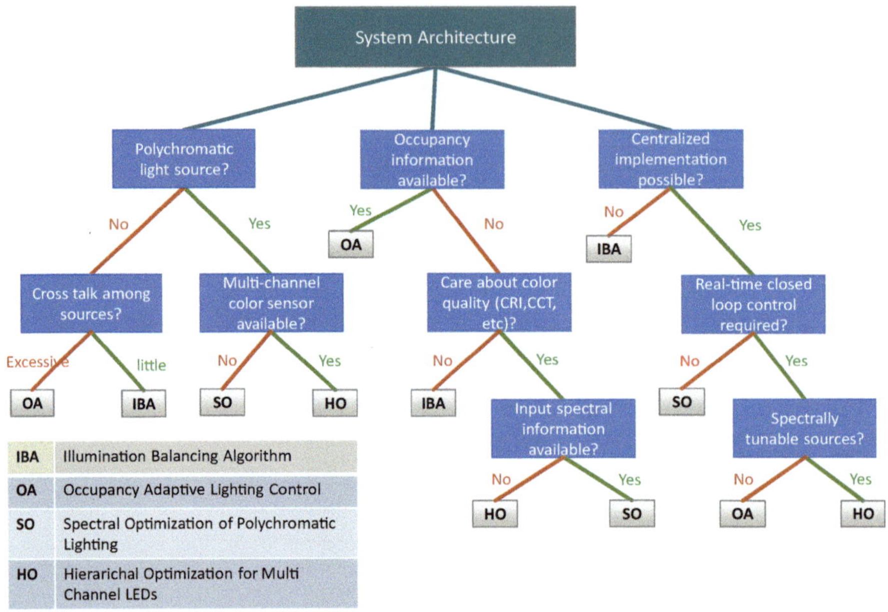


**Fig. 9.21** A decision tree showing some sample scenarios and the corresponding preferred feedback control algorithm among the four selected algorithms discussed in the article. [Reprinted from Energy and Buildings, 130 (2016), M.H. Toufiq Imam, Sina Afshari, and Sandipan Mishra, An experimental survey of feedback control methodologies for advanced lighting systems, 600–612, Copyright (2016) [2], with permission from Elsevier.]

In conclusion, each of these algorithms has its own merits and drawbacks depending upon the specific control objectives, constraint, and technical capabilities. There is no single control strategy which can be termed as the best solution for all possible scenarios. The choice among these lighting control strategies should be based on the particular control problem and business model.

Some of the questions that need to be addressed prior to the selection of the solution strategy are whether the lighting system is single-color or color-tunable, whether the multi-channel color sensing and occupancy sensing technologies are available, whether the sensors are at the ceiling or at the work surface in the illuminated space, what is the accuracy and speed of the sensors, etc. A decision tree is presented in Fig. 9.21 showing some sample scenarios and the corresponding preferred feedback control algorithm.

With the rapid progress in computation, sensing, and communication technologies, it is not even necessary to stick to a single control architecture for the new generation of lighting systems. These systems may be flexibly adapted and reconfigured to specific requirements by switching between various methodologies. These advanced smart lighting systems, interconnected with other building systems, have introduced new services for an office building beyond provision of mere illumination.

## References


1. Conti J (2013) Annual energy outlook 2013 with projections to 2040, U.S. Energy Inform. Admin., Washington, DC, USA, Technical report DOE/EIA-0383
2. Imam MT, Afshari S, Mishra S (2016) An experimental survey of feedback control methodologies for advanced lighting systems. Energy Buildings 130:600–612
3. Williams A, Atkinson B, Garbesi K, Page E, Rubinstein F (2012) Lighting controls in commercial buildings. LEUKOS 8(3):161–180
4. ul Haq MA, Hassan MY, Abdullah H, Rahman HA, Abdullah MP, Hussin F, Said DM (2014) A review on lighting control technologies in commercial buildings, their performance and affecting factors. Renew Sustain Energy Rev 33:268–279
5. Pandharipande A, Caicedo D (2015) Smart indoor lighting systems with luminaire-based sensing: a review of lighting control approaches. Energy Buildings 104:369–377
6. Rubinstein F (1984) Photoelectric control of equi-illumination lighting systems. Energy Buildings 6(2):141–150
7. Rubinstein F, Ward G, Verderber R (1989) Improving the performance of photo-electrically controlled lighting systems. J Illum Eng Soc 18(1):70–94
8. Rubinstein F, Karayel M (1984) The measured energy savings from two lighting control strategies. IEEE Trans Industry Appl IA-20(5):1189–1197
9. Rubinstein F, Siminovitch M, Verderber R (1993) Fifty percent energy savings with automatic lighting controls. IEEE Trans Industry Appl 29(4):768–773
10. Gorgulu S, Ekren N (2013) Energy saving in lighting system with fuzzy logic controller which uses light-pipe and dimmable ballast. Energy Buildings 61:172–176
11. Matta S, Mahmud S (2010) An intelligent light control system for power saving. In: Proceedings of the 36th annual conference IEEE industrial electronics society, pp 3316–3321, November 2010. https://doi.org/10.1109/IECON.2010.5675331. ISSN 1553-572X
12. Choi A, Mistrick R (1998) Analysis of daylight responsive dimming system performance. Building Environment 34(3):231–243
13. Choi A, Sung M (2000) Development of a daylight responsive dimming system and preliminary evaluation of system performance. Building Environment 35(7):663–676
14. van de Meugheuvel N, Pandharipande A, Caicedo D, van den Hof P (2014) Distributed lighting control with daylight and occupancy adaptation. Energy Buildings 75:321–329. https://doi.org/10.1016/j.enbuild.2014.02.016
15. Afshari S, Mishra S (2013) Decentralized feedback control of smart lighting systems. In: Proceedings of the ASME dynamic systems and control conference, p V001T13A006
16. Afshari S, Mishra S, Julius A, Lizarralde F, Wason J, Wen J (2014) Modeling and control of color tunable lighting systems. Energy Buildings 68:242–253. https://doi.org/10.1016/j.enbuild.2013.08.036
17. Afshari S, Jia L, Radke R, Mishra S (2014) Light field estimation and control using a graphical rendering engine. In: Proceedings of the ASME dynamic systems and control conference, p V001T07A004
18. Wen Y, Agogino A (2008) Wireless networked lighting systems for optimizing energy savings and user satisfaction. In: Proceedings of the wireless hive networks conference, pp 1–7, August 2008
19. Wen Y-J, Agogino AM (2010) Control of wireless-networked lighting in open-plan offices. Lighting Research & Technology 43(2):235–248, November 2010. https://doi.org/10.1177/1477153510382954
20. Sarkar A, Fairchild M, Salvaggio C (2008) Integrated daylight harvesting and occupancy detection using digital imaging. In: Proceedings of the electronics imaging conference, pp 68 160F–68 160F. doi: 10.1117/12.765961
21. Caicedo D, Pandharipande A (2016) Daylight and occupancy adaptive lighting control system: an iterative optimization approach. Light Res Technol 48(6):661–675
22. Koroglu M, Passino K (2014) Illumination balancing algorithm for smart lights. IEEE Trans Control Syst Technol 22(2):557–567

23. Sandhu J, Agogino A, Agogino A (2004) Wireless sensor networks for commercial lighting control: decision making with multi-agent systems. In: Proceedings of the AAAI workshop on sensor networks, vol 10, pp 131–140
24. Kurian C, Aithal R, Bhat J, George V (2008) Robust control and optimisation of energy consumption in daylight-artificial light integrated schemes. Light Res Technol 40(1):7–24. https://doi.org/10.1177/1477153507079511
25. Colaco S, Kurian C, George V, Colaco A (2012) Integrated design and real-time implementation of an adaptive, predictive light controller. Light Res Technol 44(4):459–476. https://doi.org/10.1177/1477153512445713
26. Lu J, Birru D, Whitehouse K (2010) Using simple light sensors to achieve smart daylight harvesting. In: Proceedings of the 2nd ACM workshop on embedded sensing systems for energy-efficiency in buildings, pp 73–78. https://doi.org/10.1145/1878431.1878448. ISBN 978-1-4503-0458-0
27. Singhvi V, Krause A, Guestrin C, Garrett Jr. J, Matthews H (2005) Intelligent light control using sensor networks. In:Proceedings of the 3rd international conference on embedded networked sensor systems, pp 218–229. ISBN 1-59593-054-X
28. Pan M, Yeh L, Chen Y, Lin Y, Tseng Y (2008) A wsn-based intelligent light control system considering user activities and profiles. IEEE Sens J 8(10):1710–1721. https://doi.org/10.1109/JSEN.2008.2004294
29. Pavlic T (2013) Physical stigmergy for decentralized constrained optimization: an intelligent lighting example. In: Proceedings of the 4th ACM/IEEE international conference on cyber-physical systems, pp 258–258. ISBN 978-1-4503-1996-6
30. Wen Y, Granderson J, Agogino A (2006) Towards embedded wireless-networked intelligent daylighting systems for commercial buildings. In: Proceedings of the international conference on sensor networks, ubiquitous, and trustworthy computing, vol 1, pp 326–331. ISBN 0-7695-2553-9
31. Aldrich M, Zhao N, Paradiso J (2010) Energy efficient control of polychromatic solid state lighting using a sensor network. In: Proceedings of the SPIE optical engineering + applications conference, p 778408
32. Afshari S, Mishra S (2015) An optimization framework for control of non-square smart lighting systems with saturation constraints. In: Proceedings of American Control Conference, pp 1665–1670
33. Aldrich M, Badshah A, Mayton B, Zhao N, Paradiso J (2013) Random walk and lighting control. In: Proceedings of the IEEE sensors conference, pp 1–4. https://doi.org/10.1109/ICSENS.2013.6688590. ISSN 1930-0395
34. Pandharipande A, Caicedo D (2011) Daylight integrated illumination control of LED systems based on enhanced presence sensing. Energy Buildings 43(4):944–950. https://doi.org/10.1016/j.enbuild.2010.12.018
35. Caicedo D, Pandharipande A, Leus G (2010) Occupancy-based illumination control of LED lighting systems. Light Res Technol 43(2):217–234
36. Caicedo D, Pandharipande A (2013) Distributed illumination control with local sensing and actuation in networked lighting systems. IEEE Sens J 13(3):1092–1104. https://doi.org/10.1109/JSEN.2012.2228850
37. Afshari S, Mishra S, Julius A, Lizarralde F, Wen J (2012) Modeling and feedback control of color-tunable LED lighting systems. In: Proceedings of the American control conference, pp 3663–3668
38. Jia L, Afshari S, Mishra S, Radke R (2014) Simulation for pre-visualizing and tuning lighting controller behavior. Energy Buildings 70:287–302
39. Afshari S, Woodstock T-K, Imam M, Mishra S, Sanderson AC, Radke RJ (2015) Short paper:: the smart conference room: an integrated system testbed for efficient, occupancy-aware lighting control. In: Proceedings of the 2nd ACM international conference on embedded systems for energy-efficient built environments. ACM, pp 245–248
40. Starr C, Evers C, Starr L (2010) Biology: concepts and applications, 8th edn. Cengage Learning, Independence, KY, USA. ISBN 1439046735, 9781439046739

41. Foley J, van Dam A, Feiner S, Hughes J (1995) Computer graphics: principles and practice in C, 2nd edn. Addison-Wesley Professional, Boston

# Chapter 10
# Energy Management Systems for Intelligent Buildings in Smart Grids

**Alessandra Parisio, Marco Molinari, Damiano Varagnolo and Karl H. Johansson**

## 10.1 Introduction

It is well-known that buildings contribute a large portion of the overall energy use worldwide. Buildings need to be more sustainable and environmentally friendly and the role of buildings in the next-generation energy grids has to be rethought. In fact, as shown in Fig. 10.1, *smart* buildings should be designed to integrate not just loads but also distributed generation (e.g., storage systems and renewable energy resources); to sense, compute, communicate, and actuate; to purchase, generate, and sell power to and from its energy suppliers; to take advantage of local generation to implement

This work is supported by the Swedish Foundation for Strategic Research (SSF), the Swedish Research Council (VR), and the Knut and Alice Wallenberg Foundation (KAW), the Swedish Governmental Agency for Innovation Systems (VINNOVA), the Celtic Plus project SENDATE-Extend (C2015/3-3), the Swedish research council Norrbottens Forskningsråd project DISTRACT, and the Swedish Energimyndigheten E2B2 project Green Power.

A. Parisio (✉)
School of Electrical and Electronic Engineering, The University of Manchester, Ferranti Building-C5, Manchester M13 9PL, UK
e-mail: alessandra.parisio@manchester.ac.uk

M. Molinari
School of Electrical Engineering, KTH Royal Institute of Technology, 100 44 Stockholm, Sweden
e-mail: marcomo@kth.se

D. Varagnolo
Department of Computer Science, Electrical and Space Engineering, Luleå University of Technology, Forskargatan 1, 971 87 Luleå, Sweden
e-mail: damiano.varagnolo@ltu.se

K.H. Johansson
ACCESS Linnaeus Centre, School of Electrical Engineering, KTH Royal Institute of Technology, 100 44 Stockholm, Sweden
e-mail: kallej@kth.se



J.T. Wen and S. Mishra (eds.), *Intelligent Building Control Systems*, Advances in Industrial Control, https://doi.org/10.1007/978-3-319-68462-8_10

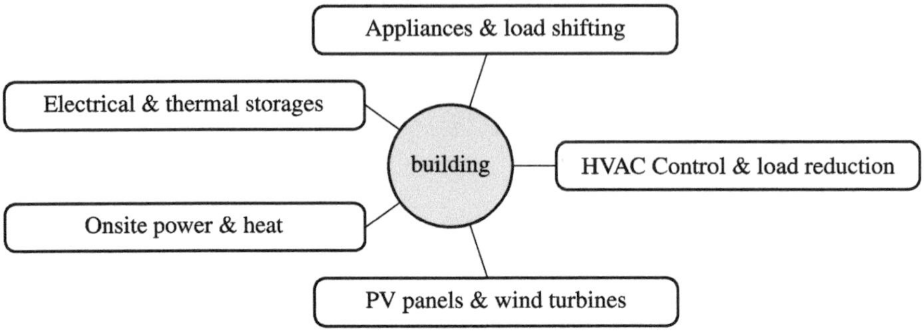


**Fig. 10.1** Technologies that can enable building-to-grid integration

demand response and optimize the energy usage. Loads of a smart building can be divided either into non-flexible ...or flexible (instead of schedulable) (critical) or schedulable loads, which implies that the decision-making algorithms should aim not only to lower the energy use without sacrificing the comfort of the occupants, but also try to support the network operator and reduce the overall environmental impact.

Hence, designing Energy Management Systems (EMS) for achieving sustainable building operation requires that several challenges are addressed. The EMS must be able to deal with the complexity such as conflicting objectives, decision-making and coordination of decisions distributed over multiple units. The increased Information and Communication Technology (ICT) capabilities can enable a wider, more sophisticated range of intelligent methods and innovative schemes to facilitate the management of energy demand and generation, including demand response, smart appliances, and price and carbon-based signals.

An EMS for smart building will: (1) respond to signals from the grid and take action on this basis (e.g., decreasing energy use when prices are high or automatically shifting consumption to times when prices are lower); (2) manage local generation facilities, such as solar panels, and feedback into the grid any energy; (3) optimally schedule storage devices, which can be used to balance out the smart grid and respond to price signals to help decrease the electricity bills [1].

Besides, this EMS needs to incorporate users into the design (to be responsive to occupants and improve their comfort) and to take advantage of the flexibility opportunities offered by storage devices (to counteract the intermittent behavior of renewable generation by providing ancillary services). These functionalities could represent new business case opportunities for various market actors such as aggregators, energy service companies, and network operators. An advanced EMS should not come with tailored configurations, but rather be cost-effective in order to be deployed by a wide range of consumers.

Such an innovative and advanced EMS would make buildings smart and smart buildings are an essential element of smart grids. Although advanced control and automation systems are becoming quite widespread, standardized and holistic solutions are still not available [1].

In this chapter, we describe the design of and scheduling frameworks for optimal management of smart buildings. In Sect. 10.2 we illustrate novel stochastic MPC schemes for HVAC systems and discuss their performance along with practical implementation issues. In Sect. 10.3, we illustrate how to integrate local generation capabilities and storage systems into a holistic building energy management framework. In Sect. 10.4, we briefly describe three demonstrators located in Sweden and UK that support ongoing research activities and experiments on an EMS for smart buildings. In Sect. 10.5, we show that data centers can play several different schemes typical of residential smart buildings data centers and they can act as bridges between three types of networks: electrical, thermal and information networks. Finally, in Sect. 10.6 we draw some conclusions.

## 10.2 Control of HVAC Systems via Scenario-Based Implicit and Explicit MPC

In this section, we discuss the design and the implementation of stochastic MPC approaches for the effective control of HVAC systems. HVAC systems are employed to maintain acceptable thermal comfort and $CO_2$ levels in buildings. A relevant share of the overall energy use in buildings is for ventilation, space heating and cooling; effective control of building HVAC, used to maintain acceptable thermal comfort and indoor air quality, is seen as an attractive approach to cost-efficiently decrease the energy use and increase the sustainability in the built environment, and has been the object of relevant research effort in recent years [2]. An improved building management can potentially lead to energy savings estimated between 5 and 30% of the total energy use [3, 4]. HVAC control systems performance can be improved by using predictive strategies.

In particular, MPC schemes are expected to become a common solution for buildings in a few years because of their capability to incorporate time-dependent energy costs, bounds on the control actions, comfort requirements, and account for uncertainties, e.g., in the models and in the forecasts [5–7]. Successful implementations will likely be based on stochastic MPC schemes with probabilistic constraints since indoor air conditions are intrinsically affected by stochastic disturbances, e.g., weather conditions and occupancy levels.

In this section, we propose MPC formulations that take into account uncertainties in the weather predictions and occupancy levels, and adopt the scenario-based distribution-free approach described in [8]. This approach does not require to assume Gaussian disturbances, as commonly done in the literature. This assumption makes the problems easier to solve, but it is often an invalid assumption in practical situations, especially for solar radiation and occupancy processes.

To address all the issues above and yet obtain an effective and computationally tractable MPC problem, we propose a stochastic MPC (SMPC) scheme that comprises two linear stochastic MPC problems and a dynamical approach for learning the statistics of the uncertainties. This SMPC scheme aims at controlling both indoor temperature and $CO_2$ concentration levels and is described in more detail in Sect. 10.2.3. The control action is computed by considering a given number of uncertainty samples and does not require any a-priori assumptions on the distributions of the uncertain variables. This implicit MPC formulation and its experimental evaluation are then discussed in Sect. 10.2.4.

The drawback associated with the proposed implicit MPC scheme is its online computational burden, whose implementation on cheap hardware platforms (such as in Programmable Logic Controllers (PLCs)) and integration in Building Automation Systems (BASs) may become prohibitive. In addition, scenario-based approaches require generation of a large number of scenarios online, further increasing the computational burden. A possible solution is to determine off-line the explicit solution of the MPC problem as a function of the current building state. This leads to explicit solutions from multiparametric programs [9, 10] where the state vector is treated as a vector of parameters, so that the optimal control profile becomes a Piecewise Affine (PWA) function of the initial state. In this way, the computational requirements of the MPC scheme reduce to a function evaluation problem, which can be implemented using simple software and cheap hardware. We formulate and describe such an explicit MPC scheme for controlling the HVAC system in Sect. 10.2.5.

Before presenting both our implicit and explicit MPC schemes, we review the relevant literature in Sect. 10.2.1. Various physical and control-oriented models of the thermal and $CO_2$ concentration dynamics are then described in Sect. 10.2.2. For further details, we refer the interested reader to [11–13] and point out that, even if we focus on one single room or thermal zone, extensions to whole building enclosures or to room networks are possible since the overall building energy use can be computed as the sum the energy usages of individual thermal zones [14].

### *10.2.1 Review of Control Strategies for HVAC Systems*

HVAC control has been extensively explored by both researchers and practitioners. Typical practice does not require coordination between HVAC control and blinds and lighting control; nonetheless, it has been experimentally shown that coordination may lead to significant energetic benefits [15].

The HVAC control problem is usually divided into a hierarchical cascade of two problems: *HVAC scheduling* and *HVAC regulation* (see [16] and references therein). More precisely,

**scheduling** means deciding both when and in which mode to operate the various HVAC subsystems (e.g., on or off), and which values should be assigned to the setpoints of subsystems (e.g., the temperature and $CO_2$ levels of thermal zones);

**regulation** means rejecting disturbances so that the actual conditions in the built environment follow closely the setpoints specified by the scheduling algorithm.

The regulation part is usually performed by means of low-level Proportional-Integral Derivative (PID) controllers, which generally provide satisfactory tracking performance, mainly because temperature and $CO_2$ dynamics are slow (see, for example, [13], where experimental evidence shows that temperature and $CO_2$ concentration have time constants of minutes in a laboratory of approximately 80 $m^2$).

The scheduling part is interesting from research perspectives, and is where the state of the art and the common practice differ. We can categorize the existing strategies as follows:

**rule-based strategies**: the control inputs are computed through rules of the type "if *condition* then *action*" where the *condition* statement typically involves thresholds and other numerical values that play the role of control parameters. Rule-based strategies may be (see also [15] and references therein):

*open-loop,* which do not make use of information to change the scheduling outcomes;

*reactive,* which use information on the current and past states of the system to change the scheduling outcomes;

*predictive,* which change the scheduling outcomes if forecasts of building usage change.

Predictive approaches are the most powerful ones; their main drawback is that they are associated to a larger set of rules and parameters, and this increases the difficulty in tuning [15];

**model-based predictive strategies**: the control strategy is determined by combining an opportune model of the building with forecasts of the disturbances to facilitate a search for the input steering the forecast system trajectory to minimize a certain objective function. This category may be divided into:

*control-oriented strategies,* where modeling of the dynamics is typically a gray- or black-box model, potentially trained using some system identification procedure;

*machine-learning-oriented strategies,* where models are typically non-parametric and data-driven (e.g., learned through opportune neural networks).

We notice that evaluations of model-based predictive strategies in smart buildings consistently show increased performance with respect to non-predictive control strategies [7, 17]. The current trend seems to add to these predictive strategies robustness with respect to uncertainties in the forecasts [18, 19].

## *10.2.2 HVAC System Modeling*

The model used in this chapter is derived by using thermodynamics laws with a thermal network approach, as described in Sect. 10.2.2.1. Compared to data-driven approaches, this approach provides a clear physical interpretation and features greater modeling generality, allowing thus a straightforward adjustment to other buildings.

Unfortunately, physical-based equations describing indoor $CO_2$ and temperature dynamics include nonlinearities. Posing then the problem of minimizing the energy use while keeping both indoor $CO_2$ and thermal comfort as one single problem results in a non-convex problem, and this would eventually lead to problems at the implementation stage. In Sect. 10.2.2.2, we thus address this issue by first deriving control-oriented models of both the $CO_2$ and thermal dynamics, and then formulating the overall HVAC control problem as a cascade of two linear (and thus more easily implementable) problems.

### 10.2.2.1 Physical Modeling of HVAC Systems

To model the energy flows in a thermal zone, we consider the flows through its envelope (i.e., walls and windows), the flows generated within the zone (e.g., occupants, lights, and equipment), and the flows provided by the heating and cooling systems via the ventilation and the heating emission systems. Our main assumptions are then that:

- there are no air latent heat, e.g., from vapor production, is neglected infiltrations, so that the inlet airflow in the zone equals the outlet airflow;
- the air in the zone is well mixed;
- latent heat, e.g., from vapor production, is neglected;
- there are no thermal bridges.

Given the assumptions above, the dynamics of the room temperature can be calculated by computing the energy balance equations of the zone, the latter modeled as a lumped node, i.e.,

$$\begin{aligned} m_{\text{air,zone}} c_{\text{pa}} \frac{dT_{\text{room}}}{dt} &= Q_{\text{venting}} + Q_{\text{int}} + \textstyle\sum_{\text{j}} Q_{\text{wall,j}} \\ &\quad + \textstyle\sum_{j} Q_{\text{win,j}} + Q_{\text{heating}} + Q_{\text{cooling}}. \end{aligned} \tag{10.1}$$

In (10.1), the left-hand side represents the heat stored in the air of the zone, $Q_{\text{venting}}$ is the heat flow due to ventilation, $Q_{\text{int}}$ are the internal gains (the sum of the heat flows due to occupancy, equipment, and lighting), $Q_{\text{wall,j}}$ and $Q_{\text{win,j}}$ represent the heat flows exchanged between walls and room and windows and room, respectively, $Q_{\text{cooling}}$ and $Q_{\text{heating}}$ are the cooling and heating flows necessary to keep the within thermally comfortable conditions.

The explicit dependence between room the variation of the temperature $T_{\text{room}}$ and the heat flows is then obtained by manipulating (10.1):

$$\begin{aligned}\frac{dT_{\text{room}}}{dt} = \frac{\dot{m}_{\text{venting}}\Delta T_{\text{venting}}}{m_{\text{air,zone}}} &+ \sum_j \frac{h_i A^j_{\text{wall}}\left(T^j_{\text{wall,i}} - T_{\text{room}}\right)}{m_{\text{air,zone}}c_{\text{pa}}} \\ &+ \sum_j \frac{\left(T_{\text{amb}} - T_{\text{room}}\right)}{R^j_{\text{win}} m_{\text{air,zone}}c_{\text{pa}}} + \frac{cN_s}{m_{\text{air,zone}}c_{\text{pa}}} \\ &+ \frac{\sum_j G^j A^j_{\text{win}} I^j}{m_{\text{air,zone}}c_{\text{pa}}} + \frac{A_{\text{rad}}h_{\text{rad}}\Delta T_{\text{h,rad}}}{m_{\text{air,zone}}c_{\text{pa}}}\end{aligned} \tag{10.2}$$

where

$$\begin{aligned} Q_{\text{venting}} &= \dot{m}_{\text{venting}}c_{\text{pa}}\Delta T_{\text{venting}} = \dot{m}_{\text{venting}}c_{\text{pa}}\left(T_{\text{air,sa}} - T_{\text{room}}\right), \\ Q_{\text{int}} &= cN_{\text{people}}, \\ Q_{\text{heating}} &= A_{\text{rad}}h_{\text{rad}}\Delta T_{\text{h,rad}} = A_{\text{rad}}h_{\text{rad}}\left(T_{\text{mr}} - T_{\text{room}}\right). \end{aligned}$$

The parameters involved in (10.2) are described in Table 10.1, reported in the appendix and presenting the parameters in alphabetical order for reading convenience. The indoor wall temperature $T^{\text{j}}_{\text{wall,i}}$ in the j-th surface is calculated with a further energy balance on the outdoor wall surface and on the indoor wall surface, with walls modeled as three resistance and two capacitance (3R2C) systems [20, 21].

The air mass flow for ventilation $\dot{m}_{vent}$ in (10.2) is determined by the $CO_2$ concentration in the room, calculated after the model proposed in [22] as:

$$V\frac{dC_{\text{CO}_2}}{dt} = \left(\dot{m}_{vent}C_{\text{CO}_2,i} + g_{\text{CO}_2}N_{people}\right) - \dot{m}_{vent}C_{\text{CO}_2}. \tag{10.3}$$

The models above capture the main features of the dynamics of the temperature and $CO_2$ concentrations. Their constants determine then the physical characteristics of a generic room, and this allows a straightforward adaptation of the models to other rooms. For example, the surface of a heating emission system may be conveniently modeled with a parameter $A_{rad}$ accounting for different sizes of the heating units. We notice that, however, heating systems with a relevant thermal mass, like floor heating, are characterized by delays that should not be neglected. In this case, the models should be adapted to account for ad-hoc time delays.

**Table 10.1** Summary of the parameters involved in the building model

| | | |
|---|---|---|
| $A_{rad}$ | $[m^2]$ | Emission area of the radiators |
| $A^j_{wall}$ | $[m^2]$ | Wall area on the j-th surface |
| $A^j_{win}$ | $[m^2]$ | Area of the window on the j-th surface |
| $c$ | $[W]$ | Constant related to equipment and occupants activity |
| $C_{CO_2,i}$ | [ppmV] | Inlet air $CO_2$concentration, assumed equal to outdoor $CO_2$ concentration |
| $C_{CO_2}$ | [ppmV] | Concentration of $CO_2$ within the room |
| $c_{pa}$ | $[J/kg°C]$ | Specific heat of the dry air |
| $g_{CO_2}$ | $[m^3_{CO_2}$/pers.] | Generation rate of $CO_2$ per person |
| $G^j$ | $[-]$ | G-value (SHGC) of the window on the j-th surface |
| $h_i$ | $[W/m^2°C]$ | Indoor heat transfer coefficient |
| $h_{rad}$ | $[W/m^2°C]$ | Heat transfer coefficient of the radiators |
| $I^j$ | $[W/m^2]$ | Solar radiation on the j-th surface |
| $m_{air,zone}$ | $[kg]$ | Air mass in the room |
| $\dot{m}_{vent}$ | $[kg/s]$ | Ventilation mass flow |
| $N_{people}$ | $[-]$ | Number of occupants in the room |
| $R^j_{win}$ | $[°C/W]$ | Thermal resistance of the window on the j-th surface |
| $T_{air,sa}$ | $[°C]$ | Supply air temperature |
| $T_{amb}$ | $[°C]$ | Outdoor temperature |
| $T^j_i$ | $[°C]$ | Indoor surface temperature of the wall on the j-th surface |
| $T_mr$ | $[°C]$ | Mean radiant tempcrature of the radiators |
| $V$ | $[m^3]$ | Volume of the air inside the room |

#### 10.2.2.2 Control-Oriented Modeling of HVAC systems

As mentioned above, we aim at rewriting the nonlinear equations. (10.2) and (10.3) as linear equations and at developing control-oriented models of both the $CO_2$ and thermal dynamics. We start by pointing out that: *(i)* the $CO_2$ concentration dynamics are independent of the thermal ones; *(ii)* $CO_2$ comfort has priority, entailing that the ventilation level cannot be lower than the one ensuring an acceptable $CO_2$ level in the room.

This allows us to address two separated subproblems: *(i)* the $CO_2$-SMPC problem, which aims at minimizing energy use while keeping $CO_2$ levels in given comfort bounds; *(ii)* the T-SMPC problem, controlling instead the indoor temperature and deciding the additional ventilation level guaranteeing the thermal comfort. The output of the sequence of the $CO_2$-SMPC problem is a sequence of air flow rates coming from the ventilation system over the whole prediction horizon, which is then integrated into the T-SMPC problem to account for the corresponding heat flow.

In the following, we present the control-oriented models incorporated into the $CO_2$-SMPC and T-SMPC problems.

**Control-oriented model of the $CO_2$ concentration dynamics**

The model state $x_{CO_2}$ and the model output $y_{CO_2}$ are equal to the nonnegative difference between the $CO_2$ concentration in the room and the inlet air $CO_2$ concentration (the latter one assumed to be equal to the outdoor $CO_2$ concentration levels). The model disturbance $w_{CO_2}(k)$ represents instead the number of occupants.

We derive an equivalent linear model of the $CO_2$ concentration dynamics by introducing the auxiliary input $u_{CO_2} := \dot{m}^{CO_2}_{\text{venting}} \cdot x_{CO_2}$ representing the reduction in the indoor $CO_2$ concentration levels induced by $\dot{m}^{CO_2}_{\text{venting}}$. To satisfy the physical bounds on the original control input $\dot{m}^{CO_2}_{\text{venting}}$, we introduce the following additional constraints on the auxiliary input $u_{CO_2}$,

$$\dot{m}^{\min}_{\text{venting}} \cdot x_{CO_2}(k) \leq u_{CO_2}(k) \leq \dot{m}^{\max}_{\text{venting}} \cdot x_{CO_2}(k). \tag{10.4}$$

The $CO_2$ concentration dynamics can eventually be described by the discrete-time Linear Time Invariant (LTI) system

$$\begin{aligned} x_{CO_2}(k+1) &= a x_{CO_2}(k) + b u_{CO_2}(k) + e w_{CO_2}(k) \\ y_{CO_2}(k) &= x_{CO_2}(k). \end{aligned} \tag{10.5}$$

**Control-oriented model of the thermal dynamics**

The state of the model is the vector of the temperatures of the room, walls, floor, and ceiling. The model disturbances represent the outdoor temperature, the incident solar radiation, the internal gains, and the heat flows due to occupancy, equipments, and lighting. The control inputs required to actuate the HVAC system are the temperature of the supplied air, $T_{\text{sa}}$, the mean radiant temperature of the radiators, $T_{\text{mr}}$, and the additional air flow rate required for guaranteeing the thermal comfort, $\Delta\dot{m}_{\text{venting}}$, defined as $\Delta\dot{m}_{\text{venting}} := \dot{m}_{\text{venting}} - \dot{m}^{CO_2}_{\text{venting}}$.

We derive an equivalent linear model of the thermal dynamics by introducing the nonnegative variables $\Delta T_{\text{h}}$, $\Delta T_{\text{c}}$, $\Delta u_{\text{h}}$ and $\Delta u_{\text{c}}$ as auxiliary inputs s.t.

$$\begin{aligned} \Delta T_{\text{h}} - \Delta T_{\text{c}} &:= T_{\text{sa}} - T_{\text{room}} \\ \Delta T_{\text{h}} + \Delta T_{\text{c}} &:= \left|T_{\text{sa}} - T_{\text{room}}\right| \\ \Delta u_{\text{h}} - \Delta u_{\text{c}} &:= \Delta\dot{m}_{\text{venting}}\left(T_{\text{sa}} - T_{\text{room}}\right) \\ \Delta u_{\text{h}} + \Delta u_{\text{c}} &:= \Delta\dot{m}_{\text{venting}}\left|T_{\text{sa}} - T_{\text{room}}\right|. \end{aligned}$$

By introducing the auxiliary inputs above, the heat flows in (10.2) can be rewritten as

$$\begin{aligned} Q_{\text{venting}} &= \dot{m}^{CO_2}_{\text{venting}} c_{\text{pa}}\left(\Delta T_{\text{h}} - \Delta T_{\text{c}}\right) + c_{\text{pa}}\left(\Delta u_h - \Delta u_c\right) \\ Q_{\text{heating}} &= A_{\text{rad}} h_{\text{rad}} \Delta T_{\text{h,rad}}. \end{aligned}$$

We consider the following additional constraints to meet physical bounds on the original control inputs

$$
\begin{aligned}
0 \le \Delta T_{\mathrm{h}}(k) &\le T_{\mathrm{sa}}^{\max} - T_{\mathrm{room}}(k) \\
0 \le \Delta T_{\mathrm{c}}(k) &\le T_{\mathrm{room}}(k) - T_{\mathrm{sa}}^{\min}
\end{aligned}
\tag{10.6}
$$

$$
\begin{aligned}
0 \le \Delta u_{\mathrm{h}}(k) &\le \Delta \dot{m}_{\mathrm{venting}}^{\max}(k) \Delta T_{\mathrm{h}}(k) \\
0 \le \Delta u_{\mathrm{c}}(k) &\le \Delta \dot{m}_{\mathrm{venting}}^{\max}(k) \Delta T_{\mathrm{c}}(k).
\end{aligned}
\tag{10.7}
$$

where $\Delta \dot{m}_{\mathrm{venting}}^{\max}(k) := \dot{m}_{\mathrm{venting}}^{\max} - \dot{m}_{\mathrm{venting}}^{\mathrm{CO_2}}(k)$. Constraints (10.6) and (10.7) rule out the possibility of simultaneous heating and cooling modes, which can happen when $T_{\mathrm{room}}$ is greater than $T_{\mathrm{sa}}^{\min}$ or smaller that $T_{\mathrm{sa}}^{\max}$ or when it is convenient to require a stronger but unfeasible cooling/heating action. Bounds on $T_{\mathrm{sa}}$ have to be carefully defined and the physical bounds on the supply air temperature should be considered.

With the newly introduced variables, the dynamics of the indoor temperature can be modeled with the discrete-time LTI system

$$
\begin{aligned}
x_{\mathrm{T}}(k+1) &= A_{\mathrm{T}} x_{\mathrm{T}}(k) + B_{\mathrm{T}}(k) u_{\mathrm{T}}(k) + E_{\mathrm{T}} w_{\mathrm{T}}(k) \\
y_{\mathrm{T}}(k) &= C_{\mathrm{T}} x_{\mathrm{T}}(k),
\end{aligned}
\tag{10.8}
$$

where the state $x_{\mathrm{T}}(k)$ contains the temperatures of the room and of the inner and outer parts of the walls, $u_{\mathrm{T}}(k)$ is the input vector, as defined above, and $w_{\mathrm{T}}(k)$ is the vector of random disturbances (outdoor temperature, solar radiation, and internal heat gains). The output $y_{\mathrm{T}}(k)$ is the indoor temperature at time $k$. We notice that the input matrix $B_{\mathrm{T}}(k)$ is time-varying since it depends on $\dot{m}_{\mathrm{venting}}^{\mathrm{CO_2}}(k)$.

### 10.2.3 SMPC Problem Formulation for HVAC Systems

In Sect. 10.2.3.1, we formulate the stochastic MPC problem for optimizing HVAC operation and illustrate how a scenario-based approximation can be derived in Sect. 10.2.3.2.

#### 10.2.3.1 Chance-Constrained MPC formulation

We start considering a generic discrete-time LTI systems with uncertainty of the form

$$
\begin{aligned}
x(k+1) &= Ax(k) + B(k)u(k) + Ew(k) \\
y(k) &= Cx(k),
\end{aligned}
\tag{10.9}
$$

where $x(k) \in \mathcal{R}^n$ is the state, $u(k) \in \mathcal{R}^m$ is the control input, $w(k) \in \mathcal{R}^r$ is the stochastic disturbance and $y(k) \in \mathcal{R}^p$ is the output. Notice that, depending on the controller under consideration ($CO_2$-SMPC or T-SMPC), (10.9) can represent either (10.5) or (10.8).

We define $\boldsymbol{x} := \left[x(1)^{\mathrm{T}}, \ldots, x(N)^{\mathrm{T}}\right]^{\mathrm{T}}$ the vector representing the state evolution over the prediction horizon $N$. Similarly, we define $\boldsymbol{u}, \boldsymbol{y}$ and $\boldsymbol{w}$. Prediction dynamics matrices $\boldsymbol{C}_A, \boldsymbol{C}_B, \boldsymbol{C}_E$ easily follow from (10.9) and the definitions of $\boldsymbol{x}$, $\boldsymbol{u}, \boldsymbol{y}$ and $\boldsymbol{w}$. Hence, we can express the output as a function of the initial state $x(0)$, i.e.,

$$\boldsymbol{y}^0 = \boldsymbol{C}_A x(0) + \boldsymbol{C}_B \boldsymbol{u} + \boldsymbol{C}_E \boldsymbol{w}, \tag{10.10}$$

with $x(0)$ the current measured value of the state.

Constraints (10.4) and (10.6) can be written in a compact form as mixed constraints on inputs and outputs. Using $\boldsymbol{y}^0$, we can write both comfort constraints and mixed constraints on inputs and outputs over the whole prediction horizon as $\boldsymbol{G}^u\boldsymbol{u}+\boldsymbol{G}^w\boldsymbol{w} \leq \boldsymbol{g}$, with the left-hand side being bi-affine functions in the vector of decision variables $\boldsymbol{u}$ and random variables $\boldsymbol{w}$, and $\boldsymbol{G}^u$, $\boldsymbol{G}^w$ and $\boldsymbol{g}$ being matrices of appropriate dimensions.

Constraints on inputs (e.g., (10.7) over the whole prediction horizon can be written as $\boldsymbol{F}\boldsymbol{u} \leq \boldsymbol{f}$, where $\boldsymbol{F}$ and $\boldsymbol{f}$ are matrices of appropriate dimensions.

To summarize, the linear constraints on the inputs and outputs (comfort constraints) over the prediction horizon are

$$\begin{aligned} \boldsymbol{G}^u\boldsymbol{u} + \boldsymbol{G}^w\boldsymbol{w} &\leq \boldsymbol{g} \\ \boldsymbol{F}\boldsymbol{u} &\leq \boldsymbol{f}. \end{aligned} \tag{10.11}$$

As introduced above, since it is possible to assume that these random constraints can be violated with a predefined probability $\alpha \in [0, 1]$, uncertainties can be handled by formulating the random constraints (10.11) as probabilistic constraints of the form

$$\mathbb{P}\left[\boldsymbol{G}^u\boldsymbol{u} + \boldsymbol{G}^w\boldsymbol{w} \leq \boldsymbol{g}\right] \geq 1 - \alpha.$$

To formulate the chance-constrained MPC problem we define a linear cost function over the whole prediction horizon as $J(x(0), \boldsymbol{u}) = \sum_{k=0}^{N-1} J(u(k))$. The control problem is formally stated as

$$\begin{aligned} &\min_{\boldsymbol{u}} \; J(x(0), \boldsymbol{u}) \\ &\text{s.t.} \;\; \mathbb{P}\left[\boldsymbol{G}^u\boldsymbol{u} + \boldsymbol{G}^w\boldsymbol{w} \leq \boldsymbol{g}\right] \geq 1 - \alpha \\ &\qquad \boldsymbol{F}\boldsymbol{u} \leq \boldsymbol{f}. \end{aligned} \tag{10.12}$$

#### 10.2.3.2 Scenario-Based Approximation

Unless the uncertainties follow specific distributions, e.g., Gaussian or log-concave, chance constraints problems as (10.12) are generally non-convex and thus numerically difficult to handle [23]. Uncertainties like solar radiation and occupancy do not usually follow probability distributions that allow us to formulate equivalent deterministic problems and make MPC problems more tractable.

The scenario-based optimization approach [8, 24] provides a way of approximating the solution of chance-constrained optimization problems, and is based on finding the optimal solution under a finite number of sampled outcomes of uncertainty. This approach does not require explicit knowledge of the uncertainty set, as in robust optimization, nor of its probability distribution, as in stochastic programming; the only requirement is that one should be able to extract a sufficient number of independent random samples from the distribution of uncertainty. The number of samples is selected to guarantee the feasibility of the solution so that the solution of the scenario problem has generalization properties, i.e., it satisfies with high probability unseen scenarios. Therefore, the original program is approximated with a deterministic one, called the *scenario problem*, which is obtained from the original by replacing the chance constraint with the $S$ sampled deterministic constraints. By construction, the scenario problem is deterministic and convex and thus it can be solved efficiently by standard numerical algorithms. In our specific HVAC control problem, the scenario approach leads to a simple linear problem to be solved at each time step; we follow this approach and extract a set of $S$ i.i.d. disturbances samples or scenarios, $\boldsymbol{w}_1, \ldots, \boldsymbol{w}_S$, with $\boldsymbol{w}_i := \left[w_i^{\mathrm{T}}(0), \ldots, w_i^{\mathrm{T}}(N-1)\right]^{\mathrm{T}}, i = 1, \ldots, S$. These generated scenarios correspond to a set of different realizations of exogenous disturbance variables $w$ which represent weather conditions and occupancy. These quantities are extracted from a learned multidimensional probability distribution trained from historical data by using copulas formalisms, see, e.g., [11–13]. We remark that posing Gaussianity assumptions limit the dependencies that can be captured by restricting the types of admitted behaviors in the tails of marginal distributions. References [11–13] also show that in HVAC control frameworks Gaussianity assumptions may be unrealistic, especially when modeling occupancy or solar radiation effects.

Hence, Problem (10.12) is approximated with the following scenario-based problem:

$$\begin{aligned} \min_{\boldsymbol{u}} \; & J(x(0), \boldsymbol{u}) \\ \text{s.t.} \; & \boldsymbol{G}^u \boldsymbol{u} \leq \boldsymbol{g} - \max_{i=1,\ldots S} \boldsymbol{G}^w \boldsymbol{w}_i \\ & \boldsymbol{F}\boldsymbol{u} \leq \boldsymbol{f}. \end{aligned} \tag{10.13}$$

Recently, authors in [25] have obtained useful results on the closed-loop constraint violations in a scenario-based MPC framework. In this work, these results are used to set a lower bound on the number of scenarios providing guarantees on the probability of constraint violation. Furthermore, since in this manuscript we formulate a cascade of two scenario-based optimization problems, we apply the results provided in [26] to guarantee the feasibility of the cascaded solution of the two individual problems (see Theorem 7 in [26]).

### *10.2.4 Implicit SMPC Formulation for HVAC Control*

We now describe our implicit SMPC formulation by illustrating the control architecture of our SMPC for HVAC systems in Sect. 10.2.4.1 and the two scenario-based optimization problems in Sect. 10.2.4.2 (as for the $CO_2$-SMPC problem) and in Sect. 10.2.4.3 (as for the T-SMPC problem). We then discuss some experimental results in Sect. 10.2.4.4.

#### 10.2.4.1 Control Architecture

The architecture of the proposed control system is illustrated in Fig. 10.2. Here the indoor temperature and the air $CO_2$ concentration levels (both to be considered as comfort indicators) are controlled through the ventilation system and radiators, while the latter two elements are actuated using low-level proportional-integral (PI) controllers. The proposed SMPC scheme computes then at each time instant the set points for the low-level controllers using new measurements and updated information about weather and occupancy patterns.

The inputs of our SMPC for HVAC systems are, at every time step: *(i)* occupancy levels, *(ii)* weather conditions, and *(iii)* measurements of the building current state. The output is a heating, cooling and ventilation plan for the next $N$ hours (with $N$ being a prediction horizon chosen by the user when designing the control system). Consistently with the MPC paradigm, at every time step $k$ only the first step of this control plan is applied to the HVAC system. After that, the whole procedure is repeated based on new measurements and updated forecasts and scenarios. This introduces feedback into the system, since the control action is a function of the

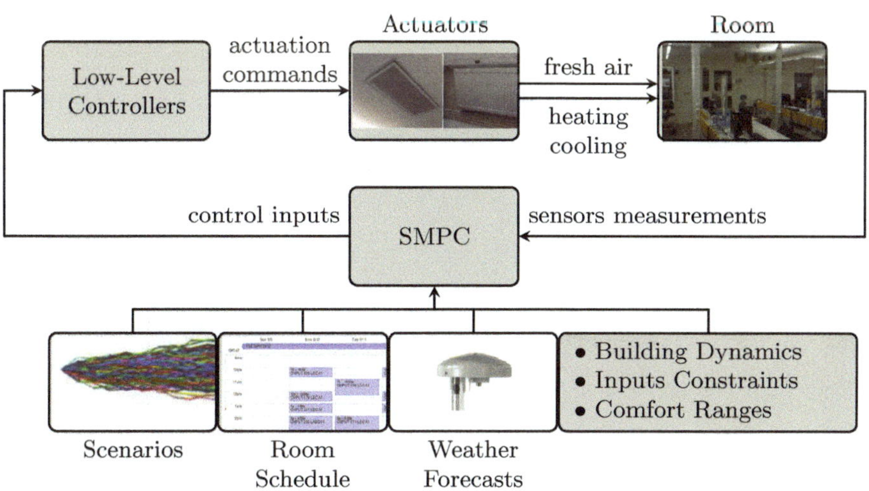


**Fig. 10.2** Architecture of the control system implemented on the testbed (reprinted with permission, from [13])

current system state and the currently acting disturbances. In our case the computed outputs are, at every time step $k$: *(i)* a mass air flow rate $\dot{m}_{\text{venting}}(k)$; *(ii)* a supply air temperature $T_{\text{sa}}(k)$; *(iii)* a radiators mean radiant temperature $T_{\text{mr}}$.

#### 10.2.4.2 $CO_2$-SMPC Problem

The generic model (10.9) is replaced in this case with (10.5). The $CO_2$ concentration over the prediction horizon, denoted by $\boldsymbol{y}_{CO_2}$, can be expressed as a function of the initial state $x_{CO_2}(0)$ as in (10.10).

The cost function is

$$J_{CO_2}(u_{CO_2}(k)) = u_{CO_2}(k)\Delta k, \tag{10.14}$$

where $\Delta k$ the sampling period. The bounds on the input $u_{CO_2}(k)$ are of the form $u_{CO_2}^{\min} \leq u_{CO_2}(k) \leq u_{CO_2}^{\max}$, while the comfort constraints on the indoor $CO_2$ concentration can be set as $0 \leq y_{CO_2}(k) \leq y_{CO_2}^{\max}$. These bounds can be expressed as polytopic constraints $F_{CO_2}u_{CO_2}(k) \leq f_{CO_2}$ and $G_{CO_2}^{y} y_{CO_2}(k) \leq g_{CO_2}^{y}$, respectively.

Constraints (10.4) can be written in a compact form as mixed constraints on inputs and outputs. By expressing $\boldsymbol{y}_{CO_2}$ as a function of $x_{CO_2}(0)$, we can write both comfort constraints and mixed constraints on inputs and outputs over the whole prediction horizon as $\boldsymbol{G}_{CO_2}^{u}\boldsymbol{u}_{CO_2} + \boldsymbol{G}_{CO_2}^{w}\boldsymbol{w}_{CO_2} \leq \boldsymbol{g}_{CO_2}$, where $\boldsymbol{G}_{CO_2}^{u}$, $\boldsymbol{G}_{CO_2}^{w}$ and $\boldsymbol{g}_{CO_2}$ are matrices of appropriate dimensions.

The $CO_2$-SMPC problem can thus be formulated as in (10.12):

$$\begin{aligned} \min_{\boldsymbol{u}_{CO_2}} \ & J_{CO_2}(x_{CO_2}(0), \boldsymbol{u}_{CO_2}) \\ \text{s.t.} \ & \boldsymbol{G}_{CO_2}^{u}\boldsymbol{u}_{CO_2} \leq \boldsymbol{g}_{CO_2} - \max_{i=1,\dots S} \boldsymbol{G}_{CO_2}^{w}\boldsymbol{w}_{CO_2,i} \\ & \boldsymbol{F}_{CO_2}\boldsymbol{u}_{CO_2} \leq \boldsymbol{f}_{CO_2}. \end{aligned} \tag{10.15}$$

After solving the $CO_2$-SMPC problem, we can easily derive $\dot{m}_{\text{venting}}^{CO_2}(k)$. The current measurement of the indoor $CO_2$ level is used to compute the control input for the current point in time.

#### 10.2.4.3 T-SMPC Problem

The system model (10.9) is in this case (10.8). The indoor temperature $\boldsymbol{y}_{\text{T}}$ over the prediction horizon can be expressed as a function of the initial state $x_{\text{T}}(0)$ as in (10.10).

The control objective is to minimize the thermal energy use

$$\begin{aligned} J_{\text{T}}(u_{\text{T}}(k)) &= \big(\rho_{\text{venting}}\big|Q_{\text{venting}}(k)\big| + Q_{\text{heating}}(k)\big)\Delta k = & (10.16) \\ &= \big(c^{\text{T}}(k)u_{\text{T}}(k)\big)\Delta k, & (10.17) \end{aligned}$$

where $u_{\mathrm{T}}(k) := \left[\Delta T_h(k), \Delta T_c(k), \Delta u_h(k), \Delta u_c(k), \Delta T_{h,\mathrm{rad}}(k)\right]$ and $c(k)$ are, respectively, the input and the cost vectors at time $k$ and $\rho_{\mathrm{venting}}$ is a factor modeling the higher energy cost of the ventilation system with respect to the radiators in the heating mode.

We define bounds on the input $u_{\mathrm{T}}(k)$ of the form $u_{\mathrm{T}}^{\min} \le u_{\mathrm{T}}(k) \le u_{\mathrm{T}}^{\max}$, and on the comfort constraints on the indoor temperature as $0 \le y_{\mathrm{T}}(k) \le y_{\mathrm{T}}^{\max}$. Bounds on inputs along with constraints (10.7) can be expressed as polytopic constraints $F_{\mathrm{T}} u_{\mathrm{T}}(k) \le f_{\mathrm{T}}$, while comfort constraints can be written as $G_{\mathrm{T}}^{y} y_{\mathrm{T}}(k) \le g_{\mathrm{T}}^{y}$.

Constraints (10.6) can be written in a compact form as mixed constraints on inputs and outputs. By expressing $\boldsymbol{y}_{\mathrm{T}}$ as a function of $x_{\mathrm{T}}(0)$, we can write both comfort constraints and mixed constraints on inputs and outputs over the whole prediction horizon as $\boldsymbol{G}_{\mathrm{T}}^{u} \boldsymbol{u}_{\mathrm{T}} + \boldsymbol{G}_{\mathrm{T}}^{w} \boldsymbol{w}_{\mathrm{T}} \le \boldsymbol{g}_{\mathrm{T}}$, where $\boldsymbol{G}_{\mathrm{T}}^{u}$, $\boldsymbol{G}_{\mathrm{T}}^{w}$ and $\boldsymbol{g}_{\mathrm{T}}$ are matrices of appropriate dimensions.

The T-SMPC problem can thus be formulated as in (10.12):

$$
\begin{aligned}
\min_{\boldsymbol{u}_{\mathrm{T}}}\ & J_{\mathrm{T}}(x_{\mathrm{T}}(0), \boldsymbol{u}_{\mathrm{T}}) \\
\text{s.t.}\ & \boldsymbol{G}_{\mathrm{T}}^{u} \boldsymbol{u}_{\mathrm{T}} \le \boldsymbol{g}_{\mathrm{T}} - \max_{i=1,\ldots S} \boldsymbol{G}_{\mathrm{T}}^{w} \boldsymbol{w}_{\mathrm{T},i} \\
& \boldsymbol{F}_{\mathrm{T}} \boldsymbol{u}_{\mathrm{T}} \le \boldsymbol{f}_{\mathrm{T}}.
\end{aligned}
\tag{10.18}
$$

After solving the T-SMPC problem, we can easily derive the control variables $T_{\mathrm{sa}}(k)$, $T_{\mathrm{mr}}(k)$ and $\Delta \dot{m}_{\mathrm{venting}}(k)$. The total air flow rate at the current point in time, $k$, can be computed as $\dot{m}_{\mathrm{venting}}^{\mathrm{CO_2}}(k) + \Delta \dot{m}_{\mathrm{venting}}(k)$. Current measurements of the temperatures in the room and in the ventilation system are used to compute control inputs for the current point in time.

#### 10.2.4.4 Experimental Evaluation

Here we describe the experimental setup, the KTH HVAC testbed, and discuss experimental results.

**Experimental setup**

The KTH HVAC testbed is hosted on the KTH main campus and is located in the ground floor of a seven-story office building with a concrete heavyweight structure. The testbed consists of four rooms: a laboratory and three student offices; the results presented in this work refer to the laboratory room, which has a limited windows surface and one external wall, facing southeast. The rooms are all equipped with a Supervisory Control And Data Acquisition (SCADA) and Programmable Logic Controllers (PLCs), a wireless sensor network, an actuator network, and a weather station.

The installed sensors enable continuous monitoring of the status of the system, i.e., $CO_2$ temperatures, humidity, and external weather conditions as in Fig. 10.3. The implemented platform also gathers data from weather forecasts services and is integrated with the rooms web-based scheduling services of the occupancy (calendars). Occupancy is measured through a photoelectric based people counter.

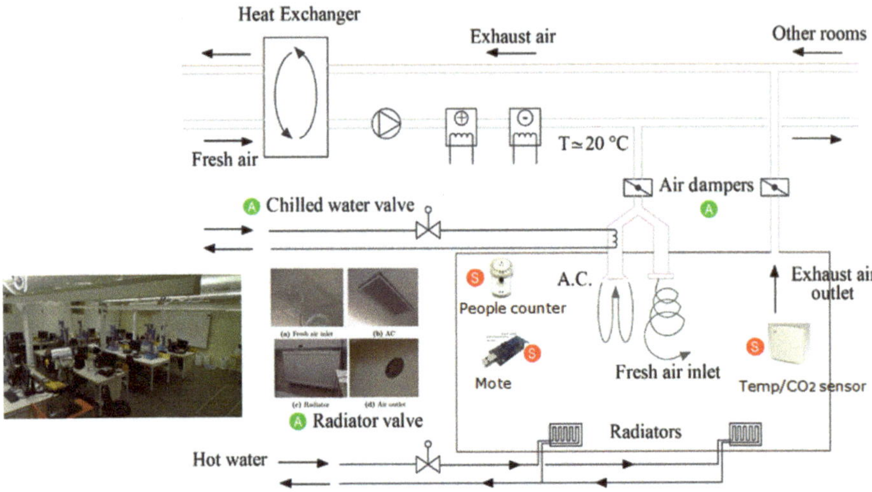


**Fig. 10.3** Scheme of the HVAC system of one of the rooms of the KTH HVAC testbed (reprinted, with permission, from [13])

The HVAC system of the rooms consists of a ventilation system supplying fresh air plus a radiator heating system. Air is ventilated into the rooms by a central fan (not controllable in the current experimental setup) running by default between 8:00 and 15:00 during weekdays. A central balanced ventilation system with heat recovery preconditions fresh air from the outdoor environment, distributing it at a temperature of about 20–21°C. Approximately 70% of the total air flow is directly conveyed into the rooms from the central air handling unit, while the remaining part can be further cooled by a local cooling coil. The ventilation airflow is controlled via two dampers that regulate the opening of the inlet and outlet ducts; the ventilation air temperature is controlled actuating on a cooling coil. When the central fan is on, a minimum air flow is distributed into the rooms due to building regulations independently from their occupancy. The heating system uses standard waterborne radiators as heat emission units. Heat is provided by district heating, with supply temperature of the water dependent on the outdoor temperature. Heat emission from radiators is controlled acting on the valve that regulates the hot water flow.

**Experimental results**

Here we discuss and evaluate the results of experiments performed on the KTH HVAC testbed.

We point out that the thermal model used for control has been implemented in MATLAB and verified against the results provided by IDA-ICE 4.5 [27], a commercial software program for energy and comfort calculations in buildings. Furthermore, the temperature and $CO_2$ models have been successfully validated against measurements from the testbed [11, 13].

Three different controllers have been tested on the HVAC system in the testbed main room during three different days (see Fig. 10.4):

1. the current practice, which corresponds to a simple control logic with PI control loops and switching logic, indicated in the following by the acronym AHC (the controller from Akademiska Hus, the company managing the building of the testbed);
2. a Deterministic Model Predictive Control (DMPC) neglecting information on the uncertainties in the forecasts, and computing the control inputs by solving the HVAC control problem with deterministic constraints obtained by replacing the unknown disturbances with their forecasts;
3. our SMPC introduced above.

The three controllers have been tested between April and June 2014 for 7 h each day, from 8:00 to 15:00. The sampling time for the MPC-based controllers is 10 min, while the predictions horizon is 8 h. The comfort range of the indoor temperature is [20, 22] °C. The controllers have been tested during spring and summer, when cooling was required.

Figure 10.4 depicts the results of the three controllers tested in 3 days in April and May. The horizontal axis reports the time period of the experiments. Each row

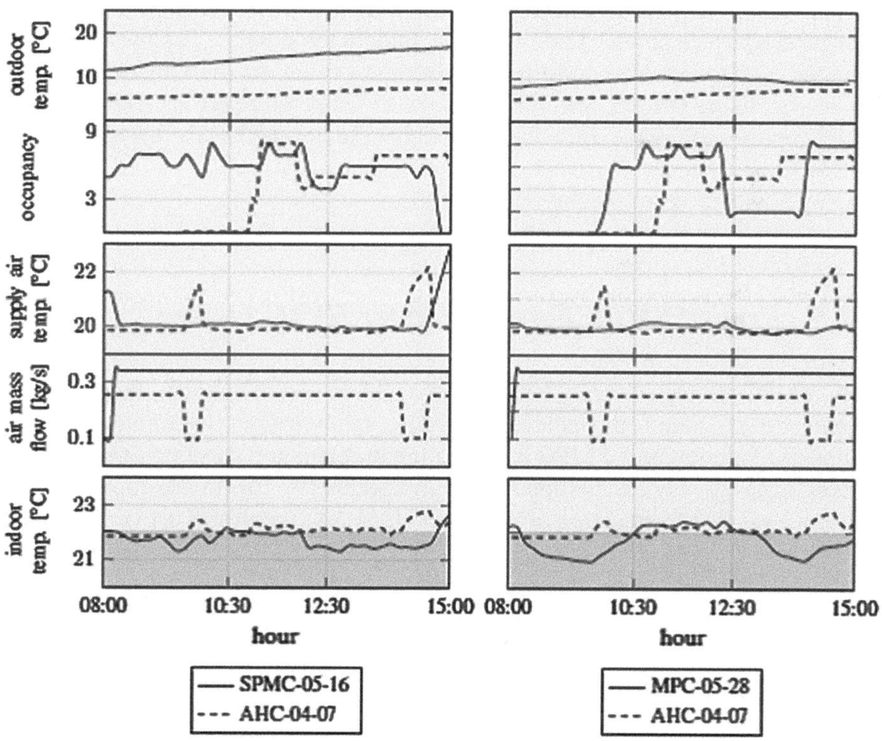


**Fig. 10.4** Disturbances, indoor temperature and control input profiles for tests on April 7, May 16 and 28

of the figure depicts the disturbances (i.e., outdoor temperature and occupancy), the control inputs (i.e., supply air temperature and air mass flow) and the controlled indoor temperature. The left column compares results between SMPC and AHC, while the right column shows the results of MPC against AHC. The solid lines of both columns represent disturbances, input and output profiles of MPC-based strategies while the dashed lines are related to the current practice AHC.

The MPC-based controllers clearly improve the thermal comfort with respect to the AHC one: indeed, despite the fact that the MPC-based controllers have to compensate for higher internal gains from occupants and higher outdoor temperatures, they also show almost no violations to the comfort constraints, while AHC yields significant violations of the thermal comfort upper bound. Notice in particular the experiment in the first column of Fig. 10.4, it can be seen that our SMPC had to compensate for a higher outdoor temperature, a more challenging situation when cooling.

The differences in the performance of the two MPC-based controllers thus emphasize the added value of incorporating information on the disturbances that affect the system: the SMPC computes its control inputs based on a worst-case scenario approach, which leads to a more robust behavior against unknown disturbances. As a consequence, the SMPC is able to keep the temperature closer to the upper bound and violates less with respect to the DMPC. The advantages of this behavior are more stressed when occupancy is higher.

In general, we can notice that, when the outdoor temperature and the occupancy are higher, SMPC has better performance in terms of violations of the thermal comfort with respect to the DMPC. See, for instance, the temperature profile resulting from DMPC, represented by the solid line in the right column and the last row of Fig. 10.4: the DMPC is not able to compensate for the occupancy peak at 11:00–12:00 am, resulting in violations of the thermal comfort.

## *10.2.5 Explicit SMPC Formulation for HVAC Systems*

Implicit MPC strategies come with an online computational burden, intensified by the scenario generation procedure. A possible solution is to determine explicitly, off-line, the solution of the MPC optimization problem as a function of the current building state by solving multiparametric programs [9, 10]. In the following, we outline the explicit formulation of the SMPC for HVAC control (10.13).

### 10.2.5.1 Computing the Explicit Solution

We adopt a two states thermal model of a single thermal zone (or room). We model a thermal zone as a network of two nodes, one accounting for the dynamics of the air within the zone, and the other one accounting for the dynamics of the walls. This model and its validation against measurements are described in detail in [12].

Problems (10.15) and (10.18) can be treated as multiparametric linear problems by considering initial state vectors, respectively, $x_{\mathrm{CO_2}}(0)$ and $x_{\mathrm{T}}(0)$ as vectors of parameters.

By solving the multiparametric $CO_2$-SMPC and T-SMPC problems, we obtain explicit state-feedback laws that are expressed as PWA functions of the initial state vectors.

Hence, the explicit $CO_2$-SMPC and T-SMPC control laws have, respectively, the form

$$\begin{aligned} u^{E\mathrm{CO_2}}(x) &= Q_i^{E\mathrm{CO_2}} x + q_i^{E\mathrm{CO_2}} && \text{if } H_i^{E\mathrm{CO_2}} x_{\mathrm{CO_2}}(0) \leq K_i^{E\mathrm{CO_2}} \\ u^{ET}(x) &= Q_j^{ET} x + q_j^{ET} && \text{if } H_j^{ET} x_{\mathrm{T}}(0) \leq K_j^{ET} \end{aligned}$$

where the polyhedral sets $\mathcal{X}_i^{E\mathrm{CO_2}} := \left\{H_i^{E\mathrm{CO_2}} x \leq K_i^{E\mathrm{CO_2}}\right\}$ and $\mathcal{X}_j^{ET} := \left\{H_j^{ET} x \leq K_j^{ET}\right\}$, with $j = 1 \ldots N^{ET}$ are partitions of the set of states, with $i = 1, \ldots, N^{E\mathrm{CO_2}}$ and $j = 1, \ldots, N^{ET}$.

To guarantee that the overall procedure satisfies the requirements on the minimum mass flow needed to maintain comfortable $CO_2$ levels, we proceed as follows: the set of admissible values of the mass flow is partitioned using standard algorithms (e.g., Lloyd algorithm [28]) and $Q$ quantized values are computed, $\dot{m}^{\mathrm{CO_2}}_{1,\mathrm{quantized}}, \ldots, \dot{m}^{\mathrm{CO_2}}_{Q,\mathrm{quantized}}$. Then, for each $\dot{m}^{\mathrm{CO_2}}_{i,\mathrm{quantized}}$, a corresponding explicit $\mathrm{T}_i$-SMPC problem that takes $\dot{m}^{\mathrm{CO_2}}_{i,\mathrm{quantized}}$ as lower bound on the mass flow is solved.

The scenario-based controller can be then implemented online according to Algorithm 1.

**Algorithm 10.1** On-line Implementation

1: **for** $k = 1, 2, \ldots$ **do**
2: measure $x_{\mathrm{CO_2}}(k)$ and $x_{\mathrm{T}}(k)$
3: compute $u^{E\mathrm{CO_2}}(x_{\mathrm{CO_2}}(k))$ and derive $\dot{m}^{\mathrm{CO_2}}_{\mathrm{venting}}(k)$
4: identify the quantized value $\dot{m}^{\mathrm{CO_2}}_{i,\mathrm{quantized}}(k)$
5: compute $u^{ET}(x_{\mathrm{T}}(k))$ by using the solution of the explicit $\mathrm{T}_i$-SMPC problem and derive the setpoints $(\dot{m}_{\mathrm{venting}}(k), T_{\mathrm{sa}}(k), T_{\mathrm{mr}}(k))$
6: send the computed setpoints to the low-level PI controllers
7: **end for**

## 10.3 Building Generation-Side and Demand-Side Management

As highlighted in Sect. 10.1, smart buildings can integrate not just flexible loads, but also storage systems and generation capabilities. Energy management frameworks for smart buildings should support multiple performance criteria (load shaping, economic costs, comfort, power imbalances minimization), along with more standard objectives (control of the indoor air quality and thermal comfort), and optimize the flexibility provided by onsite generation (e.g., Photovoltaic (PV)), storage systems, and Demand Response (DR) policies [1] (commonly defined as changes in electricity use by consumers in response to changes in the electricity price over time [29]). Effective energy management with DR policies can help flattening the aggregated demand curve and reducing the number of expensive generation plants used for peak load periods. In recent years, more and more utilities and governments offer programs that provide incentives for residential consumers to adopt on-site distributed generators and energy storage systems [30]. However energy storage devices and renewables are not still common within homes, they are often included in energy management frameworks of future environmentally friendly homes [31].

In this section, we describe a novel MPC-based EMS for smart buildings to optimally manage and coordinate energy supply and demand in multiple houses, taking user preferences into account. The control system computes an optimal energy plan based on forecasts of weather conditions, renewable generation and thermal demand; imbalances can be compensated through the feedback mechanism integrated into our framework. The feedback mechanism introduced through the MPC receding horizon philosophy allows us to compute current power imbalances, and thus take corrective actions to guarantee power balance and user comfort. Smart appliances are included since we consider DR, which can take advantage of the additional flexibility offered by storage devices, to store energy and release it when it is more convenient [32].

We consider a system representing either a residential district made up of several single-family houses with local generation capabilities owning a shared DER, or a smart building composed of apartments with heating systems and storage devices sharing a common Distributed Energy Resource (DER).

We illustrate the modeling and MPC problem formulation for the system described above in Sect. 10.3.1. An extension of this framework, including the user behavior uncertainty in scheduling shiftable appliances, is outlined in Sect. 10.3.2. Lastly, we sketch in Sect. 10.3.3 a distributed approach to solve the problem of coordinating a set of smart appliances located in $N$ apartments sharing an Energy Storage System (ESS).

### 10.3.1 MPC-Based EMS for Smart Buildings

As mentioned above, we consider a system representing either a residential district or a smart building. The subsystems we account for are then either single-family houses or apartments owning a shared DER and comprising heat pumps, ESSs and both thermal and electrical loads (e.g., heating system, electrical appliances).

We remark that the proposed EMS framework can be easily adapted to other energy systems (e.g., networks of microgrids, industrial facilities) and extended to include other DER and control objectives (e.g., electrical vehicles, fuel cells, emission reduction).

We would like to mention that a virtual experimental testbed has been built for evaluating experimental results, the Virtual MicroGrid Lab [33], where partners from industry and academia have combined resources to develop a virtual laboratory interconnecting partner laboratory premises using secure connections on top of the public Internet. A case study of five residential microgrids is implemented and simulated in the virtual laboratory; simulation results show that the storage devices allow a 10.47% cost savings and the proposed control framework can achieve up to 58.8% cost savings.

In the following, we describe the modeling of the system components and we formulate the MPC problem.

#### 10.3.1.1 Modeling

Here we briefly describe the modeling framework of the considered system.

As depicted in Fig. 10.5, each subsystem is equipped with a Home Energy Management System (HEMS), which is responsible for operating the end-user smart appliances, the local generation devices and the interaction with the grid at the residential level according to the setpoints computed by the control system. The MPC controller is responsible for coordinating the energy sources, and deals with the long-term behavior of the system (e.g., from 10 min to 1 h). This implies that the controller

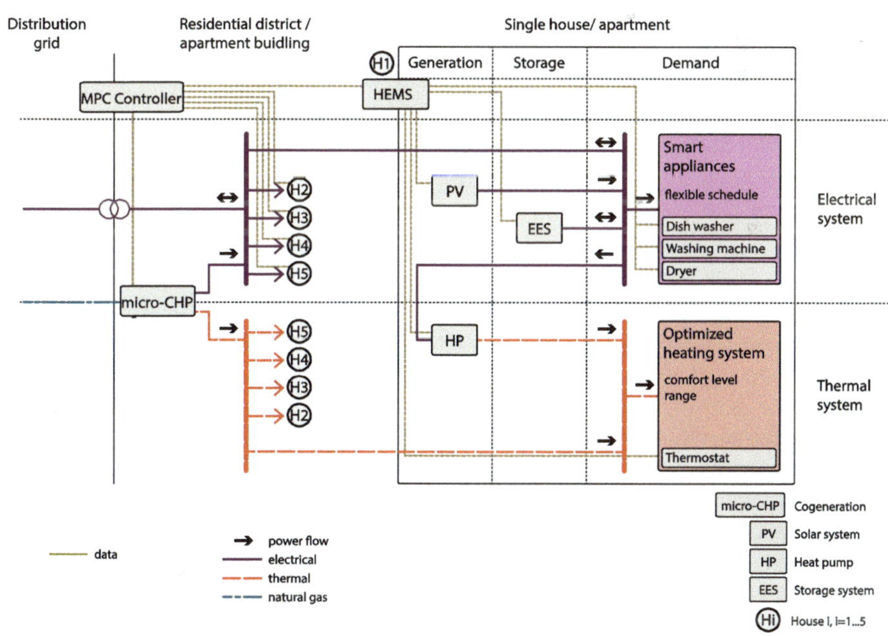


**Fig. 10.5** Schematics of the architecture of the system under consideration (reprinted, with permission, from [34])

**Table 10.2** Parameters

| Parameters | Description |
|---|---|
| $H$ | Scheduling horizon |
| $N_{\text{appliance,h}}$ | Number of home smart appliances for house h |
| $N_{\text{h}}$ | Number of houses |
| $n_i$ | Number of energy phases of appliance i |
| $E_{i,j}$ | Energy requirements for energy phase j of appliance i |
| $\underline{P}_{i,j}, \overline{P}_{i,j}$ | Bounds on energy phase power |
| $\underline{T}_{i,j}, \overline{T}_{i,j}$ | bounds on number of time slots for energy phase j |
| | of appliance i |
| $\underline{D}_{i,j}, \overline{D}_{i,j}$ | Bounds on between-phase delays in the number |
| | of time slots |
| $\text{TP}_i$ | User time preference interval for appliance i |
| $c_{\text{gas}}$ | Fuel (natural gas) cost for the micro-CHP |
| $\alpha_{\text{s,h}}$ | Storage energy degradation for house h |
| $c_{\text{s,h}}$ | Operating and maintenance cost of the power exchanged with the storage unit for house h |
| $\eta^{\text{c}}_{\text{s,h}}, \eta^{\text{d}}_{\text{s,h}}$ | Charging/discharging efficiencies of storage for house *h* |
| $\alpha_1, \alpha_0, \beta_1, \beta_0$ | Parameters of the micro-CHP model |
| $p^{\text{peak,h}}$ | "Peak signal" for house h (i.e., total slot energy upper bound) |
| $\underline{P}^{\text{heat}}_{\text{chp}}, \overline{P}^{\text{heat}}_{\text{chp}}$ | Bounds on micro-CHP thermal power |
| $\underline{P}^{\text{el}}_{\text{chp}}, \overline{P}^{\text{el}}_{\text{chp}}$ | Bounds on micro-CHP electrical power |
| $\underline{P}_{\text{gas}}, \overline{P}_{\text{gas}}$ | Bounds on micro-CHP gas power |
| $\underline{P}^{\text{heat}}_{\text{hp,h}}, \overline{P}^{\text{heat}}_{\text{hp,h}}$ | Bounds on heat pump thermal power to house h |
| $\underline{P}^{\text{el}}_{\text{hp,h}}, \overline{P}^{\text{el}}_{\text{hp,h}}$ | Bounds on heat pump electrical power to house h |
| $\underline{P}_{\text{s,h}}, \overline{P}_{\text{s,h}}$ | Bounds on the power exchanged with the storage for house h |
| $\underline{E}_{\text{s,h}}, \overline{E}_{\text{s,h}}$ | Bounds on the storage energy level for house h |

is weakly dependent on the transient behavior of the fast dynamics and a steady-state assumption for the components can be safely made without much loss of accuracy. Thus, the most relevant dynamics are the building thermal dynamics and the storage dynamics.

The forecasts, parameters and decision variables used in the proposed formulation are described, respectively, in Tables 10.2, 10.3, and 10.4 reported in the appendix. For further details on the components modeling we refer the interested reader to the technical report [33].

Forecasts of PV generation are computed by considering efficiency changes due to the given weather data and taking also other losses (e.g., inverter) into account.

We consider two types of loads:

- *thermal loads*, i.e., demand levels related to thermal indoor comfort;
- *electrical loads*, i.e., demand levels related to smart appliances.

**Table 10.3** Forecasts

| Forecasts | Description |
|---|---|
| $c_{\mathrm{tariff}}$ | Electricity tariff |
| $\underline{P}_{\mathrm{h}}^{\mathrm{heat}}, \overline{P}_{\mathrm{h}}^{\mathrm{heat}}$ | Bounds on thermal power demand for house h (corresponding to the thermal comfort range) |
| $COP_{\mathrm{hp,h}}$ | Coefficient of Performance (COP) |
| $P_{\mathrm{res,h}}$ | Power generation from renewables for house h |

**Table 10.4** Decision and logical variables

| Variables | Description |
|---|---|
| $\delta_{\mathrm{chp}}$ | Off(0)/on(1) state of the micro-CHP |
| $\delta_{\mathrm{s,h}}^{\mathrm{c}}, \delta_{\mathrm{s,h}}^{\mathrm{d}}$ | Storage charging/discharging state for house h |
| $P_{i,j}$ | Power profile for each phase |
| $x_{i,j}$ | Indicator of whether a phase is on or off |
| $t_{i,j}$ | Indicator of whether a transition is happening |
| $s_{i,j}$ | Indicator of whether a phase has been off |
| $P_{\mathrm{h}}^{\mathrm{grid}}$ | Importing(positive)/exporting(negative) power from/to the grid for house h |
| $P_{\mathrm{gas}}$ | Gas power input to the micro-CHP |
| $P_{\mathrm{chp,h}}^{\mathrm{heat}}$ | Micro-CHP thermal power to house h |
| $P_{\mathrm{chp,h}}^{\mathrm{el}}$ | Micro-CHP electrical power to house h |
| $P_{\mathrm{hp,h}}^{\mathrm{heat}}$ | Heat pump thermal power to house h |
| $P_{\mathrm{hp,h}}^{\mathrm{el}}$ | Heat pump electrical power to house h |
| $P_{\mathrm{h}}^{\mathrm{heat}}$ | Thermal power demand for house h |
| $P_{\mathrm{s,h}}^{\mathrm{c}}, P_{\mathrm{s,h}}^{\mathrm{d}}$ | Charging/discharging power exchanged with the storage for house h |
| $E_{\mathrm{s,h}}$ | Storage energy level for house h |

| *symbol* | *description* |
|---|---|
| $s$ | scenario index |
| $S$ | number of scenarios |

### Thermal Loads

Forecasts of the minimum and the maximum thermal energy required to keep the indoor temperature in the houses within a given comfort range are computed through a dynamic house model based on forecasts of the weather conditions. The house model runs on top of the Apros process simulation software [35]. The model includes the energy dynamics of building structure and indoor temperature and account for the outdoor temperature, the solar radiation and the internal gains from occupants and equipment. A further extension of the described EMS will include the HVAC controller described in the previous section.

**Electrical loads**

An appliance operation process is made up of sub-processes called energy phases. An energy phase is considered uninterruptible, and it consumes a pre-specified amount of electric energy in order to finish the physical task. Several other technical and user-specified constraints are included in the problem formulation: *(i)* both the power assigned to the energy phase at any time slot and its duration have to take values within a certain range; *(ii)* all energy phases associated with a single appliance must be run sequentially; *(iii)* delays between the energy phases are considered, but the order must be observed; *(iv)* for safety reasons, the total power assigned to all appliances at any moment cannot exceed a limit called peak signal; *(v)* user-specified time preferences are included, requiring that certain appliances should be run within some particular time intervals; *(vi)* user-specified preferences on appliances are modeled, e.g., a certain appliance cannot start before some other appliance finishes. Further details on appliance modeling and technical specifications are provided in [36, 37].

**Energy Storage System Modeling**

For a storage unit of house $h$, we consider the following discrete-time model:

$$E_{\mathrm{s,h}}(k+1) = \alpha_{\mathrm{s,h}} E_{\mathrm{s,h}}(k) + \eta^{\mathrm{c}}_{\mathrm{s,h}} P^{\mathrm{c}}_{\mathrm{s,h}}(k) \Delta T - \eta^{\mathrm{d}}_{\mathrm{s,h}} P^{\mathrm{d}}_{\mathrm{s,h}}(k) \Delta T,$$

with $0 < \eta^c, \eta^d < 1$ accounting for the energy losses and $\Delta T = t_{k+1} - t_k$ being a constant sampling time. We introduce binary variables $\delta^{\mathrm{c}}_{\mathrm{s,h}}, \delta^{\mathrm{d}}_{\mathrm{s,h}}$ to model the charging and discharging behavior and rule out the possibility of charging and discharging during the same sampling period, as expressed in the following constraints

$$\begin{aligned} \underline{P}_{\mathrm{s,h}} \delta^{\mathrm{c}}_{\mathrm{s,h}}(k) < P^{\mathrm{c}}_{\mathrm{s,h}}(k) < \overline{P}_{\mathrm{s,h}} \delta^{\mathrm{c}}_{\mathrm{s,h}}(k) \\ \underline{P}_{\mathrm{s,h}} \delta^{\mathrm{d}}_{\mathrm{s,h}}(k) < P^{\mathrm{d}}_{\mathrm{s,h}}(k) < \overline{P}_{\mathrm{s,h}} \delta^{\mathrm{d}}_{\mathrm{s,h}}(k) \\ \delta^{\mathrm{c}}_{\mathrm{s,h}}(k) + \delta^{\mathrm{d}}_{\mathrm{s,h}}(k) \leq 1. \end{aligned}$$

Bounds on the storage capacity are included in the modeling, along with limits on the total number of daily charging and discharging cycles in order to take the state of health of the ESS into account. Further details on a comprehensive storage modeling are provided in [36, 37].

**Heat pump modeling**

Heat pumps are devices able to transfer thermal energy by absorbing heat from a cold medium (heat source) and releasing it to a warmer one (heat sink). We consider an electrically operated heat pump, since they are by far the most frequently used. Since thermal efficiency of heat pump systems depends strongly on the temperature difference between heat source and sink as well as the overall operating temperature level, at each time slot, forecasts of the COP based on temperature predictions are integrated in the proposed control framework in order to predict the future heat generation from each heat pump (see [34]).

The following set of constraints model the behavior of the heat pump at each time slot $k$ and for each house $h$

$$P^{\text{heat}}_{\text{hp,h}}(k) = COP_{\text{hp,h}}(k) \cdot P^{\text{el}}_{\text{hp,h}}(k)$$
$$\underline{P}^{\text{heat}}_{\text{hp,h}} \leq P^{\text{el}}_{\text{hp,h}}(k) \leq \overline{P}^{\text{heat}}_{\text{hp,h}}$$
$$P^{\text{el}}_{\text{hp,h}}(k) \geq 0,$$

where $COP_{\text{hp,h}}(k)$ is the forecasted COP of the heat pump at time slots $k$ based on weather forecasts.

**Micro-CHP modeling**

The component represents a typical micro combined heat and power (micro-CHP) unit. The modeling of the component adopts a data-driven approach, where data from a real-world deployment was used [38], in combination with machine-learning and Big Data techniques (see [34]). The micro-CHP model is:

$$\begin{aligned} P^{\text{el}}_{\text{chp}} &= \alpha_1 \cdot P_{\text{gas}} + \alpha_0 \\ P^{\text{heat}}_{\text{chp}} &= \beta_1 \cdot P_{\text{gas}} + \beta_0. \end{aligned} \tag{10.19}$$

#### 10.3.1.2 MPC Problem Formulation

This section outlines the MPC problem formulation. The optimization problem consists of taking decisions on how to optimally schedule production by generators, storage, as well as controllable loads, to cover the system demand and minimize the generation costs and the cost of imported electricity from the distribution grid in the next hours or day. Supply and demand of electrical and thermal energy are both modeled and handled. The thermal energy is required to provide the needed thermal comfort to the house occupants, while the electrical energy is needed to run the smart appliances and the heat pump. The natural gas is required to run the shared micro-CHP.

At each MPC iteration, the problem is solved based on weather forecasts and the current system conditions. The computed optimal decision is then adjusted according to the actual values of the photovoltaic generation and of the heating requirements from the subsystems. Hence, corrective actions and the corresponding costs are taken in order to cope with potential imbalances. At the next time step, the MPC problem is solved again based on updated forecasts and system condition.

The MPC problem can be formulated as a Mixed Integer Linear Program (MILP) optimization problem. We point out that all the models and constraints described in Sect. 10.3.1.1 are constraints of the MPC problem. We next define the cost function and additional constraints included in the optimization problem to be solved at each MPC iteration. Further details can be found in [34].

**Cost function**

The aim is to minimize the cost of satisfying both the thermal and electrical loads, hence the objective function is

$$\min \sum_{k=1}^{H} \Big[ \sum_{h=1}^{N_\mathrm{h}} \big( c_\mathrm{tariff}(k) \cdot P_\mathrm{h}^\mathrm{grid}(k) + c_\mathrm{s,h} \cdot (P_\mathrm{s,h}^\mathrm{c} + P_\mathrm{s,h}^\mathrm{d}) \big) + c_\mathrm{gas} \cdot P_\mathrm{gas}(k) \Big] \Delta T.$$

**Electrical and thermal power balance**

The electrical and thermal power balances at each house need to be satisfied. At the current point in time, $k$, if an appliance is running, the power assigned to the current energy phase by the optimization problem solved at the previous time slot must be considered as a critical load for the current time slot, which cannot be rescheduled and has to be satisfied, since an energy phase is uninterruptible. We denote this amount of power for the house $h$ as $P_\mathrm{assigned,h}(k)$.

The balance between electrical energy production and consumption to be met at each time $k$ for house $h$, $\forall h, k$ is

$$\sum_{i=1}^{N_\mathrm{appliance,h}} \sum_{j=1}^{n_i} P_{i,j}(k) + P_\mathrm{assigned,h}(k) + P_\mathrm{s,h}^\mathrm{c} - P_\mathrm{s,h}^\mathrm{d} + P_\mathrm{hp,h}^\mathrm{el} + P_\mathrm{chp,h}^\mathrm{el} - P_\mathrm{res,h} = P_\mathrm{h}^\mathrm{grid}.$$

Regarding the thermal energy balance, three energy sources have to be taken into account to fulfill the thermal requirements: the heat pump, the micro-CHP and the waste heat generated by running appliances. Studies suggest that 70% of regular household electric use contributes to the household's heat demand [39].

The balance between thermal energy production and use to be met at each time $k$ for house $h$, $\forall h, k$ is

$$\underline{P}_\mathrm{h}^\mathrm{heat} \leq 0.7 P_\mathrm{assigned,h}(k) + P_\mathrm{hp,h}^\mathrm{heat} + P_\mathrm{chp,h}^\mathrm{heat} \leq \overline{P}_\mathrm{h}^\mathrm{heat}. \quad (10.20)$$

We remark that the thermal energy demand is optimized through (10.20) such that the indoor temperature in each house is within a given comfort range.

**Micro-CHP model**

The micro-CHP is driven by natural gas and generates both electric and thermal power. The following set of constraints model the behavior of the shared micro-CHP at each time slot $k$:

$$\underline{P}_\mathrm{chp}^\mathrm{el} \cdot \delta_\mathrm{chp}(k) \leq P_\mathrm{chp}^\mathrm{el}(k) \leq \overline{P}_\mathrm{chp}^\mathrm{el} \cdot \delta_\mathrm{chp}(k)$$
$$\underline{P}_\mathrm{chp}^\mathrm{heat} \cdot \delta_\mathrm{chp}(k) \leq P_\mathrm{chp}^\mathrm{heat}(k) \leq \overline{P}_\mathrm{chp}^\mathrm{heat} \cdot \delta_\mathrm{chp}(k)$$
$$\underline{P}_\mathrm{gas} \cdot \delta_\mathrm{chp}(k) \leq P_\mathrm{gas}(k) \leq \overline{P}_\mathrm{gas} \cdot \delta_\mathrm{chp}(k)$$

$$\sum_{h=1}^{N_h} P^{\mathrm{el}}_{\mathrm{chp,h}}(k) = P^{\mathrm{el}}_{\mathrm{chp}}(k)$$

$$\sum_{h=1}^{N_h} P^{\mathrm{heat}}_{\mathrm{chp,h}}(k) = P^{\mathrm{heat}}_{\mathrm{chp}}(k)$$

$$P^{\mathrm{heat}}_{\mathrm{chp,h}}(k) \geq 0$$

$$P^{\mathrm{heat}}_{\mathrm{chp,h}}(k) \geq 0,$$

where $P^{\mathrm{el}}_{\mathrm{chp}}$ and $P^{\mathrm{heat}}_{\mathrm{chp}}$ are defined according (10.19). The constraints above guarantee that the thermal and the electrical power outputs of the micro-CHP are properly shared among the houses and the bounds on the power generation and on the gas power are not exceeded.

**Interaction with the grid**

The following constraint governs the interaction with the distribution grid

$$-P^{\mathrm{peak,h}} <= P^{\mathrm{grid}}_{\mathrm{h}} <= P^{\mathrm{peak,h}}.$$

The peak signal is provided by the external power grid operator, which can be a demand response signal. The houses have the possibility to sell power to the grid (i.e., $P^{\mathrm{grid}}_{\mathrm{h}}$ can be negative).

## *10.3.2 Modeling User Behavior Uncertainty*

The MPC problem described in Sect. 10.3.1 can be extended by taking into account the user behavior uncertainty in scheduling shiftable appliances. The idea is to map the uncertainty on the decision variables (e.g., starting time of appliances) to an equivalent uncertainty in the weighted sum tariff, which is illustrated in Fig. 10.6. Deviating from starting times $x_1$ and $x_2$ by at most $M$ time slots ($M\Delta t$) turns into a variability of the tariff by at most $\Delta y_1$ and $\Delta y_2$, respectively. The parameter $M$ can then be defined based on empirical models of the users, e.g., trained from historical data.

Deviating from the optimal start time of appliances can be considered as variability in the tariff curve. A reasonable expression for defining this uncertainty for each time slot $k$,i.e., $\varepsilon^k$, is

$$\varepsilon^k = \frac{\max(C^k_\lambda, \sum_{i=k-M}^{k+M} \frac{C^i_\lambda}{2M+1}) - C^k_\lambda}{\max(C^k_\lambda, \sum_{i=k-M}^{k+M} \frac{C^i_\lambda}{2M+1})}$$

which is a function of the tariff curve within an interval of $\pm M\Delta t$ minutes in the neighborhood of time slot $k$.

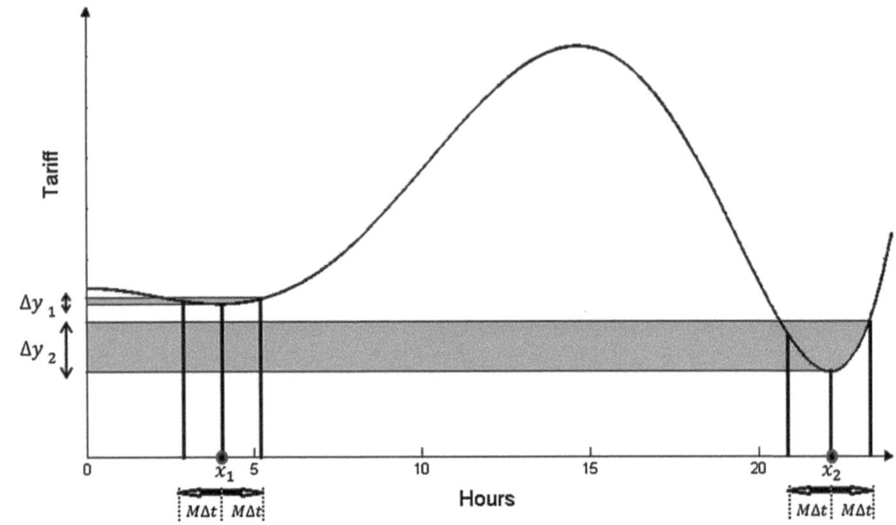


**Fig. 10.6** Converting the user behavior uncertainty to tariff uncertainty (reprinted, with permission, from [36])

When considering uncertain parameters in the cost function, a robust approach can be applied in order to obtain schedules of shiftable loads that are less sensitive to uncertainty in the user preferences. A robust solution should be feasible in all scenarios that uncertain parameters variations could define, i.e., all the possible tariff curves described by $\varepsilon^k$. This comes at the cost of a degradation of the objective value, which could be excessive as some of the uncertain scenarios rarely occur. This increase in cost over the nominal solution is the so-called *price of robustness* [40]. In order to prevent overly conservative solutions, the robust optimization approach described in [41, 42] is applied, which considers a tunable degree of robustness. In this formulation, the degree of uncertainty and the level of conservatism of the solution, and the increase in cost, can be regulated by a parameter. The authors in [40, 41] prove that the obtained robust solution will be feasible with high probability. Details on this robust scheduling framework for shiftable loads can be found in [36].

We remark that the proposed framework can be generally applied to other scenarios where different sources of uncertainty and different optimization criteria must be considered. For instance, $\varepsilon^k$ can represent the variation from the day-ahead price at time slot $k$ in the real-time energy market, while different optimization criteria can account for the user comfort or the demand peak reduction.

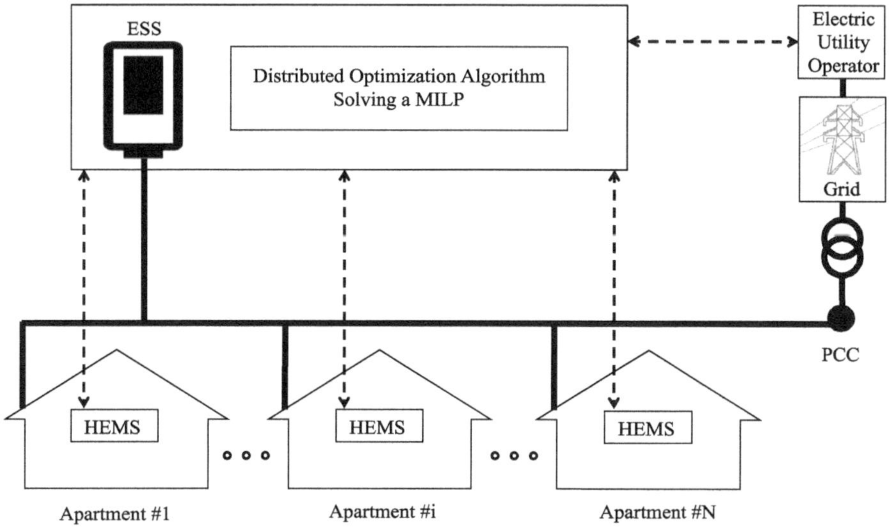


**Fig. 10.7** Schematic of interconnected apartments and aggregator (reprinted, with permission, from [37])

### *10.3.3 A Distributed Approach for Coordinating Shiftable Loads and Storage Systems*

When the number of subsystems increases, the centralized approach described so far can be prohibitive and a reasonable solution is to design distributed control architectures, which allow more flexibility in operation and simplified design and tuning.

In this section, we outline a distributed approach to solve the problem of coordinating a set of smart appliances located in $N$ apartments sharing an ESS, as shown in Fig. 10.7. Each apartment can profit from the use of the shared storage while technical and operational constraints, as well as user preferences, are satisfied. Since storage devices are still expensive, a reasonable solution to afford the expense and benefit from the use of an ESS would be to share it among several consumers. Therefore, households should be coordinated by an aggregator, which acts as mediators between users and the utility operator [43]. In order to manage a large set of appliances, we propose an iterative hierarchical approach. We distribute the decision-making mechanism by formulating problems at apartment and aggregator levels. Each apartment is equipped with an HEMS for scheduling the shiftable loads by taking the user preferences and all the technical and operational constraints into account. The local HEMSs are coordinated by the aggregator in order to provide economical incentives to the users on reshaping their demand profiles, guaranteeing that the benefits of using a shared resources are fairly allocated to the users. The framework includes the possibility of buying/selling from/to the distribution grid. The steps of the proposed algorithm are detailed in Algorithm 2. In the initialization step, the aggregated demand profile is computed by the aggregator considering the overall energy requirements from the users, the shared ESS technical and operational constraints, and the interaction with the grid, while each user sends to the aggregator the preferred individual energy profile based on their own needs and preferences, without accounting for the shared resources. The maximum profit obtainable by optimally utilizing the shared resource is then calculated. At subsequent iterations, the aggregator requires each user to shift their energy demand taking also their preferred profile into account, while each individual user aims at optimizing their shiftable load schedule accounting for the requests from the aggregator, based on pre-arranged incentives and penalties. The profit obtainable by the use of the shared resource and the interaction with the grid is calculated at each iteration and equally allocated to the users. Problems at user levels aim at minimizing the energy costs, while the problem at aggregator level maximizes the overall profits. The algorithm terminates when either all the users accept the energy shift requests, or when a maximum number of iterations is reached. The solution computed at each iteration $l \geq 2$ is then guaranteed to be feasible and to ensure fair allocation of profits.

**Algorithm 10.2** Distributed algorithm

1: Initialization and computation of the aggregated demand profile
2: **for** $l = 1, 2, \ldots,$ MaxIteration **do**
3: each user level HEMS computes and sends its preferred power profile to the aggregator
4: the aggregator calculates the optimal power shift requests aiming at maximizing the overall profits
5: the aggregator calculates the individual user profits
6: based on these individual profits, each user adjusts its own power profile
7: if all the users accept the power shifts requested by the aggregator then stop, otherwise compute a new aggregated power profile and repeat the cycle
8: **end for**

Preliminary simulation results considering four apartments with three appliances and one shared storage device show that the electricity cost resulting from the final iteration of Algorithm 2 is only 1.3% higher than the centralized optimal solution. On the other hand, the computational time has decreased by two orders of magnitude with respect to the solution time of the centralized formulation.

We point out that the proposed algorithm is suitable for model predictive control frameworks and it can be adapted to provide DR services to the grid operator. Further details can be found in [37].

## 10.4 Smart Building Demonstrators

In this section, we briefly describe three demonstrators located in Sweden and UK that support ongoing research activities and experiments on EMS for smart buildings:

1. the *Live-In Lab*, a platform for research, development and education in the building sector;
2. the *Manchester Corridor*, the Manchester's cosmopolitan hub and world-class innovation district;
3. the *SICS ICE*, a large-scale testing and experimentation data center facility that supports fully flexible experiments.

### 10.4.1 KTH Live-In Lab

The KTH Live-In Lab is a platform for research, testing and education to promote innovation in the building sector and it consists of both virtual and physical test environments.

The Live-In Lab consists of a set of three residential buildings, currently under construction, for approximately 300 studio apartments; it is located in the main campus at KTH, Stockholm, Sweden. Heating and cooling power to the buildings are provided by ground-source heat pumps. Heat is distributed airborne to the apartments through a TermoDeck™ [44] system that provides ventilation and heat distribution at the same time. In the TermoDeck™ system, air for ventilation is delivered through pipes embedded in the concrete slabs, allowing pre-heating or pre-cooling of the slabs before air is introduced in the living spaces; these features make it crucial for efficient energy management to effectively control the delays introduced by the thermal masses. Electricity is generated locally with PV panels installed on the flat roof, and the installation of storage systems, in particular batteries for electricity, is under discussion.

The buildings comprise so-called Passive and Active parts. The Passive part accounts for the majority of the floor area and is designed to be extensively equipped with state-of-the-art sensor devices to accurately log indoor and outdoor environmental parameters (e.g., temperature, humidity, light, etc.) for continuous and real-time monitoring of indoor comfort and energy use. In the initial phase of the project, the Passive part will be used only for monitoring. The Active part accounts for approximately 300 $m^2$ and will be used for more active testing: the experimental setup, including the layout of the apartments comprised in this area, will be periodically changed, allowing a holistic approach to research in buildings. The Active part has a dedicated heating and cooling system and the energy is provided with a separate heat pump and boreholes; advanced monitoring and control will be tested and fine-tuned there to be then applied to the rest of the building.

### 10.4.2 Manchester Corridor

The Corridor is Manchester's innovation district, a unique business location at the heart of the knowledge economy, generating around 20% of the city's economic output. It covers some 243 hectares, with a 60000-strong workforce (over 50% are employed in knowledge-based sectors), 72000 students and is the location of the largest clinical academic campus in Europe. It is home to a dense clustering of innovative science and technology businesses, world-class research centers, universities, leading NHS trusts, with a large-scale and transformational infrastructure investment committed to 2020. It provides an ideal testbed for innovation as a microcosm of city activity, driven by the thought leaders on its doorstep and in its institutions.

Corridor Manchester is working toward becoming one of the most sustainable urban locations in Europe and an exemplar for other cities. One of the main objective is to reduce the carbon footprint through the more sustainable management of energy and waste. The Corridor Manchester area is currently one of the demonstrator sites for the Triangulum project [45], a Horizon 2020 Lighthouse project running until 2019, to demonstrate smart green growth across energy, mobility, and ICT. The Innovate UK Internet of Things City Demonstrator project, CityVerve [46], will also be taking place in the Corridor Manchester area with Environment & Energy being a key theme within this project.

### *10.4.3 SICS ICE*

The Swedish Institute for Computer Science (SICS) Infrastructure and Cloud data center test Environment (ICE) research data center project [47] aims at supporting universities and industries with a large-scale data center infrastructure and cloud facility to be used for research, testing and demonstration purposes, especially on topics related to energy efficient software, hardware and infrastructure management (thus also its HVAC systems). The data center can be used by anybody with the formula pay-per-usage to perform tests and experimentation, and enables us to perform large-scale tests to generate realistic data.

The ICE data center is composed of different modules located in neighboring rooms and dedicated to different types of potential tests. Upon completion, forecasted in 2017, ICE will host around 5000 servers plus all the infrastructure needed to support running these devices (e.g., batteries for Uninterruptible Power Supply (UPS) purposes, Computer Room Air Conditioning (CRAC) units, Supervisory Control And Data Acquisition (SCADA) systems, and data visualization facilities) for a total of approximately 2MW peak electrical power consumption.

The facility is located in the municipality of Luleå, in northern Sweden, a favorable location for placing data centers due to the extensive optic fiber capacity, the positive local electrical energy balance (mainly based on renewable energies), and the extreme stability and redundancy of the local electrical network.

As specified in Sect. 10.5, this facility can be used for performing tests related to the coordination of the electrical loads of different buildings. The facility supports remote experimentation, in the sense that researchers can deploy control software and perform experiments remotely.

## 10.5 Coordinating Data Centers and Buildings for a More Effective Load Management

Data centers are "facilities that centralize an organization's Information Technology (IT) operations and equipment, and where they store, manage, and disseminate their data" [48]. In practice, they are facilities hosting massive amount of computer systems running IT applications, and they can be placed in containers, in dedicated rooms, or even in dedicated buildings.

We can schematize a data center as in Fig. 10.8. IT requests are processed by the various servers, and this produces heat that must be rejected to prevent overheating and thus failure of the servers. An essential component in the control of a data center is played by the IT scheduling algorithm, that decides which server should process a certain IT load. The importance of the IT scheduler can be appreciated noticing the usual structure of an air-cooled data center[1] in Fig. 10.9a and b: allocating IT loads in a place that is currently overheated or difficult to cool may induce hot-spots (and thus failures) and thermal inefficiencies. The IT allocation should thus account also for the thermal dynamics in a data center.

Single data centers may reach electrical power consumptions of more than 100 MW [49] and have industrial-scale operations. They thus have big impacts worldwide: for example, in Europe in 2013 data centers accounted for 3% of the total electric energy consumption (for corresponding emissions of 38.6 million metric tons of $CO_2$) [50], and it is moreover forecast that by 2020 in Europe 60 other large data centers with more than 1000 servers racks will be built.

Depending on the data center structure and technology, cooling of the server halls accounts for up to 40% of the electrical consumptions. Heat recovery is not yet a widespread solution in data centers management, since the currently dominant cooling strategy is indeed air-based with exhaust coolants having a quite low exergy (air up to 45 °C and with a dew point between 5.5 and 15 °C). A noticeable exception to this trend is the *open district heating* project in Stockholm [51], where data centers from different companies (e.g., Ericsson) are currently providing heat to the district heating system; the exergy of the exhausts is in this case topped up by means of heat pumps. Forecasts are that in the next decade more and more data centers will switch to liquid cooling technology (due to higher energy density servers, that will eventually demand liquid cooling) and thus be integrated into the district heating networks.

### *10.5.1 Data Centers as Smart Buildings*

Data centers can play several different schemes typical of residential smart buildings. Indeed:

---

[1] Same considerations can be drawn in liquid-cooled data centers.

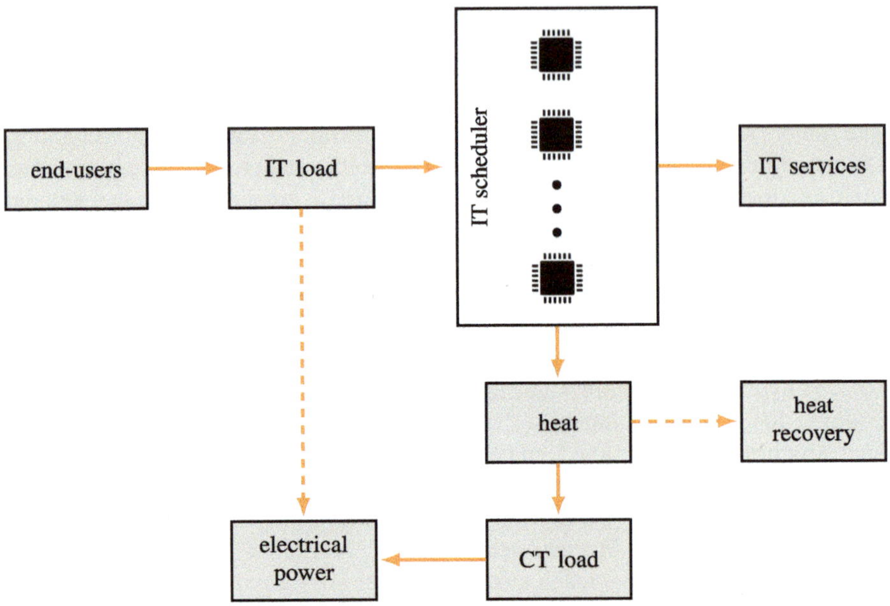


**Fig. 10.8** Schematic representation of the operations involved in a data center. Users require some IT services, that are executed by the servers. The corresponding IT load is sorted by the IT scheduler that associates a certain load to a certain physical server. The server thus generates heat that may be recovered or rejected by the Cooling Technologies devices. Both servers and CT devices use electrical power

*when considered as elements of the electrical grid,* data centers can provide ancillary services like load shifting and load shedding. Indeed data centers' electrical consumptions are mainly due to running the servers and the infrastructure for cooling the servers. Given that it is possible to play pre-cooling schemes like in other smart residential buildings, it is thus possible to perform load shifting and load shedding in data centers in a similar way one does it in other smart buildings;

*when considered as elements of the thermal grid,* data centers can sometimes act as deferrable prosumers: data centers may figure in this case either in the production or in the consumption sides of the network, especially when connected to cooling grids. In any case pre- and post- cooling, plus temporal shift of deferrable IT loads means to defer thermal production/consumption loads;

*when considered as elements of the information grid:* moving photons is cheaper than moving electrons. Under favorable electricity price conditions, networks of data centers may use the Internet to communicate and potentially exchange tasks among them, an operation commonly referred to as *geographical load balancing*. Geographical load balancing thus does not just shift IT loads from one location to an other, but also electrical load requests and potential thermal generation capability.

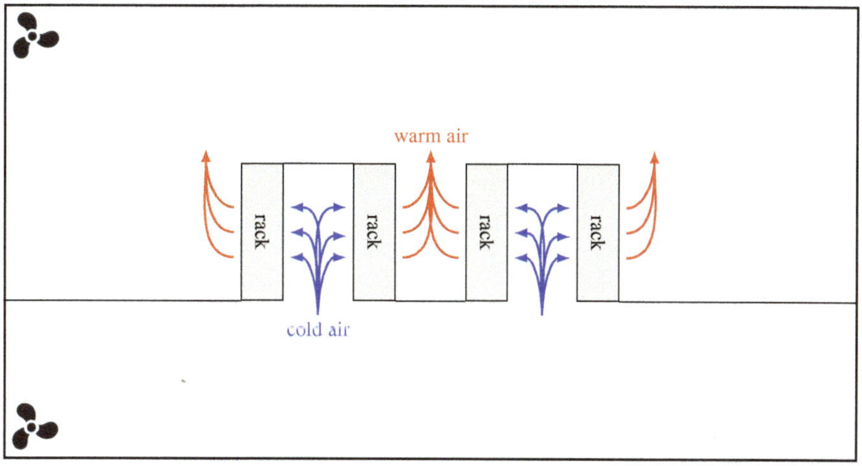


(a) Schematic representation of the typical organization of an air-cooled computer room: warm and cold aisles alternate each other so to diminish disturbances in the cold-air generation unit.

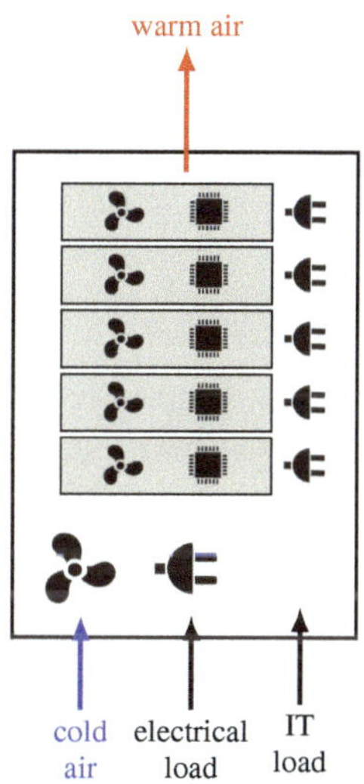


(b) Schematic representation of the typical organization of an air-cooled computer rack: servers are piled on each other so to diminish disturbances in the cold-air generation unit.

**Fig. 10.9** **a** Schematic representation of the typical organization of an air-cooled computer room: warm and cold aisles alternate each other so to diminish disturbances in the cold-air generation unit **b** Schematic representation of the typical organization of an air-cooled computer rack: servers are piled on each other so to diminish disturbances in the cold-air generation unit

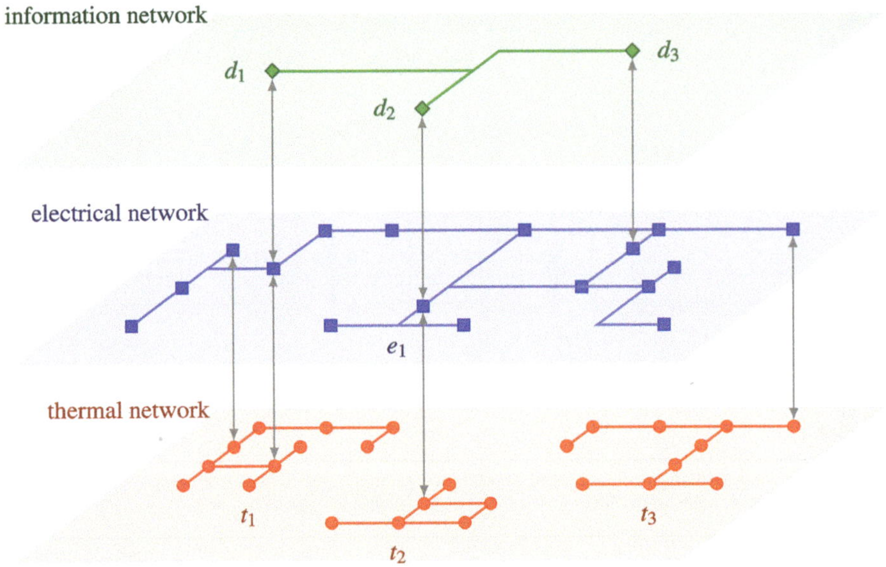


**Fig. 10.10** Data centers are consumers in the electrical network, are potential producers in the thermal network, and are prosumers in the information network

The schemes listed above explain why data centers can act as bridges between three types of networks: electrical, thermal and information networks, as graphically shown in Fig. 10.10. This means that data centers are viable players in the smart city environment: obviously, depending on the data center type and electrical/thermal network conditions, one may interplay different types of coordination and synergies. The simplest one is when a single building contains a data center room, as in [52], where heat recovery and electric consumptions can be coordinated directly at a single building's level. More complicated situations may instead require networked control strategies.

## 10.6 Conclusions and Future Directions

Implementing the next-generation smart energy grids requires buildings that can manage flexible loads, multiple performance criteria, local generation, storage systems and DR policies. These buildings, moreover, shall embed user comfort and preferences during their daily operations. All these requirements entail the need of control schemes that are predictive and allow seamless coordination among different entities. Buildings can indeed help flatten the aggregated demand curve in the electrical network and thus reduce the number of expensive generation plants used for peak load periods.

A natural scheme for optimally operating buildings while accounting for all these aforementioned requirements is MPC, since it allows for flexibility in the design of control objectives.

In this chapter we showed how the basic idea of using a MPC strategy for the management of a smart building can be further refined following some specific directions: the first is taking into account uncertainties, i.e., making the control scheme robust against imprecise models or imprecise knowledge about the future. Numerical experiments performed in daily-life conditions then showed that our approach of following a scenario-based SMPC approach led to effective results in terms of energy savings versus keeping comfortable conditions in the building.

The second direction is formulating the controller so that its implementation complexity is sufficiently low to be implementable in real-world situations. Also in this case, we performed numerical experiments and found that converting the implicit MPCs mentioned above into their explicit versions was leading to effective control schemes.

Furthermore, we described how local generation and storage systems can be integrated into a building energy management framework.

Ongoing and future studies aim at extending the proposed EMS framework to other energy systems and additional DER and control objectives. In particular, it is possible to design distributed, integrated and holistic approaches for the management of networks of thermal and electrical energy systems in coordination with network operators. With this approach the overall energy management problem can be solved in a distributed fashion, aiming to compute feasible solutions to the energy management problem addressing the issue of scalability and taking advantage of opportunities afforded by flexibility services and distributed generation.

## References


1. Annaswamy AM, Amin M (2013) IEEE vision for smart grid controls: 2030 and beyond. In: IEEE vision for smart grid controls: 2030 and beyond, pp 1–168, June 2013
2. European Commission (2008) Debate Europe-building on the experience of plan D for democracy, dialogue and debate, European Economic and Social Committee and the Committee of the Regions, COM 158/4, Brussels
3. Costa A, Keane MM, Torrens JI, Corry E (2013) Building operation and energy performance: monitoring, analysis and optimisation toolkit. Appl Energy 101:310–316
4. Chua K, Chou S, Yang W, Yan J (2013) Achieving better energy-efficient air conditioning - a review of technologies and strategies. Appl Energy 104:87–104
5. Morari M, Lee J, Garcia C (2001) Model predictive control. Prentice Hall
6. Treado S, Chen Y (2013) Saving building energy through advanced control strategies. Energies 6(9):4769–4785
7. Sturzenegger D, Gyalistras D, Gwerder M, Sagerschnig C, Morari M, Smith RS (2013) Model predictive control of a Swiss office building. In: Clima-RHEVA world congress, pp 3227–3236
8. Calafiore G, Campi M (2008) The scenario approach to robust control design. IEEE Trans Autom Control 51(5):742–753
9. Alessio A, Bemporad A (2009) A survey on explicit model predictive control. Springer, Berlin. ISBN 9783642010934

10. Bemporad A, Borrelli F, Morari M (2002) Model predictive control based on linear programming~ the explicit solution, Automatic Control Laboratory, ETH, Zurich, Switzerland, Technical report
11. Parisio A, Varagnolo D, Risberg D, Pattarello G, Molinari M, Johansson KH (2013) Randomized model predictive control for HVAC systems. In: Proceedings of the 5th ACM Workshop on Embedded Systems For Energy-Efficient Buildings, ser. BuildSys'13. ACM, New York, NY, USA, pp 19:1–19:8. ISBN 978-1-4503-2431-1
12. Parisio A, Fabietti L, Molinari M, Varagnolo D, Johansson KH (2014) Control of HVAC systems via scenario-based explicit mpc. In: 53rd IEEE conference on decision and control, pp 5201–5207, December 2014. ISSN 0191-2216
13. Parisio A, Varagnolo D, Molinari M, Pattarello G, Fabietti L, Johansson KH (2014) Implementation of a scenario-based MPC for HVAC systems: an experimental case study. IFAC Proc 47(3):599–605
14. Gwerder M, Toedtli J (2005) Predictive control for integrated room automation. In: Clima - RHEVA world congress
15. Gwerder M, Boetschi S, Gyalistras D, Sagerschnig C, Sturzenegger D, Smith R, Illi B (2013) Integrated predictive rule-based control of a swiss office building. In: 11th REHVA world congress clima
16. Haniff MF, Selamat H, Yusof R, Buyamin S, Ismail FS (2013) Review of HVAC scheduling techniques for buildings towards energy-efficient and cost-effective operations. Renew Sustain Energy Rev 27:94–103
17. Široký J, Oldewurtel F, Cigler J, Prívara S (2011) Experimental analysis of model predictive control for an energy efficient building heating system. Appl Energy 88(9):3079–3087
18. Ma Y, Matusko J, Borrelli F (2015) Stochastic model predictive control for building hvac systems: complexity and conservatism. IEEE Trans Control Syst Technol 23(1):101–116
19. Maasoumy M, Razmara M, Shahbakhti M, Vincentelli AS (2014) Handling model uncertainty in model predictive control for energy efficient buildings. Energy Build 77:377–392
20. Dong B, Lam KP, Neuman C (2011) Integrated building control based on occupant behavior pattern detection and local weather forecasting. In: Building simulation, pp 193–200
21. O'Neill Z, Narayanan S, Brahme R (2010) Model-based thermal load estimation in buildings. In: SimBuild, pp 474–481
22. Aglan HA (2003) Predictive model for $CO_2$ generation and decay in building envelopes. J Appl Phys 93(2):1287–1290
23. Kall P, Mayer J (2005) Stochastic linear programming: models, theory, and computation. Springer
24. Calafiore G (2010) Random convex programs. SIAM J Opt 20(6):3427–3464
25. Schildbach G, Fagiano L, Frei C, Morari M (2014) The scenario approach for stochastic model predictive control with bounds on closed-loop constraint violations. CoRR abs/1307.5640v2
26. Margellos K, Prandini M, Lygeros J (2014) On the connection between compression learning and scenario based optimization. CoRR abs/1403.0950
27. Equa Simulation AB, IDA-ICE. http://www.equa.se/en/, August 2016
28. Sabin MJ, Gray R (1986) Global convergence and empirical consistency of the generalized Lloyd algorithm. IEEE Trans Inf Theor 32(2):148–155
29. Singh S, Ostergaard J (2010) Use of demand response in electricity markets: An overview and key issues. In: 2010 7th international conference on the European energy market (EEM), pp 1–6
30. Bird L, Reger A, Heeter J (2012) Distributed solar incentive programs: recent experience and best practices for design and implementations, NREL/TP-6A20-56308, National Renewable Energy Laboratory, Technical report
31. Beaudin M, Zareipour H (2015) Home energy management systems: a review of modelling and complexity. Renew Sustain Energy Rev 45:318–335
32. Khan AA, Razzaq S, Khan A, Khursheed F, Owais (2015) HEMSs and enabled demand response in electricity market: an overview. Renew Sustain Energy Rev 42:773–785

33. EIT - ICT Labs, Virtual micro grid laboratory. http://www.eitictlabs.eu/evsgl/areas-of-application/supporting-sub-projects/virtual-micro-grid-lab/
34. Parisio A, Wiezorek C, Kyntäjä T, Elo J, Johansson KH (2015) An mpc-based energy management system for multiple residential microgrids. In: 2015 IEEE international conference on automation science and engineering (CASE), pp 7–14, August 2015
35. Apros. http://www.apros.fi/en/
36. Paridari K, Parisio A, Sandberg H, Johansson KH (2016) Robust scheduling of smart appliances in active apartments with user behavior uncertainty. IEEE Trans Autom Sci Eng 13(1):247–259
37. Paridari K, Parisio A, Sandberg H, Johansson KH (2015) Demand response for aggregated residential consumers with energy storage sharing. In: 2015 54th IEEE conference on decision and control (CDC), pp 2024–2030, December 2015
38. Sviluppo T. http://www.trentinosviluppo.com/Contenuti-istituzionali/News/News/2012/Crisalide-a-Roncegno-nasce-l-isola-cogenerativa
39. Levin P (2009) Brukarindata för energiberäkningar i bostäder. Technical, report, SVEBY
40. Bertsimas D, Sim M (2004) The price of robustness. Oper Res 52:35–53
41. Bertsimas D, Sim M (2003) Robust discrete optimization and network flows. Math Program 98:49–71
42. Floudas C (1995) Nonlinear and mixed-integer programming-fundamentals and applications. Oxford University Press, Oxford
43. Gkatzikis L, Koutsopoulos I, Salonidis T (2013) The role of aggregators in smart grid demand response markets. IEEE J Sel Areas Commun 31:1247–1257
44. Termodeck, TermoDeck$^{TM}$. http://www.termodeck.com/, August 2016
45. Triangulum project. http://triangulum-project.eu/, Triangulum
46. Cityverve project. http://www.cityverve.org.uk/, CityVerve
47. S.I. for Computer Science (2015) SICS infrastructure and cloud data center test environment. TODO
48. Google (2016) Definition of a data center, retrieved from the web. TODO
49. Facebook. www.datacenterknowledge.com/inside-facebooks-lulea-data-center/
50. The Boston Consulting Group (2014) Digital Infrastructure and Economic Development: An impact assessment of Facebook's data center in Northern Sweden, The Boston Consulting Group, Technical report. http://www.bcg.se/documents/file164478.pdf
51. Fortum (2016) Open district heating. http://www.opendistrictheating.com/
52. N.R.E.L. (NREL), Energy system integration facility (esif). https://hpc.nrel.gov/datacenter

# Chapter 11
# Controlling the Internet of Things – from Energy Saving to Fast Evacuation in Smart Buildings

**Qing-Shan Jia, Yuanming Zhang and Qianchuan Zhao**

## 11.1 Introduction

There is an increasing interest in connecting things into network to improve the performance and to provide novel services. This change is reshaping many of the systems in our daily lives, such as smart buildings. Generally speaking, the Internet of Things consists of things, mainly physical devices, connected through the Internet. This network-based connectivity enables different devices across different locations to sense all the other devices or to be sensed by all the other devices. Then the information and data sensed or collected by each device are shared through the Internet infrastructure. Devices connected into the network can extract and analyze the data depending on their different purposes. The Internet of things connects almost all the parts that make up a city. These objects are not merely connected but with information flowing through the network which facilitates the management and the operation of the whole society.

The Internet of Things provides a basic framework for the operation and management of different systems in the real world. When the sensing system is integrated

This work is support in part by the National Key Research and Development Program of China (No. 2016YFB0901900), National Natural Science Foundation of China (Nos. 61673229, 61425027, U1301254, 61222302, 61174072, and 91224008), the Tsinghua National Laboratory for Information Science and Technology (TNLIST) Funding for Excellent Young Scholar, and the Program for New Star in Science and Technology in Beijing (No. xx2014B056).

Q.-S. Jia (✉) · Y. Zhang · Q. Zhao
Center for Intelligent and Networked Systems, Department of Automation, Tsinghua National Laboratory for Information Science and Technology, Tsinghua University, Beijing 100084, China
e-mail: jiaqs@tsinghua.edu.cn

Y. Zhang
e-mail: zhang-ym12@mails.tsinghua.edu.cn

Q. Zhao
e-mail: zhaoqc@tsinghua.edu.cn



J.T. Wen and S. Mishra (eds.), *Intelligent Building Control Systems*, Advances in Industrial Control, https://doi.org/10.1007/978-3-319-68462-8_11

with actuators, a classic feedback-based closed-loop system is formed. The original connected system becomes a controllable system with information flow and physical devices. Then a wider application area is explored like smart grid, smart transportation, and smart buildings. The purpose of applying the Internet of Things technology to buildings is saving energy as well as improving the energy efficiency. Nowadays, the energy consumption of buildings has exceeded that of industry and transportation and made up almost 50% of the end user energy consumption. In the United States, buildings are responsible for 38% of the $CO_2$ emission, 71% of the electricity consumption, 39% of the energy use, 12% of the water consumption, and 40% of non-industrial waste. According to the statistics, almost 90% of people's time is spent indoors. Then the energy utilization is closely related to people's everyday life. How to improve the energy efficiency and at the same time satisfy the occupant's comfort requirement becomes an interesting subject.

The Internet of Things technology brings the chance to connect all the energy devices in a building and to manage them in a systematic view. The information sharing from different sensors makes the activation of different energy devices more efficient. As reported, the operational cost of buildings in China makes up about 30% of end energy consumption, in which the heating, ventilation and air conditioning system (HVAC) makes up 20%. If no actions are taken, China will become number 1 $CO_2$ emitter by 2020. Furthermore, in the coal-fired electrical energy in China, only 8% of the power reaches the end user [1]. So optimizing the end energy utilization, especially the building energy consumption, has great potential in saving energy, alleviating energy crisis, cutting $CO_2$ emission, protecting the environment, and sustaining further development of human beings.

On the other hand, with the thriving of renewable energy, distributed renewable energies like solar power, wind power, biomass energy and geothermal energy are combined with buildings for diversified energy supply and for reducing traditional power (like fossil fuel power) utilization. As the renewable energy sources are usually intermittent and fluctuant, energy storage devices are often applied for system stability and security. These energy systems bring great chances for improving building energy efficiency while at the same time increase the difficulty of managing the building operation. There are mainly three reasons as follows. First, there exists randomness in both the renewable energy supply and building energy demand. Take solar power as an example for energy supply, the power generation is mainly influenced by the position of the sun and the power may vary sharply when the solar panel is shaded by moving cloud. For the energy demand, the occupant's behavior directly influences the power consumption which is difficult to predict. Second, the operation of different energy systems is coupled which increases the management of the building system. Third, the operation and control of the system is a non-stationary decision process. In one day's operation, as the energy supply and demand vary with time advancing, the decisions in each decision window are different even with the same energy supply or energy demand situation. Facilitated by the Internet of Things technology, these various energy suppliers can be monitored and the sensed energy supply and demand information are shared by the whole energy system. Taking advantage of the stored historical data and the real-time sensed data, the distributed energy systems become "things" of an energy Internet or called building cyber-physical system and the control decisions can be made more efficiently for the renewable energy utilization.

The Internet of Things technology presents us an innovative picture of the smart building, where the consumers of the building and the energy suppliers are connected through the network. The smart thing is that the status and the behavior of the energy system become transparent through the sensing system. What's more, based on their different purposes, each device is activated with the information acquired from the network and all parts of the system can operate cooperatively and harmoniously. The final objective is an energy saving building with satisfaction of comfort the requirement for each occupant.

## 11.2 Literature Review

A brief literature review of the building sensing system and the control and optimization of the Internet of Things-based building energy management is made as follows.

For the existent building sensing, the occupant localization and counting in an indoor environment attract much attention. The methods are classified into two groups. The first group includes RADAR [2], SpotON [3], LANDMARC [4], Ekahau [5], and UWB [6] which are based on the strength measurements of Radio Frequency (RF) signals. These methods can be significantly affected by indoor electric and magnetic conditions. The second group includes Active Badge [7], Cricket [8], and Easy Living [9] which are based on the infrared information [10], ultrasound, or video cameras, and the estimation accuracy varies with the environment [11]. Li et al. [12] developed an occupant monitoring system based on an RFID. Liao et al. [13] proposed an agent-based method with graphical modeling for building occupancy. Casas et al. [14] used face detection in video image to estimate the density of occupants. Lin et al. [15] estimated the occupant density in crowds by detecting the featured areas of the head contours, and then classified these areas by Support Vector Machine (SVM) method. Guo et al. [16] estimated crowd density based on Markov random field. The problem with the above methods is that the estimation accuracy based on single sensing sources can be easily affected by the measurement environment. Then many information fusion methods are developed to improve the accuracy of the estimation, such as the Bayesian method [17], the Bayesian network [18], the Dempster–Shafer evidence combination rule [19], the fuzzy logic [20], and the neural network [18], SVM [21]. Multiple methods can be combined for further improvement [20]. For example, Alvarez-Alvarez et al. [22] proposed a method for human activity recognition in indoor environment using the information from WIFI and accelerations. Meyn et al. [23] developed a sensor-utility-network model under the criterion of maximum a posterior (MAP) probability to improve the estimation accuracy for occupant distribution. These fusion methods generally require the joint distribution of the measurements of the multiple sensing sources, which are usually not available.

For the building equipment operation and management, many efforts have been made to improve the energy efficiency of buildings through integrated/individual control of energy supply devices and energy consumption equipment. On the one hand, some studies focus on reducing the energy consumption (or cost) on the demand side. Djuric et al. [24] and Asadi et al. [25] developed some simulation-based optimization methods which consist of a simulation program (such as TRNSYS, EnergyPlus, etc.) and an optimization program (such as GenOpt and BuildOpt). Hazyuk et al. [26], [27] developed a building model and used model-based predictive control (MPC) to ensure the thermal comfort with the minimal energy consumption. On the other hand, some studies focus on improving energy efficiency of building energy supply systems through utilizing the renewable energy sources and storage devices. Figueiredo and Martins [28] presented a method based on the integration of the building automation system, demand side management, and energy production system management.

The comfort model of the occupant in most of the aforementioned work is assumed to be a fixed setpoint. In practice, this set point could be obtained according to statistical models. The PMV model [29] is one such example, which uses a regression model to predict the probability of mean vote under different indoor temperatures. This model then outputs a specific set point under which over 95% of the occupants are predicted to be comfortable. These models do not distinguish the difference between occupants. Such a difference leads to the concept of personalized comfort, which is usually learned through human–building interaction in practice. Yang and Wang [30] developed a building–human interaction system based on a multiagent system. The personalized temperature set point is generalized into a set range in [31]. As long as the temperature is within the range, the occupant is assumed to be comfortable. Different comfortable ranges are used to capture the difference among the thermal comfort of different occupants. Peschiera et al. [32] and Peterson et al. [33] found that occupant behavior can be affected by providing the electrical energy consumption back to the occupant.

The Internet of Things-based smart building management and operation is considered and several key problems are investigated. The following topics are discussed in the following sections: heterogeneous information fusion-based building occupant level estimation, energy management of smart microgrid of buildings, building mounted wind turbines for electric vehicle charging and fast evacuation.

## 11.3 Problem Formulation

### *11.3.1 Heterogeneous Information Fusion for Occupants Level Estimation*

Estimating the number of occupants at a zonal level in buildings is useful for energy-efficient control of air conditioning and lighting systems. In this section, the occupant level estimation problem is first formulated as an information fusion problem with

| Systems | | Cost (RMB) | Approach | Accuracy | Disadvantage |
|---|---|---|---|---|---|
| Active sensors | WSN/RFID | 200 | Localization | 2m-5m | Multipath effect |
| | RFID | 3,000(r) 200(t) | Localization | 2m-5m | Multipath effect |
| | UWB | 30,000(r) 200(t) | Localization | 0.30m | High cost |
| | Cricket | 2,000 | Localization | 0.10m | Ultrasonic, orientation constrained |
| Passive sensors | Video | 5,000~ 200 | Image Processing | 90%~95% | Privacy, cost |
| | CO2 | 2,000~ 100 | Signal Processing | 50%~80% | Drift |
| | Infrared | 100 | Counting | 80%~95% | No identity |

**Fig. 11.1** Occupant-Oriented Sensors

heterogeneous information sources with the criterion of the minimum mean square error (MMSE). Then two fusion methods are developed where the first method assumes independent observation noises and the second method exploits the correlation among the multiple information sources to improve the estimation accuracy. Different kinds of cheap sensors used in the fusion to get high accuracy are shown in Fig. 11.1. Consider a large office divided into multiple zones according to the positions of air outlets and lights. Suppose there are $m$ localization systems, each of which can estimate the number of occupants in the zone. Note that it is possible that the observed occupants in one system may be unequal to the estimated occupants. The question is how to find an estimate of occupants of minimize the mean square error as Eq. (11.1).

$$\min E[(\text{estimated occupants-real occupants})^2|\text{observation}] \qquad (11.1)$$

The true number of occupants is a random variable in the Bayesian viewpoint. Calculating the minimum square error requires the a priori knowledge of the distribution of the true number of occupants and the conditional joint distribution of the observations. For the former, the random arrival and departure of the occupants to a zone can be reasonably approximated by a nonhomogeneous Poisson process. For the latter, since the total number of objects that would be counted as occupants cannot be accurately estimated a priori, we assume that the number of objects that are wrongly observed as occupants by the system has Poisson distribution.

With these two assumptions about occupants' arrival process and the error detection, two approximate fusion methods for solving Eq. (11.1) are briefly presented as follows. Interested audience can refer to our publications [34, 35] for the proof and detailed mathematical formulations and deductions.

1. Assume that the observations of different systems are independent. We prove that the expected objective is a convex function. Through the derivative of the objective, the minimization problem can be solved.
2. If there are correlations among the observations of different sensing systems, assume that the solution of the fusion problem has the same form as the estimation that minimizes the mean square error of an individual system. We first obtain the optimal estimation for a single sensing system, and then the optimal estimation of multiple sensing systems by selecting the best parameters.

### *11.3.2 Energy Management for Smart Microgrid of Buildings*

The total energy consumed during building operations is determined by the building energy efficiency and the total demand. In this section, we consider a building that consists of multiple rooms. There is an occupant in each room and with an individual comfort profile. We develop a method that explores human–building interaction to improve the energy efficiency of the building, while satisfying the comfort requirement of individual occupants. To be specific, we allow each individual to adjust his/her set range of temperature according to the real-time average price of electricity (APE). Since the occupant is charged according to APE, this cost structure motivates him/her to do such an adjustment. The multiple devices in the supply side are then scheduled jointly considering this potential adjustment in the demand side. The objective function of the joint scheduling problem is to minimize the total generation cost while providing the required indoor environment for all the occupants as Eq. (11.2).

$$\min \sum_{\text{K periods}} (\text{cost of electricity +natual gas cost+penalty of battery }) + \text{tiered electricity cost} \tag{11.2}$$

This optimization problem can be transformed into a mixed-integer linear programming form which can be solved by a centralized method that uses CPLEX to solve directly with a near-optimal global optimum. However, In a commercial high rise, there are usually hundreds or thousands of rooms. The scale of the problem could be very large. In order to find a faster solution method, we decompose the optimization problem into two subproblems, as shown in Fig. 11.2.

The first subproblem is the demand side optimization problem (for a single room). There are two inputs to this problem, the comfort requirement and the APE. The various terminal devices are controlled to minimize an estimated cost for this single room with the objective function as Eq. (11.3).

$$\min \sum_{\text{K periods}} \text{price} \times (\text{total electricity consumption at each period}) \tag{11.3}$$

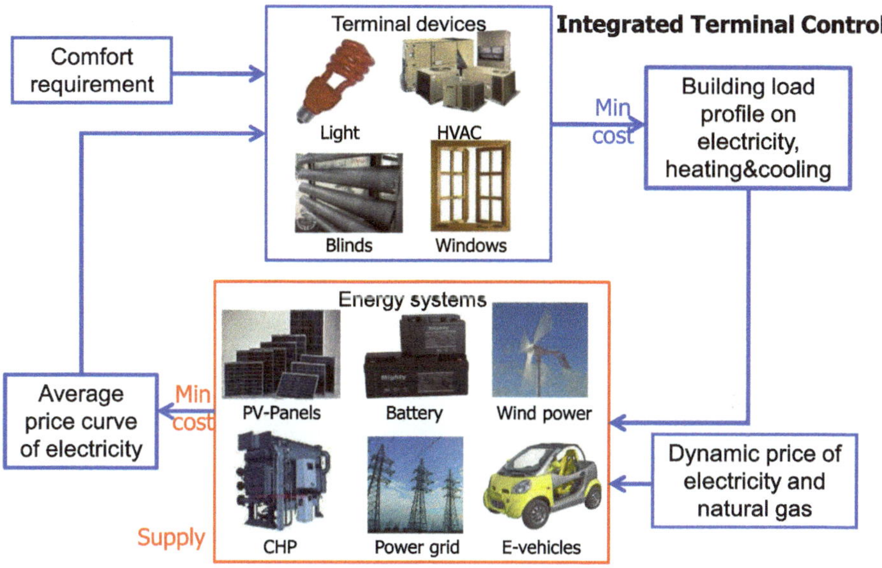


**Fig. 11.2** Supply–Demand Coordination for Efficiency and Comfort

The second subproblem is the supply-side optimization problem (for all the energy generation and storage devices in the building). There are two inputs to this problem, namely, the demand profile that is obtained from solving the demand side subproblem in the last iteration, and the dynamic prices of electricity and natural gas. The various supply side devices are controlled to minimize an estimated cost for the supply side as Eq. (11.4).

$$\min \sum_{\text{K periods}} (\text{supply side estimated cost}) \tag{11.4}$$

The interested audience can refer to [36, 37] for more detailed mathematical formulation and solving method description.

### *11.3.3 Building Mounted Wind Turbines for Electric Vehicle Charging*

The matching between random wind supply and electric vehicle (EV) charging demand can reduce the requirement of traditional power sources and the emission of $CO_2$. A bilevel Markov decision process is developed to formulate the problem of matching the stochastic electric vehicle charging demand and the volatile wind supply. At the upper level, the charging demand of EVs in different locations is aggregated into multiple aggregators. The system operator dispatches power among the aggregators in a coarse timescale to maximize the wind power utilization. At the lower level, the aggregator schedules the charging process of individual EVs at a finer timescale to minimize the charging cost. The system architecture of the bilevel and multi-timescale scheduling problem and timescales for the upper level and lower level decisions are shown in Fig. 11.3.

The upper-level objective function is formulated a weighted form with two kinds of costs. The first kind is the cost of normalized mismatch gap between the EV charging load and the wind power. The small mismatch gap means the large wind power utilization for EV charging which results in a low cost. The second kind is the cost of charging power. In the Markov decision process framework, the upper level objective is an expected cost as Eq. (11.5)

$$\min \sum_{\text{K periods}} E(\text{cost of mismatch+cost of charging aggregated EVs}) \tag{11.5}$$

In the lower level, as the EV drivers in the aggregator care about the charging cost, the responsibility of each aggregator is to minimize the charging cost of its EVs. Then the expected form of the charging cost is as Eq. (11.6).

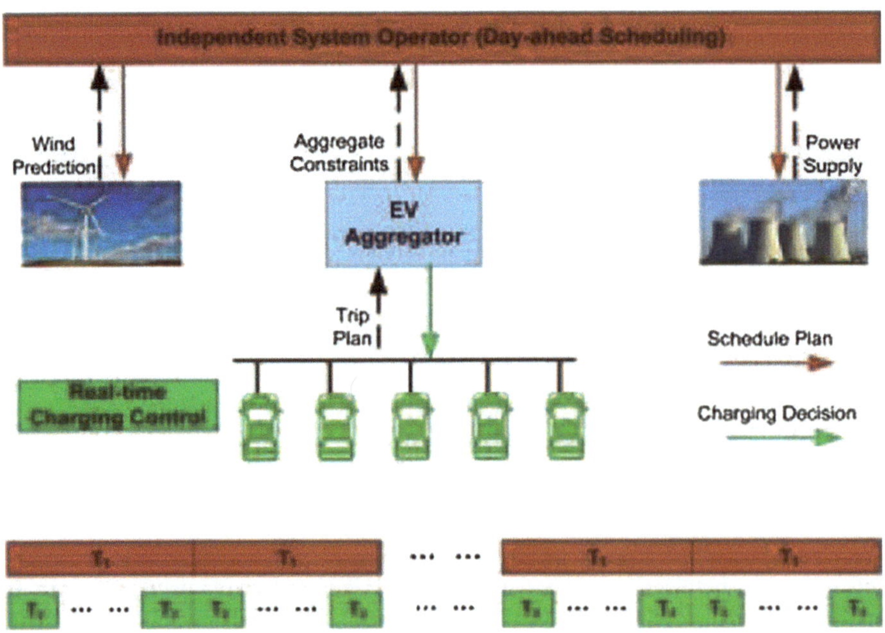


**Fig. 11.3** Bilevel Scheduling of Wind Power

$$\min \sum_{\text{K periods}} E(\text{total cost of charging EVs}) \tag{11.6}$$

For a large-scale problem, in order to solve it efficiently, bilevel simulation-based policy improvement (SBPI) method is developed. In the bilevel architecture, the lower level collects the charging requests and passes them to the upper level. Then, the upper level makes the decision and passes it to the lower level. The algorithm is summarized briefly as follows.

1. At one stage, the EVs and the wind farm generate some future sample paths of EV parking events and wind power. The number and the energy of the arrival and departure EVs in the future will be computed based on these sample paths and passed to the system operator.
2. The upper level constructs the base policy based on the base policy provided by the lower level.
3. The upper level obtains an improved policy from the constructed base policy by using SBPI.
4. Then, the lower level updates the base policy.
5. Based on the updated base policy, the lower level obtains the improved policy at each decision epoch by using SBPI and the newly generated sample paths of wind power and EV parking events.
6. Go to next stage and repeat the process.

The optimization for each EV can be parallelly computed to reduce the computation time. Furthermore, the action space in the lower level can be largely reduced by implementing the following steps. First, group the parked EVs by their remaining parking time. Second, obey the following rule to select the EVs to charge in each group, i.e., pick the EVs by their remaining required charging energy in descent order. These two steps can reduce large amounts of suboptimal actions. Interested audience can refer to [38, 39] for more detailed mathematical formulation of the bilevel EV charging problem.

## 11.4 Main Results

### *11.4.1 Results of Heterogeneous Information Fusion*

The performance of methods 1 and method 2 are evaluated by through a real system with RFID and video camera. Physical experiments are used to show how the aforementioned two fusion methods improve the accuracy of the estimation of occupant level.

The experiments were conducted on Fridays from 9:15AM to 9:45AM for four consecutive weeks. The true occupant number and the estimations given by the RFID, the video cameras, and the two approximate fusion methods are given in Fig. 11.4.

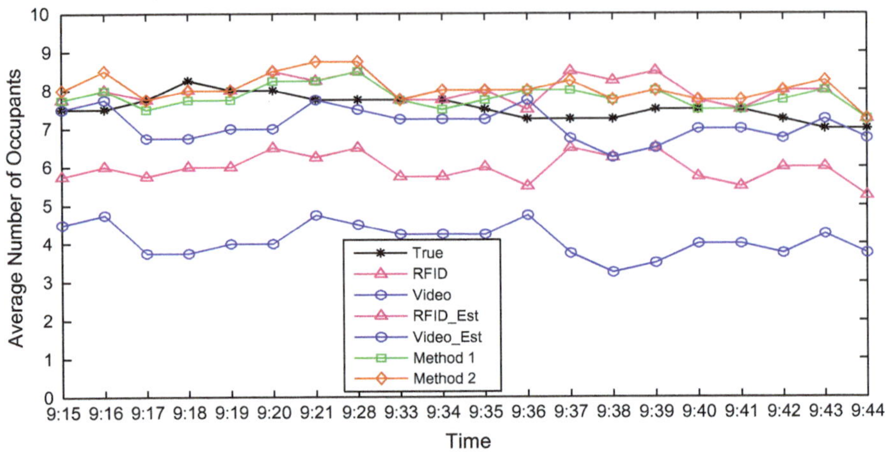


**Fig. 11.4** Number of occupants averaged over four days

From Fig. 11.4, we can see that at each time moment, there are 7 or 8 occupants in the zone on the average. Both the observation of the RFID system and that of the video camera system are less than the true number, while both estimations of the two approximate fusion methods are closer. For comparison, the linear least mean square error (LLMSE) method is applied with the assumption of Gaussian and independent observations. Let X be the true occupant level, which is unknown. The standard deviation of the observation errors and the estimation errors of the RFID system, the video camera system, Methods 1 and 2, and the LLMSE method are shown in Fig. 11.5. We can see that on the average the RFID system counts two less occupants than the true number, while the video camera systems counts four less occupants. The LLMSE method counts two less occupants. Both fusion methods only count one occupant incorrectly. In particular, from Fig. 11.5, we can see that the average standard deviation over the entire period of time, is 1.9130 for RFID, 3.9743 for video cameras, 1.3642 for LLMSE method, 1.0131 for Method 1, and 1.0872 for Method 2. Method 1 improves the accuracy of the RFID system by 47.04% and that of the video camera system by 74.51%. Similarly, we can calculate that Method 2 improves the accuracy of the RFID system and that of the video camera system by 43.17% and 72.64%, respectively. The LLMSE method only improves the accuracy of the RFID system and that of the video camera system by 28.69% and 65.67%, respectively.

## *11.4.2 Results of Smart Microgrid of Buildings*

Consider three cases where a single room in Beijing is occupied by a frugal occupant (referred as occupant 1), a neutral occupant (referred as occupant 2), and a profligate

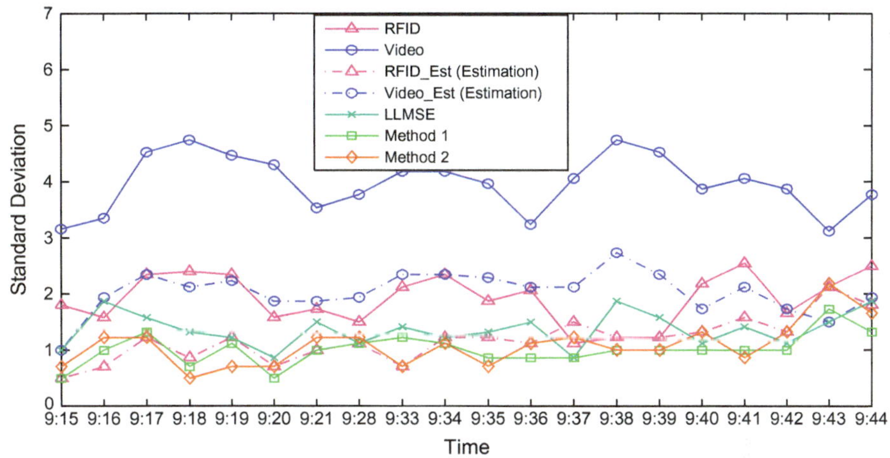


**Fig. 11.5** The standard deviations of the two approximation methods

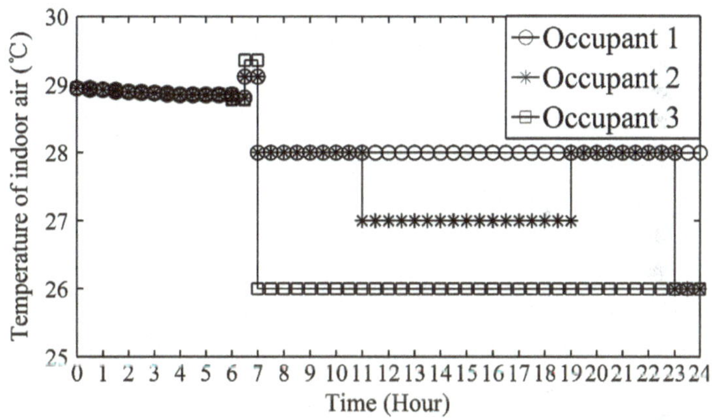


**Fig. 11.6** The indoor air temperature

occupant (referred as occupant 3), respectively. The room is 8 m long, 5 m wide, and 4 m high. The comfort models of these occupants are shown in Table I. Assumed that the room is occupied during 7:00–24:00. The energy supply systems consist of the power grid, the solar PV panels (7 cells in series), one HVAC (3 kWh), one battery (0.4 kWh), one ice storage device (18 kWh), and a CCHP (rated power of 5 kW). In this paper, the power generation of the PV panels and the battery are assumed to be cost-free. The test is performed on a typical summer day, which is hot and humid in Beijing.

The joint scheduling problems are solved using the CPLEX. Cases 1, 2, and 3 are corresponding to occupants 1, 2, and 3, respectively. The indoor air temperature of the above three cases is shown in Fig. 11.6. We can see that the indoor air temperature

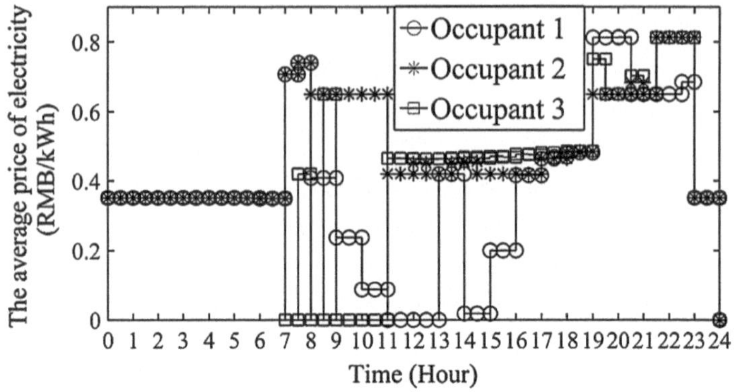


**Fig. 11.7** The APEs of the three cases

is set at the upper bound of the comfort range since that is the most energy-efficient choice in summer.

The APEs of the three cases are shown in Fig. 11.7. They are different from each other due to the different control strategies of the supply systems. The control strategies of the HVAC (including the ice storage) and the CCHP are similar in all the cases. But the control strategy of battery in case 3 is different from that in cases 1 and 2, as shown in Fig. 11.8. In all the three cases, the battery is charged before 7:00, and discharged during the TOU peak-price periods in the night. The SOC by the end of the day is set to equal to the initial value. This is achieved by charging during two consecutive stages. In case 3, there is a discharging–charging cycle of the battery from 7:00 to 16:30. But the battery is neither charged nor discharged during these stages in cases 1 and 2. This difference emerged due to the following reasons. (1) In case 1, the demand for electricity is relatively low, and the cost savings through

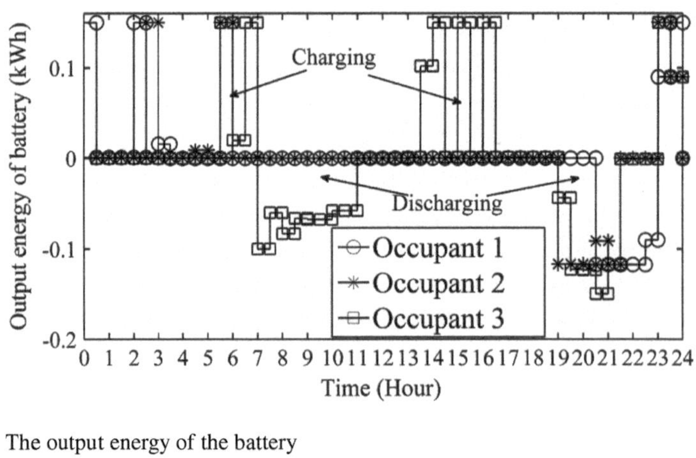


**Fig. 11.8** The output energy of the battery

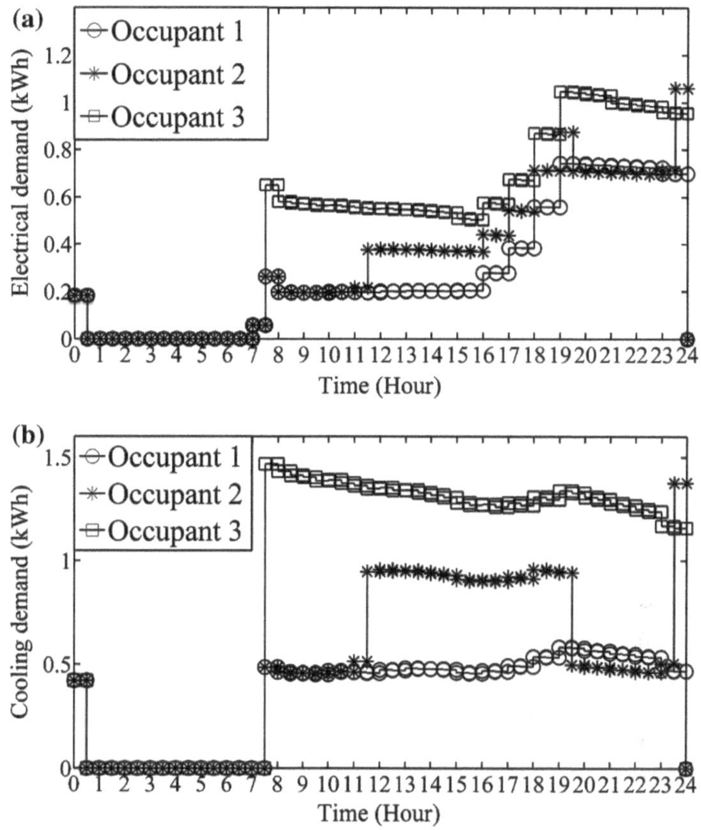


**Fig. 11.9** The electrical demand and cooling demand

using the battery during these stages is thus less than the penalty cost due to the cycle lifetime of the battery. (2) In case 2, if the battery is discharged during these periods, the APE will be decreased and fall in the range of (0.42, 0.65]. Then, the demand for electricity will increase. Because the increased cost exceeds the cost saving using the battery, the battery is not used. (3) In case 3, the demand for electricity is constant under different APEs, and is relatively high. So, the cost can be saved by using the battery during these periods.

The demand for electricity including the energy consumption of the FCU and the lighting equipment and that of the HVAC system for cooling and the cooling demand of each occupant is shown in Fig. 11.9. The demand for electricity is high in the night because more lights are used. The peak cooling loads occur from 7:30 to 24:00 due to the high outdoor temperature in summer. The HVAC system performs precooling from 0:00 to 0:30 because the TOU electricity price is low, and does not perform precooling from 0:30 to 7:00 because the outdoor temperature is low. The demands on electricity in the three cases are 13.79, 16.86, and 24.60 kWh, respectively. The total costs are 7.17, 8.71, and 14.58 RMB, respectively. In particular, the energy consumption of occupant 1 is 44% less than that of occupant 3. This shows that significant energy saving is possible through the supply–demand coordination.

### *11.4.3 Results of Electrical Vehicle Charging from Wind Turbines*

The performance of this multi-timescale and bilevel coordination approach is demonstrated through case studies in the city of Beijing. The real wind data and vehicle data are used to demonstrate the effectiveness of our model and method. The wind speed at 12 m-height is collected at a weather station in Tsinghua University. The Vestas V100-2 MW wind turbine is used. We assume the wind power follows normal distribution. The mean value is the actual generated wind power at 1-h-timescale and 15-min-timescale, which is shown in Fig. 3. The standard deviation is 10% of the mean value. The vehicle data in Beijing are used to simulate the parking event in the future. In the experiment, the scheduling problem is considered on a daily basis. The following three control policies are considered to compare.

1. Greedy Charging: Each EV will be charged to the full state as soon as possible. This is the current control strategy for the EV charging.
2. Delayed Charging: Each EV will be postponed to charge as long as possible. This policy can utilize the low electricity price at early morning and late night to save the charging cost.
3. Bilevel Charging: This is our proposed method. The charging policy is obtained by using the bilevel SBPI, where the greedy charging policy is chosen as the base policy.

The total charging power of these three control policies are shown in Fig. 11.10. In the experiment settings, each EV has a charging requirement at the initial stage and requires to finish all the charging requirements at the last stage. This causes the total charging power at the first stage and the last stage are not the same. As can be seen in Fig. 11.10, the greedy policy performs worst during the early morning. The total charging power of the greedy policy keeps decreasing within this period, while the wind power keeps increasing during this period. In contrast, the delayed charging policy and the bilevel charging policy can make the EV charging power follow the wind power. Due to this reason, the total charging costs of the greedy policy, delayed charging policy, and bilevel charging policy are 2609.3, 1237.4, and 1623.2 RMB, respectively. Note that as the system operator not only focuses on the charging cost optimization, but also focuses on the matching degree optimization. This makes that the charging cost of the bilevel charging policy is not the smallest among the three control policies. Note that, the bilevel charging policy is obtained based on the greedy policy. It can be found that the improvement of the bilevel charging policy is significantly compared with the greedy charging policy. This demonstrates the effectiveness of the bilevel SBPI method.

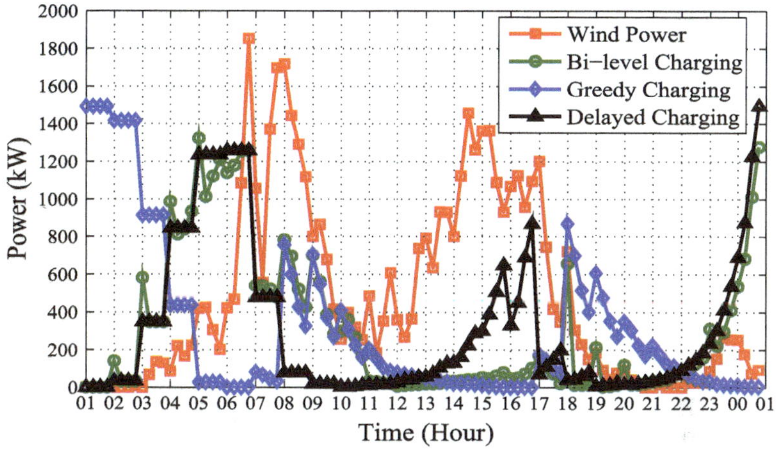


**Fig. 11.10** Wind power and total charging power in different charging policies

## *11.4.4 Fast Evacuation*

Evacuation in indoor environment has significant impact on saving human lives under emergency conditions. Comparing with the large population to evacuate, the number of guides is much smaller. Thus the evacuation policy optimization usually

**Fig. 11.11** Real-time evacuation guidance

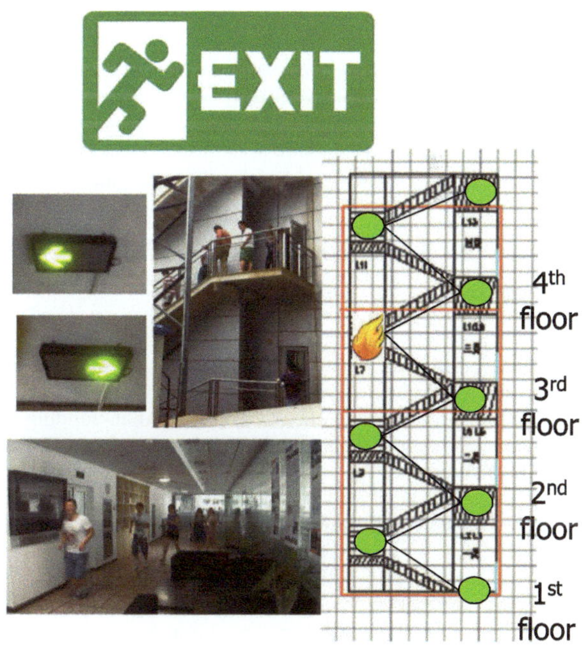

suffers from partial information, partial control, and the pervasive uncertainty in the evacuation process, and is nontrivial. The Internet of Things connects the guides as a whole system. During emergency situations, as part of the smart building system, an optimized evacuation policy is made online. Then different evacuation signals are generated when people pass through the guides which shows people the safest and shortest way to safe places. A real-time evacuation guidance example is shown in Fig. 11.11. The interested audience can refer to [40, 41] for more detailed description of IOT-based smart evacuation system in smart buildings.

## 11.5 Conclusion

The Internet of Things technology brings great opportunities to the smart building system. For the sensing, heterogeneous information fusion increases the accuracy of networked cheap sensors and then shows great potential in practical application considering the high cost of installing high-accuracy sensors. For the building energy management, the Internet of Things technology connects the energy users with the energy suppliers. The coordination among different energy systems not only improves the building energy efficiency but also decreases the fossil fuel consumption, which greatly influences people's living environment. What's more, with the shared information, the renewable energy like wind power can be better utilized by electric vehicles. For the occupants of buildings, a more secure and responsive evacuation plan can be made when facing emergency situations. These benefits brought by the Internet of Things technology have reshaped the smart building system and greatly influenced our everyday life. In the near future, combined with the date-based technology and artificial intelligence technology, the Internet of Things may present us another new blueprint of smart buildings.

**Acknowledgements** The authors would like to thank Guan Xiaohong, Xu Zhanbo, Wang Hengtao and Huang Qilong for their excellent work in applying Internet of Things to smart buildings.

## References

1. Guan X (2011) Efficient energy production, consumption and transmission facilitated by information technology. Keynote Speech of IEEE ISAS 2011, Japan, 17–19 June 2011
2. Bahl P, Padmanabham VN (2000) Radar: An in-building RF-based user location and tracking system. In: Proceedings of the IEEE InfoCom 2000, Tel Aviv, Israel, vol. 2, March 2000, pp 775–784
3. Hightower J, Want R, Borriello G (2000) SpotON: An indoor 3d location sensing technology based on RF signal strength, UW CSE 2000–02-02. Univ, Washington, Seattle
4. Ni LM, Liu Y, Lau YC, Patil AP (2004) Landmarc: Indoor location sensing using active RFID. Wirel Netw 10:701–710
5. Ekahau Inc Ekahau Positioning Engine 2.0. http://www.ekahau.com
6. Steggles P, Gschwind S (2005) The ubisense smart space platform. Tech, Rep, Ubisense Limited

7. Want R, Hopper A, Falcao V (1992) The active badge location system. ACM Trans Inf Syst 10(1):91–102
8. Priyantha NB, Chakraborty AH, Balakrishnan A (2000) The cricket location-support system. In: Proceedings of the 6th annual international conference on mobile computing and networking, August 2000. Boston, pp 32–43
9. Brumitt B, Meyers B, Krumm J, Kern A, Shafer S (2000) Easyliving technologies for intelligent environments, Lecture notes of computer science, 2000, pp 97–119
10. Wahl F, Milenkovic M, Amft O (2012) A Green autonomous self-sustaining sensor node for counting people in office environments. In: EDERC 2012: 5th European DSP education and research conference, IEEE, 2012
11. Gu Y, Lo A, Niemegeers I (2009) A survey of indoor positioning systems for wireless personal networks. IEEE Commun Surv Tutor 11:13–32
12. Li N, Calis G, Becerik-Gerber B (2012) Measuring and monitoring occupancy with an RFID based system for demand-driven HVAC operations. Autom Constr 24:89–99
13. Liao C, Lin Y, Barooah P (2002) Agent-based and graphical modelling of building occupancy. J Build Perform Simul 5(1):5–25
14. Casas JR, Sitjes AP, Folch PP (2005) Mutual feedback scheme for face detection and tracking aimed at density estimation in demonstrations. IEEE Proc Vis Image Signal Process 152(3):334–346
15. Lin SF, Chen JY, Chao HX (2001) Estimation of number of people in crowded scenes using perspective transformation. IEEE Trans SMC 31(6):645–654
16. Guo J, Wu X, Cao T, Yu S, Xu Y (2010) Crowd density estimation via Markov Random Field (MRF), In: Proceedings of the 2010 8th world congress on intelligent control and automation, Jinan, China, July 2010, pp 258–263
17. Bloch I (1996) Information combination operators for data fusion: A comparative review with classification. IEEE Trans SMC 26(1):52–67
18. Li CY, Heinemann P, Sherry R (2007) Neural network and Bayesian network fusion models to fuse electronic nose and surface acoustic wave sensor data for apple defect detection. Sens Actuators B: Chem 125(1):301–310
19. Basir O, Yuana XH (2007) Engine fault diagnosis based on multi-sensor information fusion using Dempster–Shafer evidence theory. Inf Fus 8(4):379–386
20. Tahani H, Keller JM (1990) Information fusion in computer vision using the fuzzy integral. IEEE Trans SMC 20(3):733–741
21. Banerjee TP, Das S (2012) Multi-sensor data fusion using support vector machine for motor fault detection. Inf Sci 217:96–107
22. Alvarez-Alvarez A, Alonso JM, Trivino G (2013) Human activity recognition in indoor environments by means of fusing information extracted from intensity of WiFi signal and accelerations. Inf Sci 233:162–182
23. Meyn S, Surana A, Lin Y, Oggianu SM, Narayanan S, Frewen TA (2009) A sensor-utility-network method for estimation of occupancy in buildings. In: Proceedings of the 48th IEEE conference on decision and control, Shanghai, China, December 2009, pp 1494–1500
24. Djuric N, Novakovic V, Holst J, Mitrovic Z (2007) Optimization of energy consumption in buildings with hydronic heating systems considering thermal comfort by use of computer-based tools. Energy Build 39(4):471–477
25. Asadi E, Silva MGD, Antunes CH, Dias L (2012) A multi-objective optimization model for building retrofit strategies using TRNSYS simulations. GenOpt MATLAB, Build Environ 56:370–378
26. Hazyuk I, Ghiaus C, Penhouet D (2012) Optimal temperature control of intermittently heated buildings using model predictive control: Part I–Building modeling. Build Environ 51:379–387
27. Hazyuk I, Ghiaus C, Penhouet D (2012) Optimal temperature control of intermittently heated buildings using model predictive control: Part II–Control algorithm. Build Environ 51:388–394
28. Figueiredo J, Martins J (2010) Energy production system management–Renewable energy power supply integration with building automation system. Energy Convers Manag 51(6):1120–1126

29. Fanger PO (1970) Thermal Comfort. Danish Tech Press, Denmark
30. Yang R, Wang L (2013) Development of multi-agent system for building energy and comfort management based on occupant behaviors. Energy Build 56:1–7
31. Zhao Y, Zhao Q, Wang F, Jiang Y, Zhang F (2011) On-line adaptive personalized dynamic thermal comfort (PDTC) model using recursive least square estimation. In:proceding international conference intelligent buildings and management, Singapore, pp 275–279
32. Peschiera G, Taylor JE, Siegel JA (2010) Response-relapse patterns of building occupant electricity consumption following exposure to personal, contextualized and occupant peer network utilization data. Energy Build 42(8):1329–1336
33. Petersen JE, Shunturov V, Janda K, Platt G, Weinberger K (2007) Dormitory residents reduce electricity consumption when exposed to real-time visual feedback and incentives. Int J Sustain High Educ 8(1):16–33
34. Jia Q-S, Wang H, Lei Y, Zhao Q, Guan X (2015) A decentralized stay-time based occupant distribution estimation method for buildings. IEEE Trans Autom Sci Eng 12(4):1482–1491
35. Wang H-T, Jia Q-S, Song C, Yuan R, Guan X (2014) Building occupant level estimation based on heterogeneous information fusion. Inf Sci 272:145–157
36. Guan X, Xu Z, Jia Q-S (2010) Energy-efficient buildings facilitated by microgrid. IEEE Trans Smart Grid 1(3):243–252
37. Xu Z, Jia Q-S, Guan X (2015) Supply demand coordination for building energy saving: Explore the soft comfort. IEEE Trans Autom Sci Eng 12(2):656–665
38. Huang Q, Jia Q-S, Guan X A multi-timescale and bi-level coordination approach for matching uncertain wind supply with EV charging demand, IEEE Trans Autom Sci Eng, to appear
39. Huang Q, Jia Q-S, Qiu Z, Guan X, Deconinck G (2015) Matching EV charging load with uncertain wind: A simulation-based policy improvement approach. IEEE Trans Smart Grid 6(3):1425–1433
40. Jia Q-S, Xia L (2012) Approximate event-based optimization for evacuation.In: The 11th international Workshop on Discrete Event Systems (WODES2012), Guadalajara, Mexico, 3–5 Oct. 2012
41. Jia Q-S, Guo Y (2012) Event-based evacuation in outdoor environment. In: Proceedings of the 24th Chinese Control and Decision Conference (CCDC2012), Taiyuan, China, 23–25 May, 2012, pp 33–38

# Index





J.T. Wen and S. Mishra (eds.), *Intelligent Building Control Systems*,
Advances in Industrial Control, https://doi.org/10.1007/978-3-319-68462-8

Kolibri 360 GmbH
Ferdinand-Jühlke-Straße 7
99095 Erfurt, Deutschland
produktsicherheit@kolibri360.de